Biochemical Interactions of Iron Nutrition in Plants

Biochemical Interactions of Iron Nutrition in Plants

Editor

Ferenc Fodor

MDPI

Basel • Beijing • Wuhan • Barcelona • Belgrade • Novi Sad • Cluj • Manchester

Editor
Ferenc Fodor
ELTE Eötvös Loránd
University
Budapest
Hungary

Editorial Office
MDPI
St. Alban-Anlage 66
4052 Basel, Switzerland

This is a reprint of articles from the Special Issue published online in the open access journal *Plants* (ISSN 2223-7747) (available at: https://www.mdpi.com/journal/plants/special_issues/OI9ZP0018D).

For citation purposes, cite each article independently as indicated on the article page online and as indicated below:

Lastname, A.A.; Lastname, B.B. Article Title. *Journal Name* **Year**, *Volume Number*, Page Range.

ISBN 978-3-7258-0657-7 (Hbk)
ISBN 978-3-7258-0658-4 (PDF)
doi.org/10.3390/books978-3-7258-0658-4

Contents

About the Editor . . . . **vii**

Ferenc Fodor
Iron Nutrition and Its Biochemical Interactions in Plants: Iron Uptake, Biofortification, Bacteria, and Fungi in Focus
Reprinted from: *Plants* **2024**, *13*, 561, doi:10.3390/plants13050561 . . . . **1**

Akihiro Saito, Kimika Hoshi, Yuna Wakabayashi, Takumi Togashi, Tomoki Shigematsu, Maya Katori, et al.
Barley Cultivar Sarab 1 Has a Characteristic Region on the Thylakoid Membrane That Protects Photosystem I under Iron-Deficient Conditions
Reprinted from: *Plants* **2023**, *12*, 2111, doi:10.3390/plants12112111 . . . . **4**

Joaquín Vargas, Isabel Gómez, Elena A. Vidal, Chun Pong Lee, A. Harvey Millar, Xavier Jordana and Hannetz Roschzttardtz
Growth Developmental Defects of Mitochondrial Iron Transporter 1 and 2 Mutants in Arabidopsis in Iron Sufficient Conditions
Reprinted from: *Plants* **2023**, *12*, 1176, doi:10.3390/plants12051176 . . . . **23**

Amarjeet Singh, Fruzsina Pankaczi, Deepali Rana, Zoltán May, Gyula Tolnai and Ferenc Fodor
Coated Hematite Nanoparticles Alleviate Iron Deficiency in Cucumber in Acidic Nutrient Solution and as Foliar Spray
Reprinted from: *Plants* **2023**, *12*, 3104, doi:10.3390/plants12173104 . . . . **46**

Barbara Frąszczak, Renata Matysiak, Marcin Smiglak, Rafal Kukawka, Maciej Spychalski and Tomasz Kleiber
Application of Salicylic Acid Derivative in Modifying the Iron Nutritional Value of Lettuce (*Lactuca sativa* L.)
Reprinted from: *Plants* **2024**, *13*, 180, doi:10.3390/plants13020180 . . . . **62**

Miao Zhang, Meng-Han Chang, Hong Li, Yong-Jun Shu, Yan Bai, Jing-Yun Gao, et al.
MsYSL6, A Metal Transporter Gene of Alfalfa, Increases Iron Accumulation and Benefits Cadmium Resistance
Reprinted from: *Plants* **2023**, *12*, 3485, doi:10.3390/plants12193485 . . . . **73**

Zhengtong Qu and Hiromi Nakanishi
Amino Acid Residues of the Metal Transporter OsNRAMP5 Responsible for Cadmium Absorption in Rice
Reprinted from: *Plants* **2023**, *12*, 4182, doi:10.3390/plants12244182 . . . . **89**

José María Lozano-González, Silvia Valverde, Mónica Montoya, Marta Martín, Rafael Rivilla, Juan J. Lucena and Sandra López-Rayo
Evaluation of Siderophores Generated by *Pseudomonas* Bacteria and Their Possible Application as Fe Biofertilizers
Reprinted from: *Plants* **2023**, *12*, 4054, doi:10.3390/plants12234054 . . . . **103**

Jorge Núñez-Cano, Francisco J. Romera, Pilar Prieto, María J. García, Jesús Sevillano-Caño, Carlos Agustí-Brisach, et al.
Effect of the Nonpathogenic Strain *Fusarium oxysporum* FO12 on Fe Acquisition in Rice (*Oryza sativa* L.) Plants
Reprinted from: *Plants* **2023**, *12*, 3145, doi:10.3390/plants12173145 . . . . **122**

Yi Liu, Zimo Xiong, Weifeng Wu, Hong-Qing Ling and Danyu Kong
Iron in the Symbiosis of Plants and Microorganisms
Reprinted from: *Plants* **2023**, *12*, 1958, doi:10.3390/plants12101958 . . . . **138**

Zoltán Molnár, Wogene Solomon, Lamnganbi Mutum and Tibor Janda
Understanding the Mechanisms of Fe Deficiency in the Rhizosphere to Promote Plant Resilience
Reprinted from: *Plants* **2023**, *12*, 1945, doi:10.3390/plants12101945 . . . . **150**

Irene Murgia and Piero Morandini
Plant Iron Research in African Countries: Current "Hot Spots", Approaches, and Potentialities
Reprinted from: *Plants* **2024**, *13*, 14, doi:10.3390/plants13010014 . . . . **162**

About the Editor

Ferenc Fodor

Ferenc Fodor, Ph.D., is an associate professor and head of the Department of Plant Physiology and Molecular Plant Biology at ELTE Eötvös Loránd University, Hungary. He has studied at ELTE and UCLA and received his Ph.D. in plant physiology at ELTE. He has been studying the uptake and translocation of iron and toxic heavy metals. His current research topic is the application of nanomaterials in iron nutrition.

plants

Editorial

Iron Nutrition and Its Biochemical Interactions in Plants: Iron Uptake, Biofortification, Bacteria, and Fungi in Focus

Ferenc Fodor

Department of Plant Physiology and Molecular Plant Biology, ELTE Eötvös Loránd University, 1/C Pázmány Péter. sétány, H-1117 Budapest, Hungary; ferenc.fodor@ttk.elte.hu

Microelements are vital for plant growth and development. Among the micronutrients, iron is required in relatively high concentrations, and although it is one of the most abundant elements in the environment, its availability in soils is restricted due to its low solubility, especially in alkaline conditions and calcareous soils [1]. In order to cope with its limited availability, plants have evolved special mechanisms to mobilize, complex, reduce, and take up iron. It has been a long time since Römheld and Marschner [2] postulated the two basic strategies of iron uptake in higher plants. Strategy I is based on the reduction of soluble Fe-chelates in the apoplast by plasma-membrane-localised ferric reduction oxidase (FRO) family enzymes, which is followed by the uptake of ferrous ions by a divalent metal ion transporter of the iron-regulated transporter (IRT) family. This process is aided by several complementary mechanisms. Iron solubility is increased via the acidification of the rhizosphere by H^+-ATPase enzymes located in the plasma membrane. In some plants, such as *Arabidopsis thaliana*, phenolics (mostly coumarines) are synthesized and secreted to the rhizosphere [3]. These may reduce and/or complex iron, thereby increasing its solubility. Other plants, such as *Cucumis sativus*, synthesize flavins (e.g., riboflavin) which increase iron availability by reduction and/or chelation or modifying the microbial population in the rhizosphere [4,5].

Strategy II is based on the synthesis and release of phytosiderophores (mugineic acids) by a transporter of mugineic acids (TOM). These compounds are strong iron chelators, mobilizing and chelating ferric iron in the soil; they are taken up without further chemical transformation by yellow stripe-like (YSL) transporters [6].

Strategy II was only described in graminaceous plants, whereas Strategy I was confined to the rest of higher plants. Nevertheless, combined strategies also occur, e.g., in *Oryza sativa*, both YSL and IRT1 proteins were found to facilitate iron uptake [7].

As a response to insufficient iron supply, the synthesis of proteins is upregulated in both strategies, accelerating the rate of iron uptake. After entering the cytoplasm, ferrous iron is thought to be complexed by glutathione (GSH) and/or nicotianamine; the latter is a ubiquitous amino acid derivative. Nicotianamine may have a key role in carrying iron to the sites of assimilation, storage, or retranslocation [8]. In the xylem, iron is translocated predominantly as Fe(III)-carboxylates such as Fe(III)3-(Citrate)3, which then become available in the apoplast for leaf cells [9].

The major sink of iron is photosynthetic apparatus. About 90% of tissue iron can be found in chloroplasts [10]. The mechanism of iron uptake to cellular compartments such as plastids, mitochondria, and the vacuole, has been the focus of more recent research, and despite the rapidly accumulating knowledge in this field, there are still unanswered questions. In chloroplasts and mitochondria, a reductive step by FRO enzymes seems to be necessary. Ferrous iron enters the stroma/matrix of these organelles, facilitated by another transport system: PIC1/TIC21 or nickel cobalt transporter (NiCo) in chloroplasts, and mitochondrial iron transporters (MITs) in mitochondria. In vacuoles, iron enters through vacuolar iron transporter (VIT) proteins; however, the reduction step has not yet been revealed [8].

Citation: Fodor, F. Iron Nutrition and Its Biochemical Interactions in Plants: Iron Uptake, Biofortification, Bacteria, and Fungi in Focus. *Plants* **2024**, *13*, 561. https://doi.org/10.3390/plants13050561

Received: 7 February 2024
Accepted: 9 February 2024
Published: 20 February 2024

The export of iron from the root and leaf cell cytoplasm as well as cell compartments are thought to be crucial in root-to-shoot transport, metabolic processes, loading of fruits and seeds and remobilization during senescence. Cellular export is facilitated by YSL family transporters but iron translocation in the xylem also relies on citrate loading to the xylem by ferric reductase defective 3 (FRD3) proteins. Iron export from chloroplasts may also occur through YSL proteins or a prokaryotic-type ECF ABC-transporter complex described recently [11]. Both chloroplasts and mitochondria release Fe-S proteins, which are important in cellular functions. Finally, iron is released from vacuoles through natural-resistance-associated macrophage proteins (NRAMP) [8].

Transport processes of iron may interact with divalent metal ions such as Cd, Mn, Ni, and Zn, but other elements may also influence iron metabolism. Competition or inhibition may occur in divalent transport systems, but also upon binding to phytosiderphores. Cadmium inhibits several metabolic processes in which it interacts with iron transport and assimilation. Iron acquisition in soils cannot be independent from other organisms such as fungi and bacteria. These may also release phytosiderophores or reduce iron, which can be highly advantageous for plants, especially under conditions of low iron availability. Iron deficiency in humans, known as anaemia, is one of the most critical micronutrient deficiencies worldwide; thus, it is a major goal to increase the bioavailable iron content in edible grains, fruits, and plant parts to serve as food products. Various techniques can be applied for the biofortification of iron, such as foliar sprays with Fe-chelates, iron-containing nanomaterials, or biostimulants; all these may influence iron uptake and assimilation processes.

The published papers in this Special Issue focus on several aspects of iron nutrition and its interactions during uptake, transport, and assimilation. Increasing the bioavailability and uptake/accumulation of iron is still an important topic, as is iron incorporation into and release from cellular compartments. The presence of bacteria and fungi in the rhizosphere is under developing interest as iron acquisition by plants may be aided by these organisms. Although this Special Issue does not address the question of regulation of iron homeostasis and signalling, it is another very important area of current and future research that will help to develop a holistic picture on the metabolic roles and interactions of iron.

Acknowledgments: As Guest Editor of the Special Issue, entitled "Biochemical Interactions of Iron Nutrition in Plants", I would like to thank all authors who have contributed to this valuable collection of original research and reviews, promoting the exciting world of iron nutrition and assimilation.

Conflicts of Interest: The author declares no conflicts of interest.

List of Contributions:

1. Frąszczak, B.; Matysiak, R.; Smiglak, M.; Kukawka, R.; Spychalski, M.; Kleiber, T. Application of Salicylic Acid Derivative in Modifying the Iron Nutritional Value of Lettuce (*Lactuca sativa* L.). *Plants* **2024**, *13*, 180. https://doi.org/10.3390/plants13020180.
2. Qu, Z.; Nakanishi, H. Amino Acid Residues of the Metal Transporter OsNRAMP5 Responsible for Cadmium Absorption in Rice. *Plants* **2023**, *12*, 4182. https://doi.org/10.3390/plants12244182.
3. Lozano-González, J.M.; Valverde, S.; Montoya, M.; Martín, M.; Rivilla, R.; Lucena, J.J.; López-Rayo, S. Evaluation of Siderophores Generated by Pseudomonas Bacteria and Their Possible Application as Fe Biofertilizers. *Plants* **2023**, *12*, 4054. https://doi.org/10.3390/plants12234054.
4. Zhang, M.; Chang, M.-H.; Li, H.; Shu, Y.-J.; Bai, Y.; Gao, J.-Y.; Zhu, J.-X.; Dong, X.-Y.; Guo, D.-L.; Guo, C.-H. MsYSL6, A Metal Transporter Gene of Alfalfa, Increases Iron Accumulation and Benefits Cadmium Resistance. *Plants* **2023**, *12*, 3485. https://doi.org/10.3390/plants12193485.
5. Núñez-Cano, J.; Romera, F.J.; Prieto, P.; García, M.J.; Sevillano-Caño, J.; Agustí-Brisach, C.; Pérez-Vicente, R.; Ramos, J.; Lucena, C. Effect of the Nonpathogenic Strain Fusarium oxysporum FO12 on Fe Acquisition in Rice (*Oryza sativa* L.) Plants. *Plants* **2023**, *12*, 3145. https://doi.org/10.3390/plants12173145.
6. Singh, A.; Pankaczi, F.; Rana, D.; May, Z.; Tolnai, G.; Fodor, F. Coated Hematite Nanoparticles Alleviate Iron Deficiency in Cucumber in Acidic Nutrient Solution and as Foliar Spray. *Plants* **2023**, *12*, 3104. https://doi.org/10.3390/plants12173104.

7. Saito, A.; Hoshi, K.; Wakabayashi, Y.; Togashi, T.; Shigematsu, T.; Katori, M.; Ohyama, T.; Higuchi, K. Barley Cultivar Sarab 1 Has a Characteristic Region on the Thylakoid Membrane That Protects Photosystem I under Iron-Deficient Conditions. *Plants* **2023**, *12*, 2111. https://doi.org/10.3390/plants12112111.
8. Vargas, J.; Gómez, I.; Vidal, E.A.; Lee, C.P.; Millar, A.H.; Jordana, X.; Roschzttardtz, H. Growth Developmental Defects of Mitochondrial Iron Transporter 1 and 2 Mutants in Arabidopsis in Iron Sufficient Conditions. *Plants* **2023**, *12*, 1176. https://doi.org/10.3390/plants12051176.
9. Murgia, I.; Morandini, P. Plant Iron Research in African Countries: Current "Hot Spots", Approaches, and Potentialities. *Plants* **2024**, *13*, 14. https://doi.org/10.3390/plants13010014.
10. Liu, Y.; Xiong, Z.; Wu, W.; Ling, H.-Q.; Kong, D. Iron in the Symbiosis of Plants and Microorganisms. *Plants* **2023**, *12*, 1958. https://doi.org/10.3390/plants12101958.
11. Molnár, Z.; Solomon, W.; Mutum, L.; Janda, T. Understanding the Mechanisms of Fe Deficiency in the Rhizosphere to Promote Plant Resilience. *Plants* **2023**, *12*, 1945. https://doi.org/10.3390/plants12101945.

References

1. Rout, G.R.; Sahoo, S. Role of iron in plant growth and metabolism. *Rev. Agric. Sci.* **2015**, *3*, 1–24. [CrossRef]
2. Römheld, V.; Marschner, H. Evidence for a specific uptake system for iron phytosiderophores in roots of grasses. *Plant Physiol.* **1986**, *80*, 175–180. [CrossRef] [PubMed]
3. Robe, K.; Izquierdo, E.; Vignols, F.; Rouached, H.; Dubos, C. The Coumarins: Secondary Metabolites Playing a Primary Role in Plant Nutrition and Health. *Trends Plant Sci.* **2021**, *26*, 3. [CrossRef] [PubMed]
4. Sisó-Terraza, P.; Rios, J.J.; Abadía, J.; Abadía, A.; Álvarez-Fernández, A. Flavins secreted by roots of iron-deficient Beta vulgaris enable mining of ferric oxide via reductive mechanisms. *New Phytol.* **2016**, *209*, 733–745. [CrossRef] [PubMed]
5. Gheshlaghi, Z.; Luis-Villarroya, A.; Álvarez-Fernández, A.; Khorassani, R.; Abadía, J. Iron deficient Medicago scutellata grown in nutrient solution at high pH accumulates and secretes large amounts of flavins. *Plant Sci.* **2021**, *303*, 110664. [CrossRef] [PubMed]
6. Kobayashi, T.; Nozoye, T.; Nishizawa, N.K. Iron transport and its regulation in plants. *Free. Radic. Biol. Med.* **2019**, *133*, 11–20. [CrossRef] [PubMed]
7. Vatansever, R.; Filiz, E.; Ozyigit, I.I. Genome-wide analysis of iron-regulated transporter 1 (IRT1) genes in plants. *Hortic. Environ. Biotechnol.* **2015**, *56*, 516–523. [CrossRef]
8. Sági-Kazár, M.; Solymosi, K.; Solti, Á. Iron in leaves: Chemical forms, signalling, and in-cell distribution. *J. Exp. Bot.* **2022**, *73*, 1717–1734. [CrossRef] [PubMed]
9. Rellán-Álvarez, R.; Giner-Martínez-Sierra, J.; Orduna, J.; Orera, I.; Rodríguez-Castrillón, J.A.; García-Alonso, J.I.; Abadía, J.; Álvarez-Fernández, A. Identification of a tri-iron (III), tri-citrate complex in the xylem sap of iron-deficient tomato resupplied with iron: New insights into plant iron long-distance transport. *Plant Cell Physiol.* **2010**, *51*, 91–102. [CrossRef] [PubMed]
10. Terry, N.; Abadía, J. Function of iron in chloroplasts. *J. Plant Nutr.* **1986**, *9*, 609–646. [CrossRef]
11. Vigani, G.; Solti, Á.; Thomine, S.; Philippar, K. Essential and Detrimental—An Update on Intracellular Iron Trafficking and Homeostasis. *Plant Cell Physiol.* **2019**, *60*, 1420–1439. [CrossRef] [PubMed]

plants

MDPI

Article

Barley Cultivar Sarab 1 Has a Characteristic Region on the Thylakoid Membrane That Protects Photosystem I under Iron-Deficient Conditions

Akihiro Saito, Kimika Hoshi, Yuna Wakabayashi, Takumi Togashi, Tomoki Shigematsu, Maya Katori, Takuji Ohyama and Kyoko Higuchi *

Laboratory of Biochemistry in Plant Productivity, Department of Agricultural Chemistry, Tokyo University of Agriculture, Setagaya-ku, Tokyo 156-8502, Japan; a3saito@nodai.ac.jp (A.S.); to206474@nodai.ac.jp (T.O.)
* Correspondence: khiguchi@nodai.ac.jp; Tel.: +81-354772315

Abstract: The barley cultivar Sarab 1 (SRB1) can continue photosynthesis despite its low Fe acquisition potential via roots and dramatically reduced amounts of photosystem I (PSI) reaction-center proteins under Fe-deficient conditions. We compared the characteristics of photosynthetic electron transfer (ET), thylakoid ultrastructure, and Fe and protein distribution on thylakoid membranes among barley cultivars. The Fe-deficient SRB1 had a large proportion of functional PSI proteins by avoiding P700 over-reduction. An analysis of the thylakoid ultrastructure clarified that SRB1 had a larger proportion of non-appressed thylakoid membranes than those in another Fe-tolerant cultivar, Ehimehadaka-1 (EHM1). Separating thylakoids by differential centrifugation further revealed that the Fe-deficient SRB1 had increased amounts of low/light-density thylakoids with increased Fe and light-harvesting complex II (LHCII) than did EHM1. LHCII with uncommon localization probably prevents excessive ET from PSII leading to elevated NPQ and lower PSI photodamage in SRB1 than in EHM1, as supported by increased Y(NPQ) and Y(ND) in the Fe-deficient SRB1. Unlike this strategy, EHM1 may preferentially supply Fe cofactors to PSI, thereby exploiting more surplus reaction center proteins than SRB1 under Fe-deficient conditions. In summary, SRB1 and EHM1 support PSI through different mechanisms during Fe deficiency, suggesting that barley species have multiple strategies for acclimating photosynthetic apparatus to Fe deficiency.

Keywords: barley; iron deficiency; LHCII; photosystem I; thylakoid membrane

Citation: Saito, A.; Hoshi, K.; Wakabayashi, Y.; Togashi, T.; Shigematsu, T.; Katori, M.; Ohyama, T.; Higuchi, K. Barley Cultivar Sarab 1 Has a Characteristic Region on the Thylakoid Membrane That Protects Photosystem I under Iron-Deficient Conditions. *Plants* **2023**, *12*, 2111. https://doi.org/10.3390/plants12112111

Academic Editor: Ferenc Fodor

Received: 30 March 2023
Revised: 23 May 2023
Accepted: 24 May 2023
Published: 26 May 2023

1. Introduction

A large amount of iron (Fe) is required for efficient electron transfer in the photosynthetic apparatus. Specifically, the photosystem I (PSI) reaction center harbors three 4Fe-4S clusters and is the primary target of Fe deficiency [1–5]. Two strategies to maintain photosynthesis in Fe-deficient chloroplasts are possible: a continuous supply of Fe to reaction center proteins or the reorganization of protein complexes to ensure electron transfer utilizing a smaller amount of Fe. Barley cultivar Sarab1 (SRB1) has excellent tolerance to Fe deficiency among more than 20 barley cultivars originating worldwide based on Photosynthetic Iron-Use efficiency (PIUE) developed recently in our study on the chloroplast Fe economy [6]. Using quantification of the Fe uptake rate using the live autography system, unexpectedly, SRB1 exhibited a lower rate of Fe acquisition into developing leaves with a smaller accumulation of reaction center proteins of both PSI and photosystem II (PSII) under the Fe-deficient condition than other cultivars [7]. In contrast, another Fe deficiency-tolerant cultivar Ehimehadaka 1 (EHM1) accumulated more Fe in shoots and maintained more amounts of reaction center proteins than did SRB1 under Fe-deficient conditions [7]. However, the expression of some of the genes involved in the Fe-S cluster supply pathway to photosystems, including that of sulfur utilization factor (SUF), reduced similarly under Fe deficiency in both SRB1 and EHM1 [7]. These results revealed that active Fe acquisition into shoots or regulation of the Fe-S delivery system under Fe-deficient conditions is not always the main factor contributing to Fe deficiency tolerance [7].

Several important findings have been reported regarding the adaptation of the plant photosynthetic apparatus to Fe deficiency, mainly using algae as research material. In cyanobacteria, the "iron-stress-induced" gene *isiA* is expressed, and the product IsiA protein (CP43′) forms a giant 18-subunit ring around the trimeric PSI core complex under Fe deficiency [8]. Recently, an analysis of energy transfer in the PSI-IsiA supercomplex indicated that IsiA functions as an energy donor but not as an energy quencher in the supercomplex [9]. In addition to PSI, to protect the acceptor side of PSII against Fe deficiency in cyanobacteria, an additional pigment–protein complex, IdiA (iron deficiency-induced protein), is expressed [10]. In the obligate photoautotrophic alga *Dunaliella salina*, Fe deprivation induces the expression of a chlorophyll *a*/*b*-binding protein Tidi, similar to that of IsiA protein. Tidi resembles the light-harvesting antenna complex protein of PSI (LHCI) and acts as an accessory antenna of PSI [11]. Additionally, in another eukaryote algae *Chlamydomonas reinhardtii*, Fe deprivation causes the remodeling of LHCI and decreases the antenna size of PSI to reduce the efficiency of excitation energy transfer between LHCI and PSI [12,13]. The stress-inducible light-harvesting antenna LHCSR3 protein was also expressed under conditions of Fe deficiency in *C. reinhardtii*, leading to increased non-photochemical quenching (NPQ), thereby providing protection from photoinhibition [14]. Although NPQ is most often related to PSII photoprotection, it also protects PSI through quenching of the LHCII antenna pool functionally associated with PSI. Along with the regulation of these antenna systems, the chloroplasts in *Chlamydomonas* change the Fe economy to preferentially maintain the Fe-containing enzyme Fe superoxide dismutase by balancing the rates of synthesis and degradation of many plastid Fe-proteins [15,16].

In contrast to algae, higher plants lack the Fe deficiency-induced light-harvesting antennae proteins such as IsiA, Tidi, or LHCSR3 in their genomes, making it challenging to develop a link between photosynthesis and Fe deficiency. Nevertheless, some plant species, including barley, can maintain photosynthetic function even after prolonged exposure to Fe deficiency [17]. In this context, we have shed light on the diversity of Fe-deficient responses on photosystems in Graminaceae plants and revealed the barley-specific photoprotective mechanism using light-harvesting antenna Lhcb1 isoforms during Fe deficiency [5,17,18]. Interestingly, we also found that SRB1, a barley variety with significantly higher Fe deficiency tolerance, may have a unique electron transfer function to protect downstream PSI [6]. However, the details surrounding this mechanism still need to be discovered.

In this work, we investigated the characteristics of photosynthetic electron transfer and Fe and protein distribution on thylakoid membranes in the barley cultivar SRB1. Thereafter, we compared these characteristics between SRB1 and other barley cultivars under Fe-deficient conditions. Among the barley cultivars we used, EHM1 was selected as the reference cultivar for Fe deficiency tolerance. This specific cultivar was selected to investigate the SRB1-specific tolerant mechanism and obtain physiologically meaningful data under severe Fe-limited conditions as other cultivars are Fe deficiency-susceptible and cannot maintain photosynthesis under such severe Fe-deficient conditions. Through comparative analysis, we explore the characteristics of SRB1 that contribute to the maintenance of photosynthesis under limited Fe.

2. Results

2.1. SRB1 Has Superior Ability to Suppress Electron Transfer Downstream of PSI

SRB1 maintained the photosynthetic electron transfer function downstream of PSII, including cytochrome (cyt) b_6f, and PSI, through an unknown mechanism despite having low Fe and very small amounts of reaction-center proteins in its leaves under Fe-deficient conditions [6,7]. To elucidate this, we analyzed PSII and PSI simultaneously using Dual-PAM-100 among four cultivars (SRB1 and EHM1 for tolerant cultivars, ETH2 and MSS for susceptible cultivars) with different Fe deficiency tolerance levels, as identified in a previous study [6].

The leaves of Fe-deficient and susceptible barley cultivars often exhibit highly severe chlorotic and wilt symptoms. Further, we cannot observe physiologically meaningful chlorophyll fluorescence under the same Fe-deficient condition as Fe deficiency-tolerant cultivars. We used the same cultivation techniques applied in previous reports [6,7]. Figure S1A shows a typical cultivation result: chlorophyll content in all four cultivars under Fe-sufficient conditions was 1.5 mg/gFW. The chlorophyll content under Fe-deficient conditions was 0.3–0.5 mg/gFW in all four cultivars with no severe necrosis spots (Figure S1C). Fe content in leaves was also comparable among all cultivars under Fe-deficient conditions (Figure S1B); thus, we successfully prepared plant materials exhibiting almost the same extent of Fe deficiency chlorosis and Fe content as previously reported [6].

The maximum quantum yield of PSII, denoted by Fv/Fm, was around 0.8 for all four cultivars under Fe-sufficient conditions (Figure 1A), equivalent to the theoretical value required for healthy leaves. EHM1 did not reduce Fv/Fm due to Fe deficiency among four cultivars. In contrast, Fv/Fm in the Fe-deficient leaves compared was slightly, but significantly, decreased compared to that in the Fe-sufficient leaves in SRB1, ETH2, and MSS varieties (Figure 1A). The quantum yield of regulated energy dissipation, Y(NPQ), increased with Fe deficiency compared with Fe-sufficient conditions in all cultivars (Figure 1B). This is consistent with the results of our previous report that Fe-deficient barley induces NPQ to dissipate excess light energy as heat to avoid PSII photoinhibition resulting from Fe-deficiency-mediated defects in electron transport, regardless of the barley variety [6,17]. Interestingly, SRB1 had higher Y(NPQ) than the other cultivars under both Fe-sufficient and Fe-deficient conditions. The NPQ of MSS, the most susceptible cultivar [6], was also similar to that of SRB1 under Fe-deficient conditions (Figure 1B). The quantum yields of non-regulated energy dissipation related to PSII photoinhibition, Y(NO), were not significantly different among the cultivars, irrespective of the Fe nutritional status (Figure 1C), and Fe deficiency increased the absolute Y(NO) in all cultivars. Therefore, the degree of PSII photoinhibition in all varieties under Fe-deficient conditions was comparable. Thus, consistent with the results of our previous report [6], the photoprotective mechanism of PSII through NPQ induction in Fe-deficient leaves can be considered a ubiquitous Fe-acclimation system for barley species, confirming that PSII maintenance itself is not a main factor in imparting differential Fe deficiency tolerance within barley cultivars.

Next, to clarify the status of PSI, P700 absorbance was examined; P700 maximum oxidation capacity (Pm) is generally used as an indicator of PSI quantity and function because it reflects the maximum absorbance of P700. Pm values in all cultivars were dramatically reduced under Fe-deficient conditions compared to those under Fe-sufficient conditions (Figure 1D), confirming previous findings that Fe deficiency primarily affects PSI [2]. The Fe deficiency tolerant cultivars SRB1 and EHM1 had comparable Pm values under Fe-deficient conditions and significantly higher Pm values than those of the Fe deficiency-susceptible varieties ETH2 and MSS, suggesting that functional PSI maintenance is an essential factor in Fe deficiency tolerance.

Y(ND) is a PSI donor-side limitation, a mechanism of PSI protection that inhibits electron transfer from PSII to PSI by NPQ, plastoquinone (PQ) reduction, or ΔpH expansion in the thylakoid lumen. SRB1 showed the highest Y(ND) under Fe deficiency among the four cultivars, followed by MSS with equally high Y(ND) (Figure 1E). Y(NA) is a parameter for PSI acceptor-side limitation related to PSI photoinhibition by P700 over-reduction and subsequent generation of the toxic reactive oxygen species (ROS). Interestingly, Y(NA) increased under Fe deficiency in all cultivars except SRB1. In contrast, SRB1 showed almost no increase in Y(NA), even under Fe-deficient conditions, indicating that SRB1 suppressed P700 over-reduction.

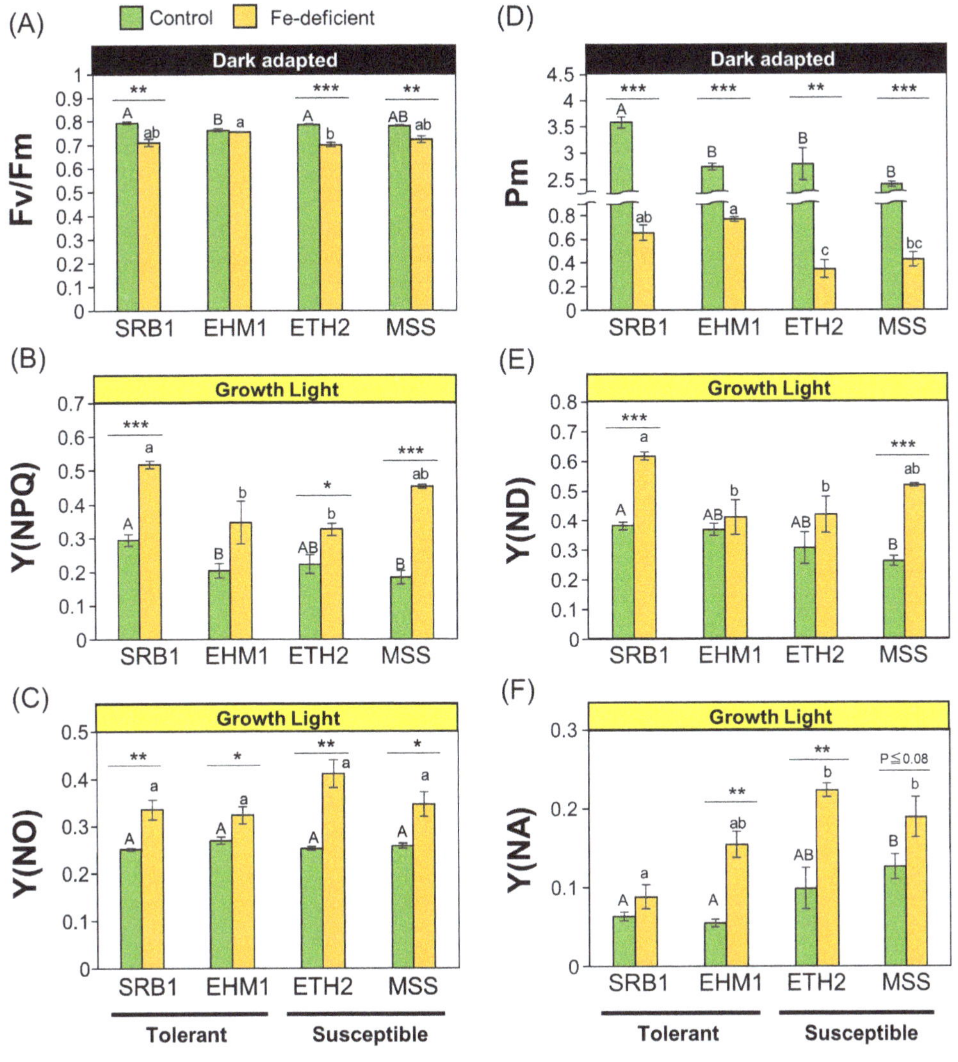

Figure 1. Comparison of the functionality of PSII and PSI in four barley cultivars with varied Fe deficiency tolerance levels: PSII maximum quantum yield, Fv/Fm (**A**); the quantum yield of light-induced non-photochemical fluorescence quenching induced for photoprotection of PSII, Y(NPQ) (**B**); the quantum yield of non-light-induced non-photochemical fluorescence quenching related to PSII photoinhibition; Y(NO) (**C**); the maximal P700 signal, Pm (**D**); PSI donor-side electron transfer limitation, Y(ND) (**E**); PSI acceptor-side electron transfer limitation, Y(NA) (**F**). Data are represented as the means ± SE of three to four independent measurements. * $p < 0.05$, ** $p < 0.01$, and *** $p < 0.001$, indicate significant differences between +Fe and −Fe treatments (according to Student's *t*-test). Different letters are shown on individual columns when p is <0.05 among the four barley cultivars based on Tukey multiple testing.

These results suggest that the mechanism of the P700 oxidation system [19] is the reason for the ability of SRB1 to maintain photosynthesis during Fe deficiency. There is a link between elevated NPQ and increased Y(ND) [20]. This linkage is due to decreased linear electron transfer from PSII to PSI by increased thermal dissipation of light energy.

Consistent with this, Y(NPQ) was highest in SRB1 under Fe deficiency (Figure 1B), and Y(ND) was concomitantly elevated under Fe deficiency (Figure 1E), suggesting that the high P700 oxidation induction of SRB1 is related to NPQ induction. To further confirm this observation, the related parameters were also taken into consideration. 1-qL, which indicates the reduced state of PQ pools, was higher in SRB1 than in the other cultivars (Figure S2A). Further, the Y(I)/Y(II) ratio, the ratio of the quantum yield of PSI to PSII, was significantly higher in SRB1 under Fe-deficient conditions (Figure S2B), indicating that the PSII electron transfer rate was kept low relative to the PSI electron transfer rate. Therefore, a safe PSII–PSI excitation balance that is less prone to PSI photoinhibition can be maintained in the Fe-deficient SRB1. Although MSS also had high Y(NPQ) and Y(ND) like SRB1 under Fe-deficient conditions (Figure 1E), the absolute Y(NA) of Fe-deficient MSS was twice that of Fe-deficient SRB1 (Figure 1F), indicating that PSI photoinhibition is unavoidable in the Fe-deficient leaves of MSS. Interestingly, the Fe-sufficient leaves of MSS exhibited higher Y(NA) than those of other cultivars (Figure 1F; Fe-sufficient SRB = 0.63, Fe-sufficient MSS = 0.136), suggesting that even before Fe deficiency, MSS was not highly effective in PSI photoprotection via the P700 oxidation system, presumably because of the low PSI structural stability associated with low Fe availability in the chloroplasts, as previously reported [6].

2.2. SRB1 Is Excellent at Maintaining Functional PSI under Fe-Deficient Conditions

To determine whether the Fe-deficient cultivar SRB1 maintains a higher accumulation of PSI proteins than the other cultivars, we quantified the contents of major photosystem proteins of both PSII and PSI. The amounts of PSII core subunit proteins D1 and D2 were generally similar among cultivars under Fe-sufficient conditions (Figure 2A,B). However, because the Western blot (Figure 2A) was based on data obtained from different electrophoresis gels for each variety, the proteins extracted from each variety grown under Fe-sufficient conditions were also electrophoresed on the same gel, confirming that the PSII amount was equivalent among the four cultivars. On the other hand, the D1 and D2 protein amounts in Fe-deficient SRB1 leaves were significantly lower (approximately 30%) compared with those in Fe-deficient EHM1 or ETH2 leaves. The other PSII core subunit, cyt b_{559} (PsbE), which contains a hem Fe, did not differ significantly among cultivars regardless of Fe sufficiency or Fe deficiency. Thus, although SRB1 contained lower amounts of D1 and D2 than those in other cultivars, the electron transfer function on PsbE within the PSII complex was maintained under Fe-deficient conditions, consistent with the fact that Fv/Fm was not greatly reduced by Fe deficiency (Figure 1A). These results substantiate that PSII is not the primary reason for the difference in Fe deficiency tolerance levels among these cultivars.

In the case of PSI core subunits, the amounts of PsaA, PsaB, and [4Fe-4S]-containing PsaC in the Fe-deficient SRB1 were unexpectedly and significantly decreased by about 10% compared to their levels in Fe-sufficient control (Figure 2A,C). This residual percentage of PSI core subunits in SRB1 was equal to or less than that of the other Fe deficiency-susceptible cultivars ETH2 and MSS, in which considerably fewer PSI reaction center proteins could be detected. In contrast, EHM1, another Fe deficiency-tolerant variety, retained about 30% of PsaA and as much as 20–25% for PsaB and PsaC under Fe-deficient conditions. As the Western blot analysis in Figure 2A was performed after adjusting the protein concentration to be the same in both Fe-sufficient and Fe-deficient samples, only a tiny amount of PSI protein could be detected in the Fe-deficient samples due to chlorosis. Because of concerns about the quantitation limit of Western blot analysis on protein content basis in Figure 2, we re-performed the same analysis on samples in which the chlorophyll concentration was adjusted to make them equal among all samples to better detect thylakoid membrane proteins in the Fe-deficient samples (Figure S3). As shown in Figure S3A,C (Western blot on a chlorophyll content basis), the reduction in PSII and PSI core subunits was significantly observed in SRB1 and the Fe deficiency-susceptible cultivars ETH2 and MSS. In contrast, the decrease in both PSII and PSI was less pronounced in EHM1, supporting the results shown in Figure 2. These results shown in Figure S3 (Western blot on a chlorophyll content basis) reconfirmed the data shown in Figure 2 (Western blot on a protein content basis); unlike SRB1, it is evident that EHM1 has a strategy to maintain the total amount of photosystem proteins during Fe deficiency.

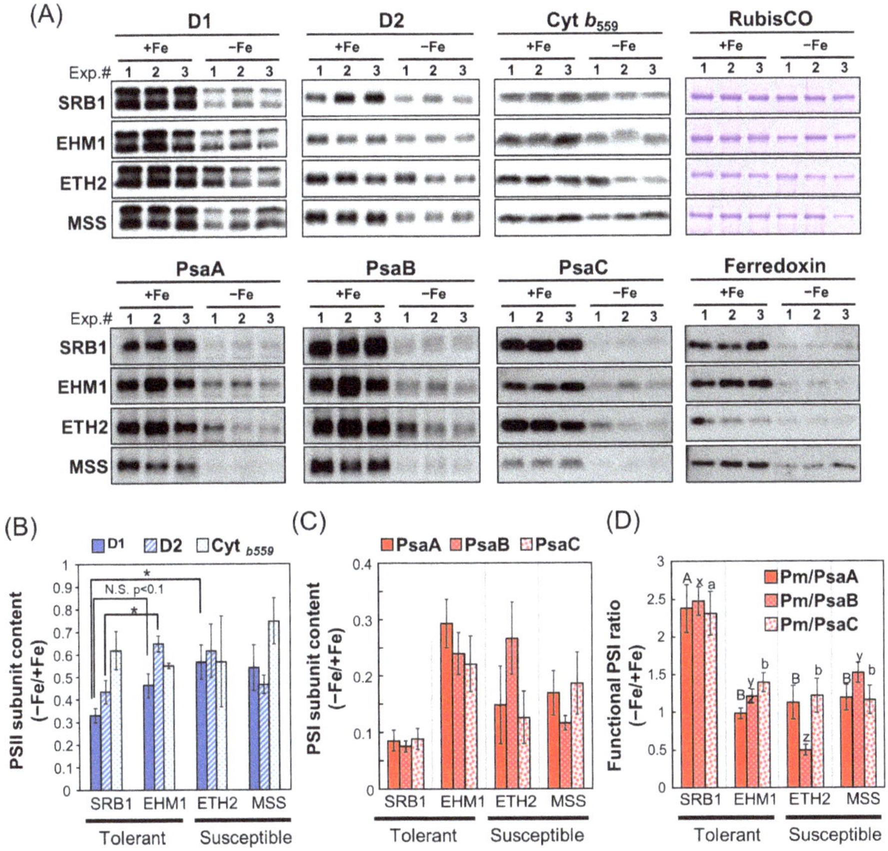

Figure 2. Western blot analysis normalized on a per total leaf protein to compare whole thylakoid proteins and functional PSI levels in four barley cultivars with different Fe deficiency tolerance: (**A**) Immunoblot analysis of PSII reaction center proteins (D1, D2, and cyt b_{559} [PsbE]) and PSI reaction center proteins (PsaA, PsaB, and PsaC), Ferredoxin, and CBB-stained RubisCO large subunits. Whole proteins extracted from leaves were separated by SDS-PAGE (1 μg protein/lane for D1, 5 μg protein/lane for the other proteins) and detected with specific antibodies for each protein; (**B**,**C**) Immunoblots detected with specific antibodies against each PSII subunit (D1, D2, and cyt b_{559}) (**B**) or PSI subunit (PsaA, PsaB, and PsaC) (**C**) in panel A was quantified by Image J software and calculated as relative values for the Fe-sufficient condition (Fe-sufficient condition = 1); (**D**) The retention rate of functional PSI under the Fe-deficient condition is expressed as the relative value of Pm per PSI subunit content under Fe-deficient conditions (value of Figure 2C) to that under Fe-sufficient conditions. Data are presented as means ± SE of three independent leaves, with different letters shown on individual columns when $p < 0.05$ among four barley cultivars based on Tukey multiple testing. * $p < 0.05$.

Because SRB1 has only a small amount of PSI reaction centers (Figures 2A,C and S3A,C), we speculate that the remaining PSI in the Fe-deficient SRB1 may be more functional than those in the other three cultivars. PSI functionality has generally been calculated based on the ratio of Pm before and after PSI photoinhibition treatments [21]. In relation to this concept, we attempted to calculate PSI functionality by using the ratio of Pm under the Fe-deficient conditions to the Pm of leaves under the control (Fe-sufficient) conditions. The problem is that in the case of Fe deficiency, Pm is considerably lower, influenced by a decrease in PSI complexes, regardless of PSI functionality. To eliminate the influence of the decline in PSI content in Fe-deficient leaves, we divided the Pm value by the relative accumulation of PSI core proteins, PsaA, PsaB, and PsaC, respectively (Figure 2C), for normalization to align PSI content computationally under Fe-sufficient and Fe-deficient conditions. As a result of this modified method from Lempiäinen et al. [21], we have successfully calculated PSI functionality under Fe-deficient conditions (Figure 2D). As shown in Figure 2D, the residual functional PSI rate of the Fe-deficient SRB1 was approximately two times higher than that of the other three cultivars, EHM1, ETH2, and MSS. This result was also confirmed in Figure S3D (Western blot on a chlorophyll content basis), i.e., the functional PSI rate of SRB1 under Fe-deficient conditions was about two-fold higher than in other varieties.

These results suggest the different strategies between the two Fe deficiency-tolerant cultivars: SRB1, which keeps a large proportion of functional PSI to overcome the decrease in PSI accumulation, and EHM1, which maintains a sufficient amount of PSI proteins to preferentially bind Fe-S clusters. Unlike the tolerant cultivars, EHM1 and SRB1, the susceptible cultivars, ETH2 and MSS, exhibited low PSI accumulation (Figure S3C) or the functional PSI ratio (Figures 2D and S3D). The low PSI stability under Fe-deficient conditions in these varieties may be related to various systems and would not be necessarily related to photosystem function. Therefore, we excluded these cultivars from further biochemical analyses focusing on the thylakoid membranes.

2.3. *Organization of Thylakoid Membrane in Fe-Sufficient Leaves*

With regard to the differences in the amount of functional PSI maintained under Fe-deficient conditions among varieties, we investigated the sequence of the *PsaC* gene, whose product harbors two 4Fe-4S clusters in PSI, based on the query whether the differences in the primary structure itself may be responsible for PSI's Fe availability. However, the gene encoding PsaC is a single-copy gene in the barley genome and no sequence differences were detected between SRB1 and EHM1 using Sanger dideoxy sequencing. Thus, differences in its structure or ability to bind the Fe cofactors are not likely to explain the different responses in photosystems to Fe deficiency between SRB1 and EHM1.

Therefore, to focus on whether thylakoid membrane structure and its protein distribution affect Fe availability, the analysis using transmission electron microscopy (TEM) was conducted to observe the organization of grana stack and stromal thylakoids in chloroplasts of SRB1 and EHM1. Although the structures of thylakoids in Fe-deficient mesophyll cells varied from rather normal to swollen abnormal thylakoids [5,18], we found that SRB1 seemed to have more stromal lamellar sheets than did EHM1, even under the Fe-sufficient condition. To confirm this characteristic quantitatively, we tried to simply estimate the stromal thylakoid/grana stacks on the TEM images.

In this analysis, the central part of the newest fully expanded leaf was used for imaging analysis as the same position used in Figures 1 and 2. It is known that there are two types of stroma lamellae membranes: one is directly linked to the grana stacks, and the other is like a large sheet that is not directly linked to the grana structure [22]. The former type, called "stroma lamella-grana (SG) structure", is responsible for the compartmentalization of PSI and the restoration of PSII [23]. Therefore, only stromal lamellae in the SG structure were selected for this analysis as structures that have a significant effect on photosynthesis.

The granal and stromal thylakoid membrane lengths were traced on the vertically visual image of thylakoids using NIH ImageJ (Figures 3 and S4). More than twenty representative SG structures (n = 21 for EHM1 and n = 24 for SRB1) in four TEM images for each cultivar were analyzed (Figure S4). As shown by the yellow line in Figure 3A, the stroma lamella (the non-appressed region) includes the top and bottom planes of the grana stacks and the sheet structure connected with the grana stack (Figure 3A). After calculating the stromal thylakoid to granal thylakoid ratio, we confirmed that the ratio for EHM1 was about 0.6, while that for SRB1 was about 0.8, which in turn showed that the ratio of SRB1 was significantly higher than that of EHM1 by about 45% (Figure 3B).

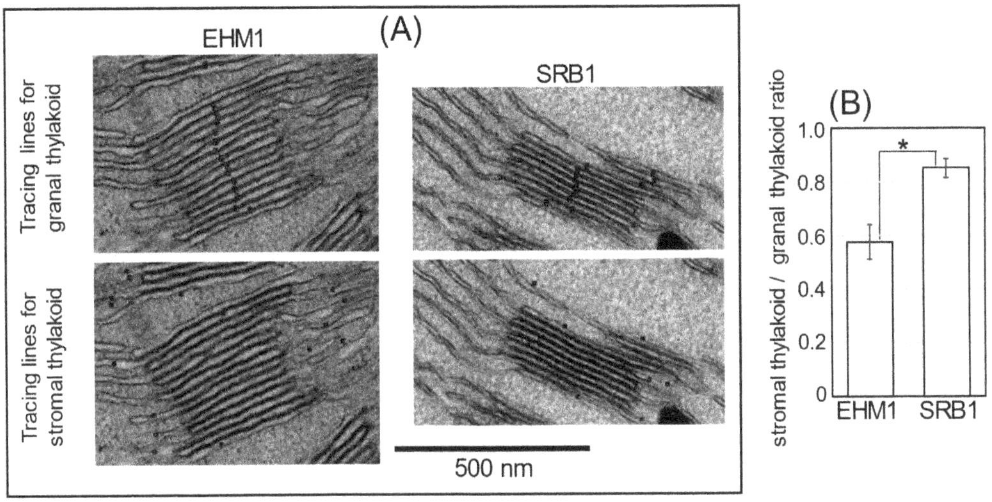

Figure 3. The ratio of stromal and granal thylakoids in Fe-sufficient chloroplasts. TEM images of Fe-sufficient chloroplasts were obtained from SRB1 and EHM1. All original images are shown in Figure S4. We sampled four independent chloroplasts images per cultivar, and selected grana stacks with stroma lamellar membranes clearly recognized in each image. Then, granal thylakoid and thylakoid membranes in stroma connected to selected grana were measured by line length using ImageJ. Typical magnified images of yellow lines tracing the membrane of grana and stroma are shown in (**A**). All tracing lines on membranes are shown in Figure S4. Summations of line lengths to grana or stroma were calculated for each image, then averages of the stromal thylakoid/granal thylakoid ratio, as shown in (**B**). Values represent the mean ± SE of the four images. * $p < 0.05$ indicates significant differences (according to Student's *t*-test) between two cultivars.

2.4. Protein Complexes Contained in Thylakoid Fractions with Different Densities in Fe-Sufficient Leaves

As shown above, we observed differences in the organization of the thylakoid membrane between the two cultivars using TEM. We also found experimentally that the supernatant obtained during the thylakoid membrane extraction procedure from SRB1 leaves always showed more chlorophyll than that from EHM1. The thylakoid membranes in the supernatant of SRB1 were probably derived from unstacked thylakoid membranes that were physically disrupted during homogenization and could not be precipitated after the normal low-speed centrifugation. Therefore, we analyzed the composition of such thylakoid membranes. We designated the slurries obtained after the 2500× *g* centrifugation as "heavy/high-density thylakoids (H-Thy)," corresponding to typical thylakoid membranes usually analyzed. In contrast, the remaining thylakoids in the supernatant were recovered as pellets after centrifugation at 10,000× *g*, designated as "light/low-density thylakoids (L-Thy)."

After obtaining H-Thy and L-Thy from Fe-sufficient leaves of SRB1 and EHM1, we applied each fraction to sucrose density gradient (SDG) centrifugation following solubilization with n-dodecyl β-D-maltoside (β-DM). The appearance of SDG tubes and the composition of proteins of H-Thy were highly similar between SRB1 and EHM1 (Figure 4). This experiment was also performed on thylakoids isolated from stored frozen leaves, and the SDG analyses were generally reproducible (Figures S5 and S6). Total proteins and Fe distribution patterns, not absolute values, were almost identical among the two cultivars. In this analysis, we excluded the eighth fraction (fr.8) from consideration of Fe since the lower fractions were less reproducible for quantitative Fe. For example, fr.8 of SRB1 had a large proportion of Fe, as shown in Figure 4, but the corresponding fr.8 obtained from snap-frozen leaves did not have such an amount of Fe (Figure S6). Such irregular Fe detected from fr.8 of SRB1 in Figure 4 may not have biological meanings. Note that PSII fractions in H-Thy had more Fe than did PSI-LHCI fractions in both SRB1 and EHM1 (Figure 4), but this is because this collected fraction contains both PSII and PSI as confirmed by Western blot analysis (Figures 4 and S6) and is not considered abnormal data.

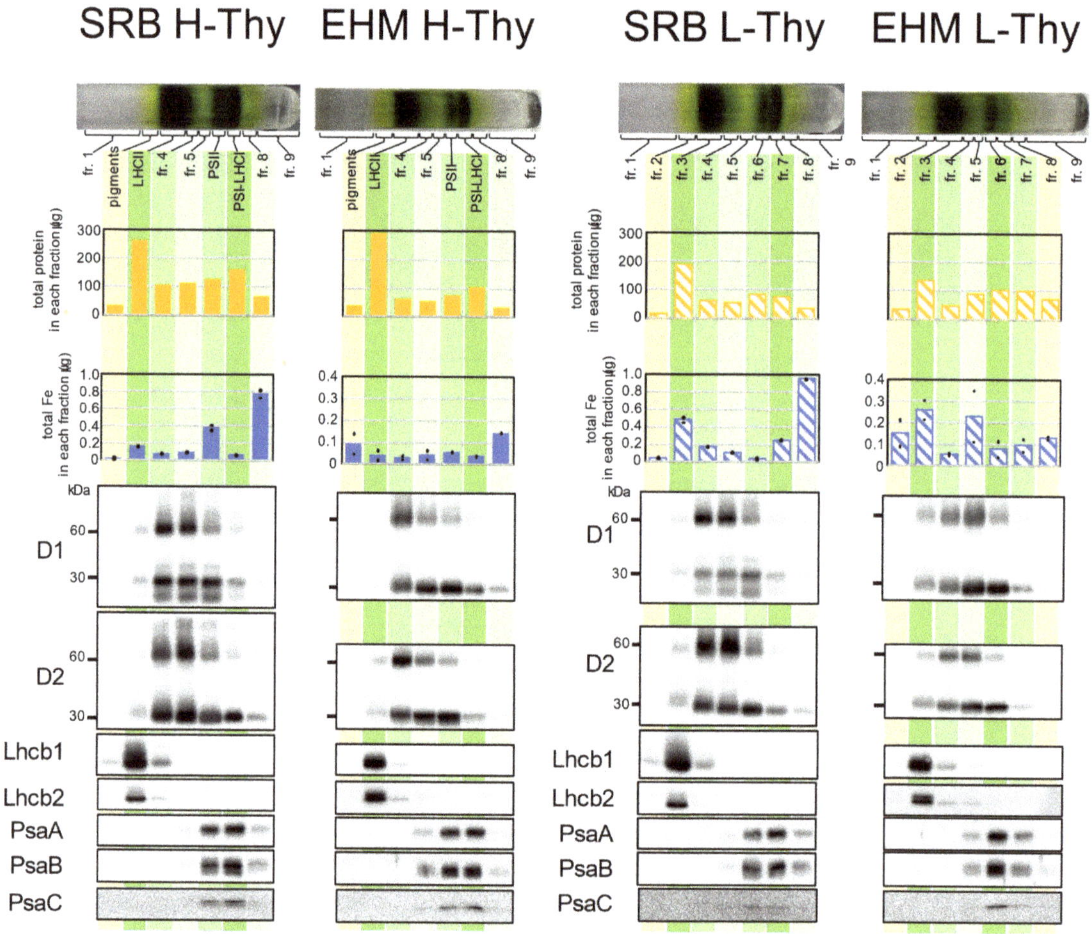

Figure 4. Analyses of fractions obtained from SDG using the thylakoid membranes derived from SRB1 and EHM1. Thylakoid membrane samples from 0.4 mg/mL of chlorophyll were solubilized with β-DM (4.8 mg β-DM for 0.1 mg chlorophyll) and loaded on the top of the SDG tube. Total amounts of proteins or Fe in each fraction are presented in the bar graphs. The amount of Fe was determined by two test solutions prepared from one fraction and measurement was conducted twice for one test solution. A dot shows the average of one test solution and a bar shows the average of two test solutions. A total of 1/1000 of each fraction was loaded for Western blot analysis, except for anti-PsaC. For anti-PsaC, 1/500 of each fraction was loaded.

The appearance and protein distribution pattern of the SDG tubes of L-Thy in SRB1 matched those of H-Thy. However, the Fe contents of fr.3 (LHCII-like fraction) and fr.7 (PSI-LHCI-like fraction) in L-Thy of SRB1 were higher than those in the corresponding fractions in H-Thy of SRB1 (Figure 4), suggesting that more Fe seemed to be allocated to the L-Thy in SRB1. In contrast, in EHM1, the distribution of green bands in L-Thy and H-Thy was different in appearance, with fewer fr.7 bands corresponding to PSI-LHCI in L-Thy than in H-Thy. Indeed, the signals of the Western blot against anti-PsaA, B, and C in fr.7 of L-Thy of EHM1 were weak when compared to the corresponding fraction (PSI-LHCI) of H-Thy of EHM1 (Figure 4).

Besides that, three differences between SRB1 and EHM1 were found in both H-Thy and L-Thy. First, the distributions of PsaA, B, and C extended to fr.8 in the case of SRB1, whereas it extended to the upper layer where fr.5 is located in the case of EHM1 (Figure 4). Thus, the thylakoid membrane of SRB1 may contain PSI complexes of larger molecular weight than those of EHM1, regardless of H-Thy and L-Thy. Second, signals of anti-Lhcb1 in both H-Thy and L-Thy from SRB1 thylakoid membranes were enlarged when compared to those from EHM1, but signals of anti-Lhcb2 were not (Figure 4). Since total protein amounts in the LHCII fractions in H-Thy and L-Thy of EHM1 were not smaller than those of SRB1 (Figure 4) and the intensities of CBB stain of LHCII proteins did not differ between the two cultivars (Figure S5), the abundance of Lhcb1 in LHCII should be higher in SRB1 than in EHM1. Third, LHCII fractions from H-Thy of SRB1 and the corresponding fr.3 from L-Thy of SRB1 had larger proportions of Fe than did adjacent fractions when compared to those of EHM1.

It is difficult to recover large quantities of thylakoid membranes from the chlorotic Fe-deficient fresh leaves due to the size scale of the experimental apparatus. Therefore, the same experiment was conducted in parallel on stocked frozen leaves to obtain the reproducibility of the data. Data of H-Thy from frozen leaves (Figure S6) were almost the same as that from fresh leaves (Figure 4). Minor differences between fresh and frozen leaves were found; a few amounts of the extra green bands appeared in fr.8 of H-Thy from EHM1 (Figure S6); also, fr.7 of L-Thy of frozen SRB1, which may correspond to PSI-LHCI, was decreased more than that of H-Thy of fresh SRB1 when comparing the appearance of the green band and the amount of PsaA, B, and C (Figures 4 and S6). Meanwhile, the following differences were well reproducible in frozen and fresh leaves: the differences in the distribution patterns of PsaA, B, and C and signal intensities of anti-Lhcb1 (Figure 4) were almost reproducible in the case of frozen leaves (Figure S6); the LHCII fraction of SRB1 H-Thy and fr.3 of SRB1 L-Thy had a relatively higher proportion of Fe than that of EHM1 (Figure S6), similar to the data of fresh SRB1 leaves (Figure 4); the relative composition of Lhcb1, which was observed to be higher in SRB1 than in EHM1, was also reproducible among frozen leaves (Figure S6).

2.5. The Influences of Fe Deficiency on Protein Distribution in the Thylakoid Membrane

The SDG centrifugation method, which requires large amounts of thylakoid, is disadvantageous for further analysis because the recovery of the L-Thy fraction is further reduced in the case of Fe-deficient leaves. Therefore, we tried to use Native-PAGE, which generally requires fewer thylakoid proteins. However, Fe contamination from the PAGE gel prevented accurate analysis of trace amounts of Fe. Instead, we quantitatively investigated the influences of Fe deficiency on the partitions of chlorophyll, Fe, and proteins between H-Thy and L-Thy.

We used twice the amounts of leaves to isolate the thylakoid membrane from chlorotic leaves compared to those from control leaves since the accumulation of protein complexes on the thylakoid membrane was remarkably decreased by Fe deficiency, even though both SRB1 and EHM1 are Fe deficiency-tolerant cultivars. The recovery rate of chlorophyll is shown in Figure S7A. Fe deficiency clearly decreased the amount of H-Thy but not L-Thy in both cultivars based on chlorophyll (Figure 5A,B); that is, Fe deficiency increased the ratio of L-Thy to H-Thy. Fe deficiency may result in the loosening and easy disintegration of the structure of thylakoid membranes by decreasing the amount of protein complexes on them. In fact, we reported abnormal thylakoid membranes in Fe-deficient leaves of EHM1 (moderate chlorosis: [18], severe chlorosis: [5]). However, the ratio of L-Thy to H-Thy differed between the two cultivars. SRB1 exhibited a higher ratio (0.10; SE 0.007) of L-Thy to H-Thy than that of EHM1 (0.05; SE 0.015), even under the control condition based on chlorophyll when values of H-Thy were 1 (Figure 5A,B, green and green-hatched bars). This difference was increased by Fe deficiency, up to 0.46 (SE 0.12) for SRB1 and up to 0.21 (SE 0.031) for EHM1 (Figure 5A,B, orange and orange hatched bars). Based on the amounts of chlorophyll, it was estimated that one-third of the thylakoid membrane derived from Fe-deficient SRB1 leaves was L-Thy. We also determined the Fe content of each fraction. The ratio of Fe in L-Thy to Fe in H-Thy was 0.11 (SE 0.005) and 0.11 (SE 0.018) for SRB1 and EHM1, respectively, when grown under control conditions; and 0.34 (SE 0.25) and 0.20 (SE 0.034) for SRB1 and EHM1, respectively (Figure 5C,D), when grown under Fe-deficient conditions. These results indicate that SRB1 had a relatively high chlorophyll/Fe ratio in L-Thy than in H-Thy under Fe-deficient conditions.

Furthermore, we evaluated the amounts of proteins on the thylakoid membranes. Since Fe deficiency drastically decreases protein complexes on the thylakoid membrane (Figure 2A), we applied four-fold amounts of Fe-deficient samples per lane for Western blot analysis to observe differences between H-Thy and L-Thy. CBB stain images are shown in Figure S8C. The signal intensity of the Western blot was evaluated using ImageJ and we calculated the ratio of L-Thy to H-Thy. We did not normalize the values obtained from Fe-deficient samples based on those from control samples because we had to load eight-fold amounts of Fe-deficient materials to obtain enough signals to compare with those from the control materials, as described above. The Fe-containing proteins, D1 and D2 of the PSII reaction center and PsaA, PsaB, and PsaC of the PSI reaction center, were detected using specific antibodies. Anti-HvLhcb1 and Lhcb2 antibodies were also used, as we previously reported the migration of HvLhcb1 under Fe-deficient conditions [18]. Allocations of all proteins detected in L-Thy were increased by Fe deficiency in both cultivars, but the proportion of each protein on H-Thy and L-Thy was largely different between the two cultivars (Figure 5E). A large proportion of all detected proteins was localized on H-Thy in EHM1 (Figures 5E and S8A). In contrast, SRB1 allocated more D1, D2, Lhcb1, and Lhcb2 belonging to PSII to L-Thy than did EHM1, regardless of the Fe nutritional status (Figures 5E and S8A). However, PsaA, PsaB, and PsaC of PSI were mainly localized on H-Thy of the control SRB1, similar to that of EHM1, but were dramatically decreased by Fe deficiency (Figures 5E and S8A), consistent with Figure 1D. The proportions of these PSI core proteins on L-Thy, however, were further increased by Fe deficiency in SRB1 than in EHM1. Typically, PSII has relatively lower Fe content than PSI. Thus, unbalanced distributions of reaction center proteins composing PSII, PSI, and LHCII proteins on L-Thy obtained from SRB1 (Figure 5E) were consistent with a higher chlorophyll/Fe ratio in L-Thy than in H-Thy extracted from Fe-deficient SRB1 (Figure 5A,C).

We tested the protein distribution on the thylakoid membrane of MSS since we observed a significant reduction in PSI core proteins in both Fe deficiency-tolerant SRB1 and Fe deficiency-susceptible MSS (Figure 2A). The proportions of H- and L-Thy in control and Fe-deficient leaves of MSS were similar to those of SRB1 (Figure S7B). PSII core proteins (D1 and D2) and PSI core proteins (PsaA, PsaB, and PsaC) were more distributed to L-Thy than that in EHM1, similar to SRB1 (Figures S8B and 5E). The distribution of LHCII proteins to H- and L-Thy in MSS exhibited an intermediate pattern between SRB1 and EHM1 (Figures S8B and 5E).

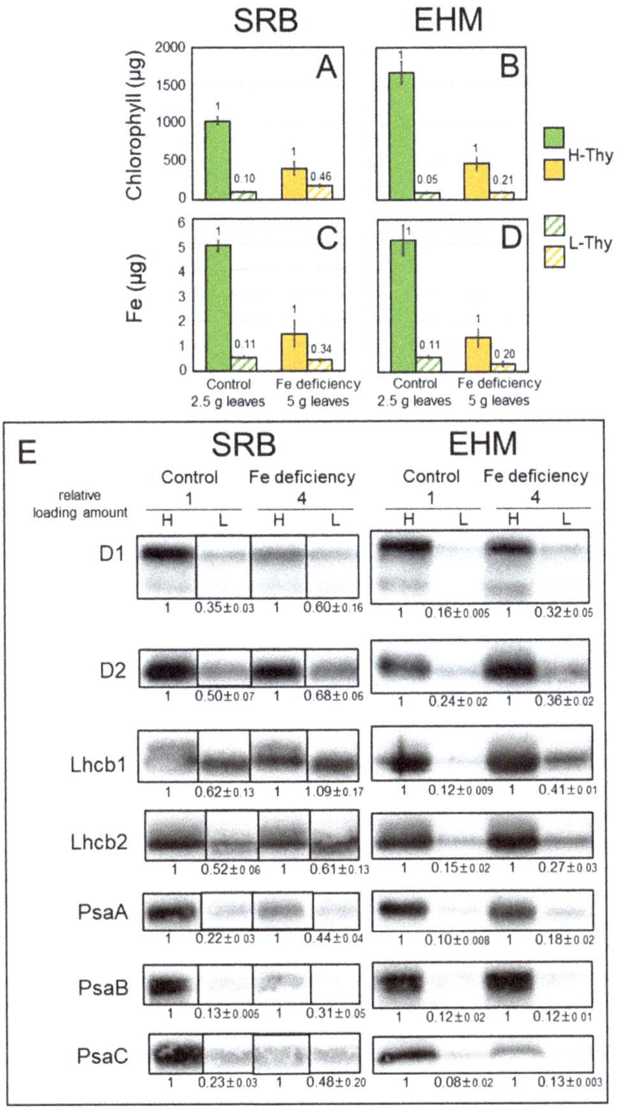

Figure 5. Chlorophyll, Fe, and proteins present on high- and low-density thylakoids from SRB1 and EHM1. (**A**,**C**) SRB1 and (**B**,**D**) EHM1. (**A**,**B**) Chlorophyll content and (**C**,**D**) Fe contents derived from 2.5 g or 5 g of control or Fe-deficient leaves, respectively. Solid bars and hatched bars represent H-Thy and L-Thy, respectively. Values represent the mean ± SE of three independent fractions. Small counts on bars indicate the ratios of L-Thy to H-Thy. (**E**) Western blot analysis. Loading amounts: 1/2000 of the fraction for the control sample and 1/500 of the fraction for the Fe-deficient sample, except anti-PsaC. For anti-PsaC, 1/1000 or 1/250 fractions were loaded for the control sample or Fe-deficient sample, respectively. Higher amounts of Fe-deficient materials than control materials were used to obtain a clear signal. Lanes of images from SRB1 were rearranged to match the order of those from EHM1. Corresponding CBB stain images are shown in Figure S8C, and original blot images of three replicates are shown in Figure S8A. Relative signal intensities of L-Thy to corresponding H-Thy were calculated and average and SE (n = 3) were shown.

3. Discussion

3.1. Electron Transfer around PSI under Fe-Deficient Conditions Is Better in Sarab1 than in Other Cultivars

We previously reported that the ability of Fe deficiency-tolerant cultivars of barley to increase PIUE is related to the optimization of the electron flow downstream of PSII, including cyt b_6f and PSI [6]. Consistent with this report, we found that the two Fe deficiency-tolerant cultivars, SRB1 and EHM1, both maintained higher Pm values (Figure 1D) than did the Fe deficiency-sensitive variety under Fe-deficient conditions. However, the mechanism of maintaining PSI function appeared to differ between the two cultivars, with SRB1 adopting a strategy dependent on the maintenance of functionality rather than the accumulation of PSI proteins, whereas EHM1 maintains more PSI protein complexes with lower functionality than SRB1.

The strategy of SRB1 for maintaining functional PSI is a P700 oxidation-inducing system [19,24] that increases $P700^+$ (Figure 1E,F). Two central mechanisms are known to be involved in the P700 oxidation induction. The first is suppression of the accumulation of reduced P700* by decreasing the electrons transferred to P700 by suppressing the PSI donor [24,25]. This mechanism includes NPQ induction in PSII, ΔpH in the thylakoid membrane lumen, and functional inhibition of the PQ pool [20]. In the present study, 1-qL, which indicates the reduced state of the PQ pool, was relatively higher in SRB1 than in other cultivars, both in Fe-sufficient and Fe-deficient plants (Figure S2A), suggesting that the PQ pool in SRB1 was likely reduced by an increase in the ΔpH prior to exposure to Fe deficiency. Such highly reduced states of the PQ pool could activate the xanthophyll cycle to increase zeaxanthin, an efficient heat-dissipating pigment [26] and STN7 kinase to enhance LHCII protein phosphorylation [5].

If LHCII phosphorylation was enhanced in SRB1, the conformation of the grana would be loosened, which could lead to a decrease in grana and an increase in unstacked stroma lamellar structures seen in the TEM images (Figure 3). Such a decrease in grana and increase in stroma lamella is also found under Fe-deficient EHM1 [18], suggesting that SRB1 can efficiently induce NPQ (Figure 1B) by retaining more mobile LHCII both under Fe-sufficient and -deficient conditions. Thus, we need to determine whether LHCII phosphorylation is more pronounced in SRB1 than in EHM1.

Another way to induce P700 oxidation is to promote electron transfer on the PSI acceptor side. The Calvin–Benson cycle, photorespiration, cyclic electron transfer flow (CEF), and the Water–Water cycle associated with ROS scavenging are essential for this in higher plants [19,24,25,27]. Our study did not investigate whether there are differences in these downstream PSI functions among cultivars. However, since ferredoxin (Fd) is depleted downstream of PSI in an Fe-deficient environment (Figure 2A), it is questionable whether Fd-mediated photorespiration and CEF are strongly induced in SRB1. The ROS scavenging pathways and the robustness and repair kinetics of PSI in SRB1 also warrant further analysis.

Despite MSS increasing the Y(NPQ) and Y(ND) under Fe deficiency similarity to SRB1, this cultivar showed the lowest Fe deficiency tolerance among barley varieties [6] and had a high Y(NA) value (Figure 1F), suggesting pronounced PSI photoinhibition during Fe deficiency. The reason PSI photoinhibition is high even though MSS could induce P700 oxidation is possibly due to its low Fe-usage efficiency within chloroplasts; MSS has the lowest photosynthetic Fe-use efficiency [6]. In fact, even under Fe-sufficient conditions, the Pm value (Figure 1D) and the accumulation of the Fe-binding protein PsaC was remarkably low in MSS (Figure 2A) even though its leaf-Fe content was higher than that of other cultivars (Figure S1). These results suggest that much Fe in MSS is not used to build PSI because the 4Fe-4S clusters are essential for de novo PSI assembly, but may be used for assembling other Fe-containing proteins, deposited in the tissue, or in a chemical form that cannot be recycled. Thus, the low efficiency of Fe supply to PSI in MSS would perturb the PSI maintenance even if the P700 oxidative system worked under Fe-deficient conditions.

Interestingly, SRB1 greatly decreased D1 and D2 accumulations than did EHM1 under Fe-deficient conditions, although the function of PSII of SRB1 was comparable to that of other cultivars (Figure 1). We calculated the ratio of signal intensity between Fe-deficient and control samples (Figures 2B, 5E and S9). The loading amount in Figure 2B was normalized by protein amount; Figure 5E by proportions of the fractions, and Figure S9 by chlorophyll amount. The results from the Western blot analysis showed that SRB1 rather decreased the accumulation of PSII core proteins. SRB1 could adopt a strategy of reserving small but functional amounts of reaction centers in the case of PSII and organizing 'economical' photosystems.

3.2. Sarab1 May Have a Characteristic Region of Thylakoid Membrane Supporting Smooth Electron Transfer under Fe-Deficient Conditions

The recovery rates of total thylakoid membranes based on chlorophyll from SRB1 leaves tended to be lower than that from EHM1 leaves regardless of Fe nutrition status, with a significant result observed in the case of Fe-deficient leaves (Figure S7A, $p = 0.16$). Additionally, the rate of L-Thy to H-Thy of SRB1 was higher than that of EHM1 (Figure 5A,B). These findings do not contradict the observations that SRB1 leaves had more stromal thylakoid membranes, which may be fragile during the extraction procedure compared with EHM1 leaves using TEM (Figure 3). The structures and numbers of grana stack and stromal thylakoid membranes change in response to short- or long-term changes in light conditions, and these structural dynamics contribute to recovery from photoinhibition [28]. Nozue et al. reported that isolated grana, which were connected to a few stroma lamellae, exhibited a slower PSII repair than did stroma–grana structures [23]. Since the ability of photosystems to convert light energy into chemical energy decrease under Fe deficiency, Fe deficiency stress may result in a similar response to that of excess light conditions for chloroplasts, even under growth light conditions. Therefore, the higher rate of stromal thylakoid membranes in SRB1 chloroplasts may support the maintenance of photochemical reactions under Fe-deficient conditions. Functional differences between stromal membranes directly connected to granal membranes and stroma lamellae not directly connected to grana stacks were assumed [22]. Since we prepared segments for TEM using chemical fixation in this work, it was difficult to obtain a clear image of the whole chloroplast and to estimate a rate of two distinct stromal thylakoid regions using sufficient replicates. Quantitative comparison of the rate of grana stack, stromal thylakoid connected to grana, and not connected stroma lamellae among barley species adopting high-pressure freezing-freeze substituted fixation [29] may identify the advantageous structures of the thylakoid membrane under Fe-deficient conditions in the future.

The composition of H-Thy, which may correspond to well-characterized thylakoid membranes from SRB1 was highly similar to that from EHM1 based on the results from SDG centrifugation (Figure 4). Overall, the composition of L-Thy was similar to that of H-Thy, even though fr.7 corresponding to the PSI-LHCI fraction of L-Thy was reduced in EHM1 (Figure 4). Components of L-Thy in this work demonstrated that supernatants, which have not been previously analyzed, may contain functional thylakoid membranes. However, the dispositions of Fe and proteins to H-Thy and L-Thy were different between two Fe deficiency-tolerant cultivars. Almost the same amount of Fe remained in H- and L-Thy from SRB1 and EHM1 when barley plants were suffering from Fe deficiency (Figure 5C,D orange and orange hatched bars), while the accumulation of PSI reaction-center proteins on SRB1 thylakoid membranes was more drastically decreased by Fe deficiency than on EHM1 thylakoid membranes (Figures 2D and 5E). Moreover, the rate of the distribution of Fe to L-thy was increased in SRB1 when compared to that in EHM1 (Figure 5C,D orange hatched bars). These data suggest that SRB1 thylakoid membranes contribute relatively larger amounts of Fe to the amounts of reaction-center proteins when compared to EHM1. However, the results from SDG centrifugation did not show the larger proportions of Fe in the fractions of PSI reaction-center proteins from SRB1 compared with those of EHM1 (Figure 4). Meanwhile, larger proportions of Fe were observed in the LHCII fractions of

SRB1 than in those of EHM1 when grown under Fe-sufficient conditions (Figures 4 and S6). We identified limited molecular species from each fraction using Western blot analysis in this work; thus, chemical species of Fe in the LHCII fraction are unknown. CBB staining of SDS-PAGE gels (Figures S5 and S8C) showed differences in the protein composition of L-Thy between SRB1 and EHM1. The changes in thylakoid membrane structure discussed in 3.1, which also affect the arrangement and complex structures of PSII and PSI within the thylakoid membrane, may well explain the differences between SRB1 and EHM1 seen in fractionated thylakoid membrane samples. Thus, analyses of both proteins and Fe in intact protein complexes obtained from native PAGE are necessary in the future.

The results from SDG centrifugation suggest that the abundance of Lhcb1 in LHCII could be higher in SRB1 than in EHM1 (Figures 4 and S5). This feature of SRB1 could be linked to a higher ratio of stromal/granal thylakoid (Figure 3). Furthermore, the rate of the distribution of Lhcb1 to L-Thy was largely increased by Fe deficiency both in SRB1 and EHM1 (Figure 5E). This data may correspond to the migration of Lhcb1 [18] and NPQ induction [6] under Fe-deficient conditions. Those Lhcb1s with high mobility in the unstacked thylakoid membranes could decrease the electron transfer from PSII in SRB1, probably forming some efficient energy quenchers around PSI, as we have suggested previously [5,17]. Indeed, the highest induction of NPQ was found in Fe-deficient SRB1 among barley cultivars (Figure 1B). Although NPQ is most often related to PSII photoprotection, NPQ also protects PSI, directly or indirectly through quenching part of the LHCII antenna pool functionally associated with PSI [30]. It is reasonable to assume that the Lhcb1-mediated NPQ induction in L-Thy of SRB1 resulted in higher 1-qL and Y(I)/Y(II) ratio (Figure S2) than in other cultivars, as well as in the strong induction of Y(ND) (Figure 1E) by Fe deficiency, leading to higher P700 oxidation levels in the Fe-deficient SRB1. Based on our data, we conclude that high P700 oxidation of Fe-deficient SRB1 could avoid the photoinhibition of PSI, allowing SRB1 to maintain PSI function even at low PSI content under prolonged Fe-deficient conditions.

Overall, L-Thy from Fe-deficient SRB1 leaves seems to have relatively high amounts of Fe, LHCII, PSII, and PSI proteins. These features are consistent with larger proportions of Fe in LHCII fractions from SDG fractionation. Details of associated proteins with the LHCII complex of SRB1 should be elucidated in the future. Based on the distribution of Fe and proteins on the thylakoid membrane described above, it is possible that SRB1 stocks some low-molecular-weight Fe polypeptide complexes fractionated to sucrose density with LHCII in L-thy, even when grown under Fe-sufficient conditions. Although many Fe proteins exist on thylakoid membranes, one of the candidate Fe proteins is PGR5, which belongs to the ferritin-like protein superfamily with Fe as a cofactor. Its primary role is CEF from PSI to the PQ pool, but its relevance to Fe partitioning on thylakoid membranes has also been discussed recently [31]. The speculation that such Fe-containing polypeptides might be important Fe reservoirs during Fe deficiency in low-density thylakoid membranes, such as stromal lamellae where PSI is localized, would be worth testing in the context of these recent discussions.

4. Materials and Methods

4.1. Plant Materials

Barley *Hordeum vulgare* L. cultivars and growth conditions were adopted for this study according to Saito et al. [6]. 'Ehime Hadaka 1' (EHM1), 'Ethiopia 2' (ETH2), 'Musashino-mugi' (MSS), 'Sarab 1' (SRB1) were kindly provided by Professor Kazuhiro Satoh (Barley Germplasm Center, Okayama University, Japan). Seedlings were grown hydroponically in a growth chamber at 24/20 °C. The growth light intensity was set at 150–200 µmol photons m^{-2} s^{-1} under 14/10 h light/dark cycles. The control nutrient solution had 30 µM Fe-EDTA, and low Fe nutrient solutions were supplemented with 0.3–3 µM depending on the demand for Fe of each cultivar since a sufficient amount of thylakoid membrane cannot be obtained from severe chlorosis leaves. We used chlorotic leaves with a SPAD value (index of total chlorophyll content in a leaf area) of 15–20 as Fe-deficient leaves. Intact plants, fresh leaves, or leaves immediately frozen and stored at −80 °C were used for further experiments.

4.2. Measurement of Chlorophyll Fluorescence and P700 Redox State

Chlorophyll fluorescence and P700 redox state were measured simultaneously using a DUAL-PAM-100 (Heinz Walz GmbH, Effeltrich, Germany) in the young fully expanded leaves of 16 to 20-day-old plants as previously described [6]. The minimal fluorescence in the dark-adapted state (Fo) was recorded after the illumination of a weak measuring light (620 nm from the Dual-DR measurement heads) at a photon flux density of 5 µmol photon m^{-2} s^{-1}. A saturating pulse (SP) light (300 ms, 14,000 µmol photon m^{-2} s^{-1}) was applied to determine the maximal fluorescence in the dark-adapted state (Fm). The actinic light intensity increased in a stepwise manner (0, 6, 14, 32, 90, 168, 210, 326, 497, 755, 1174 µmol photons m^{-2} s^{-1}), and the actinic light condition at 210 µmol photons m^{-2} s^{-1} was used for the main analysis as growth light conditions. The maximal and minimal fluorescence in the light-adapted state (Fm′ and Fo′) and steady-state chlorophyll fluorescence (Fs) were recorded during the exposure to the actinic light illumination. Prior to measuring Fm′ and Fo′, a saturating pulse light and far-red light (720 nm) were applied, respectively. The maximal quantum yield of PSII and NPQ were calculated as Fv/Fm and (Fm − Fm′)/Fm′, respectively. The index for the reduction in the primary PQ electron acceptor (QA) (1−qL) was calculated as 1 − (Fm′ − Fs)/(Fm′ − Fo′) × (Fo′/Fs). Y(II), Y(NO), and Y(NPQ) were calculated as (Fm′ − Fs)/Fm′, 1/[NPQ + 1 + qL(Fm/Fo − 1)], and 1 − Y(II) − 1/[NPQ + 1 + qL(Fm/Fo − 1)], respectively.

Simultaneous with the chlorophyll fluorescence measurements, the redox change of P700 was assessed by monitoring the changes in the absorbance of transmission light at 830 and 875 nm, according to Klughammer and Schreiber [32]. The maximal P700 signal (Pm) was determined by applying a saturated pulse light in the presence of far-red light (720 nm), while that of the oxidized P700 during actinic light illumination (Pm′) was determined by the saturated pulse-light application. The P700 signal during actinic light illumination (P) was recorded just prior to the saturated pulse light application. Y(I), Y(NA), and Y(ND) were calculated as (Pm′ − P)/Pm, (Pm − Pm′)/Pm, and P/Pm, respectively.

To determine the functional PSI from the value of Pm, the absolute Pm value, which is affected by both PSI content and PSI function in leaves, was divided by the value of the relative amounts of PSI core proteins, PsaA, PsaB, and PsaC, respectively. The obtained Pm/PsaA, Pm/PsaB, and Pm/PsaC values are the 'Normalized Pm', which is correlated to absolute functional PSI. Finally, the ratio ('Normalized Pm' of Fe-deficient leaves)/('Normalized Pm' of Fe-sufficient leaves) was calculated as the functional PSI ratio under Fe-deficient conditions shown in Figures 2D and S3D.

4.3. Transmission Electron Microscopy (TEM) and Measurement of Granal and Stromal Thylakoids

We adopted and modified the methods described in Saito et al. [18]. Barley leaves were cut with a razor to squares of lengths less than 2 mm, and fixing solution (2% paraformaldehyde, 2% glutaraldehyde in 0.05 M cacodylate buffer pH 7.4 at 4 °C) was rapidly added to the leaves for absorption in a sealed syringe by pulling the piston to reduce pressure. After incubation in the fixing solution at 4 °C overnight, the samples were washed three times with 0.05 M cacodylate buffer for 30 min each. Samples were postfixed with 2% osmium tetroxide (OsO_4) in 0.05 M cacodylate buffer at 4 °C for 3 h. After dehydration with ethanol series and infiltration with propylene oxide, samples were embedded in resin Quetol-651 (Nisshin EM Co., Tokyo, Japan) and polymerized. The polymerized resins were ultra-thin sectioned to 80 nm using an ultramicrotome Ultracut UCT (Leica, Vienna, Austria), and sections were stained with 2% uranyl acetate followed by secondary staining with a lead stain solution (Sigma-Aldrich Co., St. Louis, MO, USA). The grids were observed using a JEM-1400 plus transmission electron microscope (JEOL Ltd., Tokyo, Japan) at an acceleration voltage of 80 or 100 kV. Images were taken with a CCD camera VELETA (Olympus Soft Imaging Solutions GmbH, Münster, Germany) or EM-14830RUBY2 (JEOL Ltd.).

Granal and stromal thylakoid membranes were estimated using NIH ImageJ 1.53t software (https://imagej.nih.gov/ij/, accessed on 24 August 2022). We sampled four independent chloroplast images for each cultivar and selected all grana stacks with lamellar membranes clearly recognized in each image. Each electron-dense layer in grana stacks was traced by two lines as granal thylakoid membranes. Thylakoid membranes connected to selected grana stacks were traced as stromal thylakoid membranes. Summations of line lengths belonging to grana or stroma were calculated for each image, then stroma/grana ratios were presented.

4.4. *Thylakoid Membrane Extraction*

Fresh or frozen leaf pieces were homogenized with an isotonic solution and filtered crude thylakoid samples were corrected as pellets by centrifugation at 2500× *g* at 4 °C for 10 min, as described by Saito et al. [6]. We designated green pellets suspended in the buffer obtained by 2500× *g* centrifugation as "heavy/high-density thylakoids (H-Thy)". The supernatants after centrifugation containing considerable amounts of chlorophyll were subjected to further centrifugation at 10,000× *g* at 4 °C for 10 min, and these green pellets were designated as "light/low-density thylakoids (L-Thy)". Precipitates were resuspended in HM buffer [6] and stored at −80 °C as H-Thy and L-Thy fractions.

4.5. *Sucrose Density Gradient (SDG) Centrifugation*

Discontinuous SDG tubes were prepared by sequential layering of 1.3 M, 1.0 M, 0.7 M, 0.4 M, and 0.1 M sucrose (from bottom to top) with 0.05% (*w*/*v*) of β-DM. Thylakoid membrane samples as 0.4 mg/mL of chlorophyll were solubilized with β-DM (4.8 mg β-DM for 0.1 mg chlorophyll) at 4 °C in the dark for 2 min. The solubilized sample was loaded on the top of the SDG tube and centrifuged at 280,000× g for 20 h using a P40ST rotor (Eppendorf Himac Technologies Co., Ltd., Ibaraki, Japan) at 4 °C.

4.6. *Measurement of Chlorophyll, Proteins, and Fe*

Chlorophyll in thylakoid membranes was quantified after resuspending aliquots of samples in 80% (*v*/*v*) acetone [33]. The amounts of proteins in each sample were quantified using the BCA (bicinchoninic acid) method with Protein Assay Standard I (Bio-Rad Laboratories, Inc, Hercules, CA, USA). Dried leaves were digested in concentrated HNO_3 at 150 °C and dissolved in 1% (*v*/*v*) HNO_3. Thylakoid fractions were digested using extremely clean reagents, instruments, and atmosphere according to Saito et al. [6], due to trace amounts of Fe in thylakoid membranes. The Fe concentration was measured using an atomic absorption spectrophotometer (AA-6300, Shimadzu, Tokyo, Japan) coupled with a graphite furnace atomizer (GFA-EX7i, Shimadzu, Tokyo, Japan).

4.7. *Western Blot Analysis*

Proteins of the thylakoid membranes were solubilized in protein-solubilizing buffer and subjected to SDS–PAGE. Immunoblot analysis was performed as described previously [17]. Anti-PsbA/D1 (AS05-084), anti-PsbB/D2 (AS06-146), anti-PsbE/cyt b_{559} (AS06-112), anti-Lhcb2 (AS01-003), anti-PsaA (AS05-084), anti-PsaB (AS10-695), anti-PsaC (AS10-939), and anti-FDX1/ferredoxin (AS06-121) antibodies were obtained from Agrisera (Vännäs, Sweden). Anti-HvLhcb1 antibody was raised against the peptide derived from HvLhcb1 by Eurofins Genomics (Tokyo, Japan). Signal intensities were quantified using NIH ImageJ 1.53t software with plug-in 'BandPeakQuantification' [34].

5. Conclusions

We found that the Fe deficiency-tolerant barley cultivar SRB1 has a characteristic region on the thylakoid membrane that contains a considerable amount of Fe and specific composition of photosystem proteins when compared to the corresponding region in EHM1. Such a region on the thylakoid membrane may contribute to effective electron transfer utilizing only a small amount of Fe and reaction-center proteins under Fe-deficient conditions. The survey of the composition of photosystem proteins among various barley cultivars and QTL analysis are ongoing. The advantages of such an organization of the thylakoid membrane of SRB1 under other stress conditions will be of interest to understand the acclimation of photosynthetic apparatus with high plasticity.

Supplementary Materials: The following supporting information can be downloaded at: https://www.mdpi.com/article/10.3390/plants12112111/s1, Figure S1: Comparison of chlorophylls and Fe content of leaves of the cultivars used under Fe-sufficient and Fe-deficient conditions; Figure S2: The property of SRB1 associated with the reduction in the plastoquinone pool to a relatively high degree, a mechanism that allows photosystem I to avoid an over-reduced state; Figure S3: Data corresponding to Figure 2 when the same amount of chlorophyll were applied to each lane; Figure S4: Original TEM images and ROIs associated with Figure 3; Figure S5: CBB stain of gels applied with the same fractions in Figure 4; Figure S6: Analyses of fractions obtained from SDG using the thylakoid membranes derived from frozen leaves; Figure S7: Supplementary data for Figure 5; Figure S8: Original images of Western blot analysis and CBB stain images corresponding to Figure 5; Figure S9: Comparison of PSII core protein content among cultivars.

Author Contributions: A.S. and K.H. (Kyoko Higuchi) conceived the project, designed and performed the experiments, and prepared the manuscript. T.O. supervised experiments and prepared the manuscript. K.H. (Kimika Hoshi), Y.W., T.T., T.S. and M.K. performed the experiments. All authors have read and agreed to the published version of the manuscript.

Funding: This research was funded by JSPS KAKENHI (grant number JP20H02891).

Institutional Review Board Statement: Not applicable.

Informed Consent Statement: Not applicable.

Data Availability Statement: Not applicable.

Acknowledgments: The authors thank Professor Kazuhiro Satoh (Barley Germplasm Center, Okayama University, Japan) for kindly providing barley seeds.

Conflicts of Interest: The authors declare no conflict of interest.

References

1. Pushnik, J.C.; Miller, G.W. Iron regulation of chloroplast photosynthetic function: Mediation of PS I development. *J. Plant Nutr.* **1989**, *12*, 407–421. [CrossRef]
2. Fodor, F.; Böddi, B.; Sárvári, É.; Záray, G.; Cseh, E.; Ferenc, L. Correlation of Iron Content, Spectral Forms of chlorophyll and chlorophyll-proteins in Iron Deficient Cucumber (*Cucumis sativus*). *Physiol. Plant.* **1995**, *93*, 750–756. [CrossRef]
3. Raven, J.A.; Evans, M.C.W.; Korb, R.E. The role of trace metals in photosynthetic electron transport in O_2-evolving organisms. *Photosynth. Res.* **1999**, *60*, 111–150. [CrossRef]
4. Schmidt, S.B.; Eisenhut, M.; Schneider, A. Chloroplast transition metal regulation for efficient photosynthesis. *Trends Plant Sci.* **2020**, *25*, 817–828. [CrossRef]
5. Higuchi, K.; Saito, A. Elucidation of efficient photosynthesis in plants with limited iron. *Soil Sci. Plant Nutr.* **2022**, *68*, 505–513. [CrossRef]
6. Saito, A.; Shinjo, S.; Ito, D.; Doi, Y.; Sato, A.; Wakabayashi, Y.; Honda, J.; Arai, Y.; Maeda, T.; Ohyama, T.; et al. Enhancement of photosynthetic iron-use efficiency is an important trait of *Hordeum vulgare* for adaptation of photosystems to iron deficiency. *Plants* **2021**, *10*, 234. [CrossRef]
7. Higuchi, K.; Kurita, K.; Sakai, T.; Suzui, N.; Sasaki, M.; Katori, M.; Wakabayashi, Y.; Majima, Y.; Saito, A.; Ohyama, T.; et al. 'Live-Autoradiography' technique reveals genetic variation in the rate of Fe uptake by barley cultivars. *Plants* **2022**, *11*, 817. [CrossRef]
8. Boekema, E.J.; Hifney, A.; Yakushevska, A.E.; Piotrowski, M.; Keegstra, W.; Berry, S.; Michel, K.P.; Pistorius, E.K.; Kruip, J. A giant chlorophyll-protein complex induced by iron deficiency in cyanobacteria. *Nature* **2001**, *412*, 745–748. [CrossRef] [PubMed]
9. Akita, F.; Nagao, R.; Kato, K.; Nakajima, Y.; Yokono, M.; Ueno, Y.; Suzuki, T.; Dohmae, N.; Shen, J.R.; Akimoto, S.; et al. Structure of a cyanobacterial photosystem I surrounded by octadecameric IsiA antenna proteins. *Commun. Biol.* **2020**, *3*, 232. [CrossRef] [PubMed]
10. Michel, K.P.; Pistorius, E.K. Adaptation of the photosynthetic electron transport chain in cyanobacteria to iron deficiency: The function of IdiA and IsiA. *Physiol. Plant.* **2004**, *120*, 36–50. [CrossRef] [PubMed]

11. Varsano, T.; Wolf, S.G.; Pick, U. A chlorophyll *a*/*b*-binding Protein Homolog That Is Induced by Iron Deficiency Is Associated with Enlarged photosystem I Units in the Eucaryotic Alga *Dunaliella salina*. *J. Biol. Chem.* **2006**, *281*, 10305–10315. [CrossRef] [PubMed]
12. Naumann, B.; Stauber, E.J.; Busch, A.; Sommer, F.; Hippler, M. N-terminal processing of Lhca3 Is a key step in remodeling of the photosystem I-light-harvesting complex under iron deficiency in *Chlamydomonas reinhardtii*. *J. Biol. Chem.* **2005**, *280*, 20431–20441. [CrossRef]
13. Naumann, B.; Busch, A.; Allmer, J.; Ostendorf, E.; Zeller, M.; Kirchhoff, H.; Hippler, M. Comparative quantitative proteomics to investigate the remodeling of bioenergetic pathways under iron deficiency in *Chlamydomonas reinhardtii*. *Proteomics* **2007**, *7*, 3964–3979. [CrossRef] [PubMed]
14. Devadasu, E.; Chinthapalli, D.K.; Chouhan, N.; Madireddi, S.K.; Rasineni, G.K.; Sripadi, P.; Subramanyam, R. Changes in the photosynthetic apparatus and lipid droplet formation in *Chlamydomonas reinhardtii* under iron deficiency. *Photosynth. Res.* **2019**, *139*, 253–266. [CrossRef] [PubMed]
15. Moseley, J.L.; Allinger, T.; Herzog, S.; Hoerth, P.; Wehinger, E.; Merchant, S.; Hippler, M. Adaptation to Fe-deficiency requires remodeling of the photosynthetic apparatus. *EMBO J.* **2002**, *21*, 6709–6720. [CrossRef] [PubMed]
16. Page, M.D.; Allen, M.D.; Kropat, J.; Urzica, E.I.; Karpowicz, S.J.; Hsieh, S.I.; Loo, J.A.; Merchant, S.S. Fe sparing and Fe recycling contribute to increased superoxide dismutase capacity in iron-starved *Chlamydomonas reinhardtii*. *Plant Cell* **2012**, *24*, 2649–2665. [CrossRef]
17. Saito, A.; Iino, T.; Sonoike, K.; Miwa, E.; Higuchi, K. Remodeling of the major light-harvesting antenna protein of PSII protects the young leaves of barley (*Hordeum vulgare* L.) from photoinhibition under prolonged iron deficiency. *Plant Cell Physiol.* **2010**, *51*, 2013–2030. [CrossRef]
18. Saito, A.; Shimizu, M.; Nakamura, H.; Maeno, S.; Katase, R.; Miwa, E.; Higuchi, K.; Sonoike, K. Fe deficiency induces phosphorylation and translocation of Lhcb1 in barley thylakoid membranes. *FEBS Lett.* **2014**, *588*, 2042–2048. [CrossRef]
19. Miyake, C. Molecular mechanism of oxidation of P700 and suppression of ROS Production in photosystem I in response to electron-sink limitations in C3 plants. *Antioxidants* **2020**, *9*, 230. [CrossRef]
20. Kaiser, E.; Walther, D.; Armbruster, U. Photorespiration enhances acidification of the thylakoid lumen, reduces the plastoquinone pool, and contributes to the oxidation of P700 at a lower partial pressure of CO_2 in wheat leaves. *Plants* **2020**, *9*, 319. [CrossRef]
21. Lempiäinen, T.; Rintamäki, E.; Aro, E.M.; Tikkanen, M. Plants acclimate to photosystem I photoinhibition by readjusting the photosynthetic machinery. *Plant Cell Environ.* **2022**, *45*, 2954–2971. [CrossRef] [PubMed]
22. Anderson, J.M.; Horton, P.; Kim, E.H.; Chow, W.S. Towards elucidation of dynamic structural changes of plant thylakoid architecture. *Philos. Trans. R. Soc. Lond. B Biol. Sci.* **2012**, *367*, 3515–3524. [CrossRef] [PubMed]
23. Nozue, H.; Shigarami, T.; Fukuda, S.; Chino, T.; Saruta, R.; Shirai, K.; Nozue, M.; Kumazaki, S. Growth-phase dependent morphological alteration in higher plant thylakoid is accompanied by changes in both photodamage and repair rates. *Physiol. Plant.* **2021**, *172*, 1983–1996. [CrossRef] [PubMed]
24. Shimakawa, G.; Miyake, C. Oxidation of P700 ensures robust photosynthesis. *Front. Plant Sci.* **2018**, *9*, 1617. [CrossRef]
25. Furutani, R.; Ifuku, K.; Suzuki, Y.; Noguchi, K.; Shimakawa, G.; Wada, S.; Makino, A.; Sohtome, T.; Miyake, C. P700 oxidation suppresses the production of reactive oxygen species in photosystem I. *Adv. Bot. Res.* **2020**, *96*, 151–176. [CrossRef]
26. Ruban, A.V.; Wilson, S. The mechanism of non-photochemical quenching in plants: Localization and driving forces. *Plant Cell Physiol.* **2021**, *62*, 1063–1072. [CrossRef]
27. Asada, K. The water-water cycle in chloroplasts: Scavenging of active oxygens and dissipation of excess photons. *Annu. Rev. Plant Physiol. Plant Mol. Biol.* **1999**, *50*, 601–639. [CrossRef]
28. Kowalewska, Ł.; Bykowski, M.; Mostowska, A. Spatial organization of thylakoid network in higher plants. *Bot. Lett.* **2019**, *166*, 326–343. [CrossRef]
29. Li, M.; Mukhopadhyay, R.; Svoboda, V.; Oung, H.M.O.; Mullendore, D.L.; Kirchhoff, H. Measuring the dynamic response of the thylakoid architecture in plant leaves by electron microscopy. *Plant Direct* **2020**, *4*, e00280. [CrossRef]
30. Lima-Melo, Y.; Kılıç, M.; Aro, E.M.; Gollan, P.J. Photosystem I inhibition, protection and signalling: Knowns and unknowns. *Front. Plant Sci.* **2021**, *12*, 791124. [CrossRef]
31. Leister, D.; Marino, G.; Minagawa, J.; Dann, M. An ancient function of PGR5 in iron delivery? *Trends Plant Sci.* **2022**, *27*, 971–980. [CrossRef] [PubMed]
32. Klughammer, C.; Schreiber, U. Complementary PS II quantum yields calculated from simple fluorescence parameters measured by PAM fluorometry and the Saturation Pulse method. *PAM Appl. Notes* **2008**, *1*, 27–35.
33. Porra, R.J.; Thompson, W.A.; Kriedemann, P.E. Determination of accurate extinction coefficients and simultaneous-equations for assaying chlorophyll-a and chlorophyll-B extracted with 4 different solvents—Verification of the concentration of chlorophyll standards by atomic-absorption spectroscopy. *Biochim. Biophys. Acta* **1989**, *975*, 384–394. [CrossRef]
34. Ohgane, K.; Yoshioka, H. *Quantification of Gel Bands by an ImageJ Macro, Band/Peak Quantification Tool*; Protocols.io.: Berkeley, CA, USA, 2019. [CrossRef]

plants

MDPI

Article

Growth Developmental Defects of Mitochondrial Iron Transporter 1 and 2 Mutants in Arabidopsis in Iron Sufficient Conditions

Joaquín Vargas [1], Isabel Gómez [1], Elena A. Vidal [2,3,4], Chun Pong Lee [5], A. Harvey Millar [5], Xavier Jordana [1] and Hannetz Roschzttardtz [1,*]

1 Departamento de Genética Molecular y Microbiología, Facultad de Ciencias Biológicas, Pontificia Universidad Católica de Chile, Santiago 8331150, Chile
2 ANID-Millennium Science Initiative Program-Millennium Institute for Integrative Biology (iBio), Santiago 8331150, Chile
3 Centro de Genómica y Bioinformática, Facultad de Ciencias, Ingeniería y Tecnología, Universidad Mayor, Santiago 8580745, Chile
4 Escuela de Biotecnología, Facultad de Ciencias, Ingeniería y Tecnología, Universidad Mayor, Santiago 8580745, Chile
5 ARC Centre of Excellence in Plant Energy Biology, School of Molecular Sciences, The University of Western Australia, Bayliss Building M316, Crawley, WA 6009, Australia
* Correspondence: hroschzttardtz@bio.puc.cl; Tel.: +56-223-542-669

Abstract: Iron is the most abundant micronutrient in plant mitochondria, and it has a crucial role in biochemical reactions involving electron transfer. It has been described in *Oryza sativa* that *Mitochondrial Iron Transporter* (*MIT*) is an essential gene and that knockdown mutant rice plants have a decreased amount of iron in their mitochondria, strongly suggesting that OsMIT is involved in mitochondrial iron uptake. In *Arabidopsis thaliana*, two genes encode MIT homologues. In this study, we analyzed different *AtMIT1* and *AtMIT2* mutant alleles, and no phenotypic defects were observed in individual mutant plants grown in normal conditions, confirming that neither *AtMIT1* nor *AtMIT2* are individually essential. When we generated crosses between the *Atmit1* and *Atmit2* alleles, we were able to isolate homozygous double mutant plants. Interestingly, homozygous double mutant plants were obtained only when mutant alleles of *Atmit2* with the T-DNA insertion in the intron region were used for crossings, and in these cases, a correctly spliced *AtMIT2* mRNA was generated, although at a low level. *Atmit1 Atmit2* double homozygous mutant plants, knockout for *AtMIT1* and knockdown for *AtMIT2*, were grown and characterized in iron-sufficient conditions. Pleiotropic developmental defects were observed, including abnormal seeds, an increased number of cotyledons, a slow growth rate, pinoid stems, defects in flower structures, and reduced seed set. A RNA-Seq study was performed, and we could identify more than 760 genes differentially expressed in *Atmit1 Atmit2*. Our results show that *Atmit1 Atmit2* double homozygous mutant plants misregulate genes involved in iron transport, coumarin metabolism, hormone metabolism, root development, and stress-related response. The phenotypes observed, such as pinoid stems and fused cotyledons, in *Atmit1 Atmit2* double homozygous mutant plants may suggest defects in auxin homeostasis. Unexpectedly, we observed a possible phenomenon of T-DNA suppression in the next generation of *Atmit1 Atmit2* double homozygous mutant plants, correlating with increased splicing of the *AtMIT2* intron containing the T-DNA and the suppression of the phenotypes observed in the first generation of the double mutant plants. In these plants with a suppressed phenotype, no differences were observed in the oxygen consumption rate of isolated mitochondria; however, the molecular analysis of gene expression markers, *AOX1a*, *UPOX*, and *MSM1*, for mitochondrial and oxidative stress showed that these plants express a degree of mitochondrial perturbation. Finally, we could establish by a targeted proteomic analysis that a protein level of 30% of MIT2, in the absence of MIT1, is enough for normal plant growth under iron-sufficient conditions.

Keywords: mitochondria; iron transporters; MIT; developmental defects; RNA-seq

Citation: Vargas, J.; Gómez, I.; Vidal, E.A.; Lee, C.P.; Millar, A.H.; Jordana, X.; Roschzttardtz, H. Growth Developmental Defects of Mitochondrial Iron Transporter 1 and 2 Mutants in Arabidopsis in Iron Sufficient Conditions. *Plants* **2023**, *12*, 1176. https://doi.org/10.3390/plants12051176

Academic Editor: Ferenc Fodor

Received: 3 January 2023
Revised: 25 January 2023
Accepted: 22 February 2023
Published: 4 March 2023

1. Introduction

Iron is an essential nutrient, and it is well known that it is an integral constituent of many metalloproteins, primarily as part of heme groups and iron-sulfur clusters. As such, iron is essential for oxygen transport, electron transfer (redox), and catalytic reactions [1]. The biological versatility of iron is based on its capacity to be coordinated by proteins and to act as an electron donor and acceptor. Thus, iron can readily convert between its two common oxidation states, Fe^{2+} and Fe^{3+}, by the loss or gain of one electron. However, iron is also potentially toxic due to its redox reactivity. Indeed, free iron acts as a catalyst for oxidative stress via Fenton reactions, which yield hazardous radicals with the capacity to attack cellular macromolecules and cause tissue damage. Consequently, a tight control of iron homeostasis is imperative to satisfy metabolic needs for iron and prevent the accumulation of toxic iron concentrations. Iron homeostasis involves all the processes that regulate the balance between iron uptake, its intracellular storage, and utilization [2].

In soil, Fe^{2+} undergoes spontaneous aerobic oxidation to Fe^{3+}, which is virtually insoluble at physiological pH. This makes the acquisition of iron by cells and organisms challenging, despite its high abundance. The mechanism of iron uptake in the roots of *Arabidopsis thaliana* is now well-described and involves an acidification reduction-transport mechanism [3]. Under iron deficiency, ferric chelates are solubilized by local rhizosphere acidification caused by the release of protons by the Arabidopsis Plasma Membrane H^+-ATPase2 (AHA2; [4]). Solubilized Fe^{3+} ions are then reduced to Fe^{2+} by the Reductase Ferric Reduction Oxidase2 (FRO2) [5] and, finally, transported into the cell by the Iron Transporter Iron Regulated Transporter1 (IRT1; [6,7]). The mechanisms governing the distribution of iron to specific organs, cells, and organelles are still very poorly understood.

In plants, in addition to its role in the mitochondrial electron transport chain, common to eukaryotes, iron is essential for chloroplast photosynthesis, as shown by the chlorosis of plants grown under iron-deficient conditions [8–10]. A total of twenty-two iron atoms are required per photosynthetic electron transport chain [11]. Iron import into the chloroplast was proposed to be performed by the Permease in Chloroplast1 (PIC1) localized in the inner envelope of this organelle [12]. *PIC1* knockout mutations result in dwarf plants with altered iron homeostasis. Before being transported into the chloroplast, iron is thought to be first reduced by the Ferric Reductase7 (AtFRO7), also localized in the chloroplast envelope [13]. Indeed, chloroplasts isolated from At*fro7* loss-of-function mutants have significantly reduced iron content and altered photosynthetic complexes [13]. Iron remobilization from leaf chloroplasts seems to be mediated by YSL4 and YSL6 [14]. Recently, it has been suggested that FPN3 has a role in the iron export from Arabidopsis chloroplasts and mitochondria [15].

In plant mitochondria, iron is more abundant than other transition metals such as Cu, Zn, and Mn, consistent with its crucial role as a component in electron transfer reactions [16]. It has been suggested that iron is transported to the mitochondria through the outer membrane by voltage-dependent anion protein channels (VDACs) and then to the mitochondrial matrix by the Mitochondrial Carrier Family (MCF) transporters. The MCF gene family, with more than fifty members in *Arabidopsis thaliana*, encodes membrane proteins containing six transmembrane domains [17]. Mitochondrial Iron Transporter (MIT), a member of the MCF in *Oryza sativa*, was the first mitochondrial iron transporter identified in plants [18]. Complementation studies using OsMIT demonstrated that it is able to transport iron into yeast mitochondria, and its function is essential in *Oryza sativa*. Knockdown plants for *OsMIT* showed a decrease in iron content in mitochondria and in aconitase activity, an iron-sulfur protein [18]. In *Arabidopsis thaliana*, it has been described that two genes encode MIT proteins [19]. Fusions with fluorescent proteins demonstrated that *AtMIT1* and *AtMIT2* localized to mitochondria, and plants knockout for *MIT1* (homozygous mutant) and knockdown for *MIT2* (heterozygous mutant) showed mitochondrial defects when plants were grown in iron deficiency conditions [19]. Both mitochondrial iron deficiency and excess seem to provoke oxidative stress in a mammalian model [20]. In plants, mitochondrial iron deficiency or excess also affects mitochondrial function [15,21], indicating that plant mitochondria have a crucial role in cellular metal homeostasis [22].

In this article, we characterize *Atmit1 Atmit2* double mutant plants (knockout for *MIT1* and knockdown for *MIT2*, using the same alleles used by Jain et al., 2019 [19]) grown in iron-sufficient conditions. These plants showed pleiotropic developmental defects, some of which strikingly resemble those found in auxin transport and sensing mutants. Transcriptomic data revealed a misregulation of genes involved in iron acquisition, synthesis of coumarins, formation of the Casparian strip, suberization, and root hair development. Furthermore, we demonstrate unambiguously, by crossing knockout mutants for MIT1 and MIT2, that MIT function is essential in Arabidopsis.

2. Results

Isolation of mutants in the two Arabidopsis genes encoding mitochondrial iron transporters (MIT).

In Arabidopsis, two genes (At1g07030 and At2g30160) encode proteins with high similarity to the rice Mitochondrial Iron Transporter (MIT) [18,23], and have recently been characterized [19]. The Arabidopsis MIT isoforms share 82% of their peptide sequence identity and similarity, including the putative mitochondrial targeting peptide. Both are 66–67% identical (77–78% similar) to the rice MIT protein (excluding the putative mitochondrial targeting peptides). To evaluate the potential role of the Arabidopsis proteins as mitochondrial iron transporters, At1g07030 (*MIT2*) and At2g30160 (*MIT1*) were used to transform the *MRS3-MRS4* knockout yeast (Δ*mrs3*Δ*mrs4*). *Saccharomyces cerevisiae* Mrs3 and Mrs4 are members of the MCF responsible for transporting Fe into mitochondria under low-Fe conditions, and the double knockout Δ*mrs3*Δ*mrs4* mutant grows poorly when Fe availability is low [24,25]. Each Arabidopsis MIT isoform was able to complement the growth defect of Δ*mrs3*Δ*mrs4* yeast cells (Supplementary Figure S1), indicating that they can act as mitochondrial iron carriers. Our results are largely in agreement with previous observations by [19], with the exception of the lack of evidence for a significant difference in the efficiency of complementation by MIT1 and MIT2 (Supplementary Figure S1).

To explore MIT1 and MIT2 function in Arabidopsis, we identified two and three T-DNA insertion mutant lines for *MIT1* and *MIT2*, respectively (Figure 1A, Supplementary Figures S2 and S3). The T-DNA insertion is located in exon 1 for both *mit1* mutants, causing an interruption in gene expression 200 bp and 53 bp downstream of the start codon in *mit1-1* and *mit1-2*, respectively. In *mit2-1* and *mit2-3* mutants, the T-DNA is located in the intron (419 and 500 bp downstream of the 5′ splice site), while *mit2-2* contains an insertion in the first exon (239 bp downstream of the start codon). We analyzed the progeny of selfed heterozygous *mit1-1*, *mit1-2*, and *mit2-1* plants and found that the progeny did not deviate significantly from 1:2:1 (wild type: heterozygous: homozygous). Furthermore, homozygous mutant plants for each of the *mit1* and *mit2* alleles did not show any phenotypic alteration when compared with wild type plants (data not shown).

Next, RT-PCR analysis of *MIT1* and *MIT2* expression was carried out to ascertain that the homozygous mutant plants obtained for all five mutants (Figure 1B) were truly null mutants (Figure 1C). Results show clearly that *mit1-1*, *mit1-2*, and *mit2-2* plants are knockout mutants. Unexpectedly, *mit2-1* and *mit2-3* accumulate *MIT2* transcript and thus are not knockout mutants. Sequencing of the two *MIT2* RT-PCR products obtained from *mit2-1* RNA demonstrated that the intron is correctly spliced (Supplementary Figure S3). However, the *MIT2* transcript level as determined by RT-qPCR is significantly decreased (Supplementary Figure S6), confirming that both *mit2-1* and likely *mit2-3* are knockdown mutants.

Given that *mit1-1*, *mit1-2*, and *mit2-2* are knockout mutants, their normal growth showed that neither *MIT1* nor *MIT2* are essential per se and that they may be redundant. These results led us to perform *mit1* x *mit2* crosses.

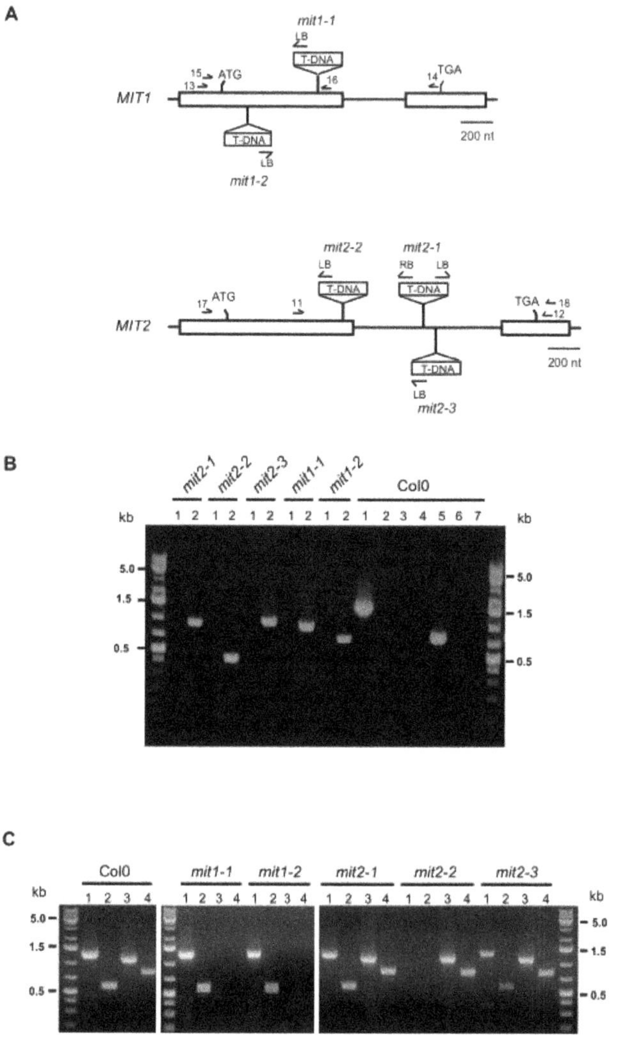

Figure 1. Identification of mit1 and mit2 individual mutants. (**A**) Exon-intron structure of *MIT* genes and T-DNA insertion sites. Exons are represented by boxes, and T-DNA insertion sites in the different mutant lines are shown (T-DNA inserts are not drawn to scale) (details in Supplementary Figures S2 and S3). Horizontal arrows indicate the positions of primers. (**B**) Genotyping of individual *mit1* and *mit2* plants showing they are homozygous mutants. Lanes 1: PCR of the wild type allele (primers 11 and 12 for *MIT2*, 15 and 16 for *MIT1*); lanes 2: amplification of mutant alleles (primers LBb1.3 and 12 for *mit2-1*, 11 and LB1sail for *mit2-2*, 11 and LBb1.3 for *mit2-3*, 15 and LBb1.3 for *mit1-1*, and LBb1.3 and 16 for *mit1-2*). PCR reactions with DNA from wild type Col0 plants were performed with all seven primer pairs (lanes 1 and 5: primers for wild type *MIT2* and *MIT1* alleles; lanes 2 to 4: primers for mutant *mit2-1*, *mit2-2*, and *mit2-3* alleles; lanes 6 and 7: primers for mutant *mit1-1* and *mit2-2* alleles). (**C**) *MIT1* and *MIT2* expression in individual WT, *mit1*, and *mit2* plants. cDNAs from homozygous *mit1-1*, *mit1-2*, *mit2-1*, *mit2-2*, *mit2-3* mutants, and WT plants were amplified (40 cycles) with two primer pairs for each (position indicated in Figure 1A). *MIT* transcript: primers 17 and 18 (lanes 1) or primers 11 and 12 (lanes 2) for *MIT2*, primers 13 and 14 (lanes 3) or primers 15 and 16 (lanes 4) for *MIT1*. Size standards correspond to the GeneRuler 1kb Plus DNA ladder from Thermo Scientific.

2.1. MIT Function Is Essential in Arabidopsis

Given that *mit1-1*, *mit1-2*, and *mit2-2* are knockout mutants, their normal growth showed that neither *MIT1* nor *MIT2* are essential per se. To determine whether MIT function is essential, we crossed the knockout mutants *mit1-1* and *mit2-2*. F2 seeds from selfed double heterozygous plants (*MIT1mit1-1 MIT2mit2-2*) were sown on 0.5X MS plates, and a plant carrying three mutated alleles was identified (*MIT1mit1-1 mit2-2mit2-2*). Visual inspection of F3 seeds in three siliques from this selfed plant showed that they contain 77.5 ± 7.0% normal seeds and 22.5 ± 7.0% aborted seeds. Furthermore, when F3 seeds were allowed to develop on soil under iron-sufficient conditions, we were unable to identify a plant carrying a double homozygous mutation (65 plants analyzed). Altogether, these results confirm that MIT function is essential, that the absence of MIT1 and MIT2 is embryo-lethal, and that *MIT1* and *MIT2* genes are redundant.

2.2. Crosses Using mit1-1 and mit2-1 Alleles Show Segregation Defects and Produce Abnormal Seeds

Next, we crossed homozygous *mit1-1* plants with homozygous *mit2-1* plants. One hundred and seventy-seven F2 plants grown from the seeds of three selfed F1 double heterozygous plants (genotype *MIT1mit1 MIT2mit2*) were genotyped (Supplementary Table S2). No double homozygous mutants were identified, and we noted a bias against plants homozygous for *mit1* and heterozygous for *mit2* (4 plants *mit1-1mit1-1 MIT2mit2-1*) that is apparent but not observed for plants heterozygous for *mit1* and homozygous for *mit2* (22 plants *mit1-1MIT1 mit2-1mit2-1*). This may be due to *mit1-1* being a knockout mutation and *mit2-1* being a knockdown mutation (see below). The plants carrying three mutated alleles did not show visible phenotypic alterations when compared with wild type plants, at least under standard growth conditions.

Then F3 seeds from selfed F2 plants carrying three mutated alleles and one wild type allele (*MIT1* or *MIT2*) were sown directly on the soil, and the grown plants were genotyped. Again, no double homozygous mutant plants were obtained (73 and 65 plants analyzed, Table 1). These results suggest that MIT function is essential and that the *MIT1* and *MIT2* genes are redundant. Furthermore, instead of the expected ratio of 2:1 for heterozygous: wild type plants, *MIT1mit1*:*MIT1MIT1* in the *mit2mit2* background, and *MIT2mit2*:*MIT2MIT2* in the *mit1mit1* background, ratios of 0.8 and 1.1 were observed. These ratios suggest a gametophytic defect, i.e., a defect in gametes carrying only mutated alleles of the mitochondrial iron transporters (*mit1mit2* gametes).

Table 1. Segregation analysis in the progeny of F2 plants carrying three mutated alleles.

F2 plant genotype: *MIT1mit1mit2mit2*.			
Analyzed F3 Plants	**Genotype of F3 Plants**		**Ratio**
	MIT1mit1mit2mit2	*MIT1MIT1mit2mit2*	
73	32	41	0.8
F2 plant genotype: *mit1mit1MIT2mit2*.			
Analyzed F3 Plants	**Genotype of F3 Plants**		**Ratio**
	mit1mit1MIT2mit2	*mit1mit1MIT2MIT2*	
65	34	31	1.1

F3 seeds obtained from selfed F2 plants carrying three mutated alleles were sown directly on soil, and inheritance of *mit1* and *mit2* alleles were analyzed by genotyping 73 and 65 F3 plants, respectively. Uppercase letters indicate wild type alleles, and lowercase letters indicate mutated alleles. Seedling genotypes were determined by PCR, as described in Methods. Expected ratio for Mendelian inheritance is 2.0.

Visual inspection of F4 seeds from F3 plants carrying three mutated alleles showed that in addition to "normal" seeds that resemble those from wild type plants, plants carrying three mutated alleles produced seeds with altered phenotypes: (i) smaller, irregular seeds ("abnormal" seeds), and (ii) shrunken, collapsed seeds ("aborted" seeds) (Figure 2A). Sets of normal, abnormal, and aborted seeds per silique were quantitatively determined using four *MIT1mit1 mit2mit2* plants, four *mit1mit1 MIT2mit2* plants, and two wild type plants as controls (Figure 2B). In siliques from plants with three mutated alleles, 26% (plants with one *MIT1* wild type allele) and 20% (plants with one *MIT2* wild type allele) of total seeds were considered "abnormal". Only a minor proportion of seeds were aborted (7% and 9%, respectively). There are no significant differences in the total number of seeds per silique between wild type plants and plants with three mutated alleles, nor are there significant numbers of non-fertilized ovules. Furthermore, these numbers were evenly distributed along the inflorescence axis, indicating that seed phenotypes are not due to flower heterogeneity (Supplementary Figure S4).

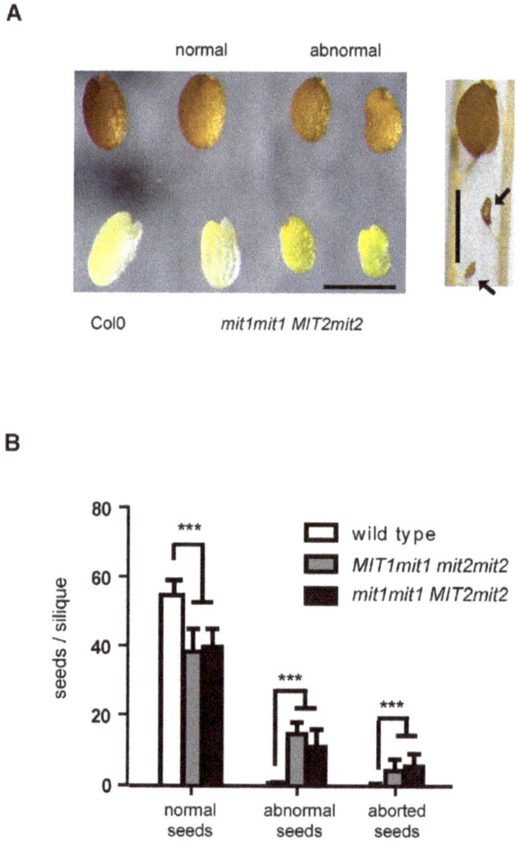

Figure 2. Seed abnormalities in plants carrying three *mit* mutated alleles. (**A**) Mature seeds and manually extracted embryos from wild type Col0 and *mit1mit1 MIT2mit2* plants. "Normal" seeds are those with a wild phenotype; "abnormal" seeds and embryos are somewhat smaller and have an irregular surface. To the right is shown a silique fragment with two shrunken, collapsed seeds, indicated by arrows. Bars = 500 µm. (**B**) Seed (normal, abnormal, and aborted) set per silique was scored along the main inflorescence for four F3 *MIT1mit1 mit2mit2* plants, four F3 *mit1mit1MIT2mit2* plants, and two wild type plants. Twelve siliques per plant (from the 5th to 16th in appearance) were scored. Error bars are SD. Asterisks indicate values that were determined by the *t*-test to be significantly different from wild type (***, $p < 0.001$).

When these "abnormal" seeds were sown on plates containing half-concentrated MS medium, almost all were able to germinate (Table 2) and expand their green cotyledons, although at a lower rate than "normal" and wild type seeds (Supplementary Figure S5). However, only around 50% of seedlings were established with true leaves and elongated roots (Table 2). A significant proportion of these abnormal seeds possessed embryos with three cotyledons (Table 2, Figure 3). We were able to identify double homozygous mutants (*mit1-1mit1-1 mit2-1mit2-1*) among plants grown from "abnormal" seeds, in addition to plants with three mutated alleles (Figure 4). Furthermore, there is a strong correlation between growth rate and genotype: all double homozygous mutants presented a delayed growth rate when compared with plants with either one or two wild type *MIT1* alleles. These results showed that seed morphology was not a clear-cut criterion to identify genotype, and, alternatively, that double homozygous mutants are viable. This unexpected result led us to determine that *mit2-1* is not a knockout mutant (see above, Figure 1C) and that *MIT2* is expressed at a lower level (12.2%) in the double homozygous mutants (Supplementary Figure S6).

Table 2. Characterization of "abnormal" seeds obtained from plants with three mutated *mit* alleles.

	"Abnormal" Seeds From	
	MIT1mit1mit2mit2	*mit1mit1MIT2mit2*
% germination	94.4 ± 7.8 [1]	97.2 ± 2.6 [2]
% establishment	57.4 [3]	53.9 ± 11.9 [4]
% 3-cotyledon embryos	12.1 ± 9.3 [5]	14.8 ± 8.4 [6]

Abnormal seeds of *MIT1mit1mit2mit2* and *mitmit1MIT2mit2* plants were sown on 0.5xMS plates, and different parameters evaluated. Germination was recorded as radicle protrusion and establishment as appearance of leaves and root growth. Replicate numbers: [1] 5 replicates (20–30 seeds each), [2] 11 replicates (30–40 seeds each), [3] 1 experiment with 29 seeds, [4] 8 replicates (29–37 seeds each), [5] 5 replicates (15–33 seeds each, 15 embryos with 3 cotyledons out of 122 total abnormal seeds), [6] 10 replicates (30–40 seeds each, 48 3-cotyledon embryos out of 310 total abnormal seeds).

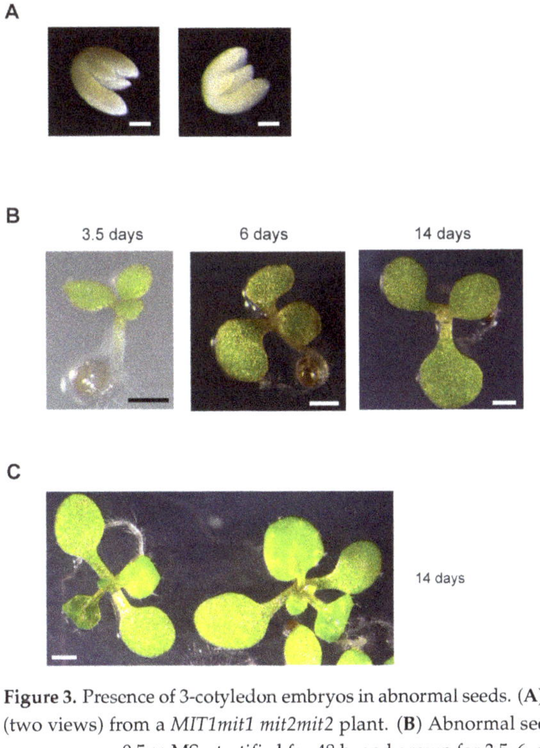

Figure 3. Presence of 3-cotyledon embryos in abnormal seeds. (**A**) Manually extracted mature embryo (two views) from a *MIT1mit1 mit2mit2* plant. (**B**) Abnormal seeds from *mit1mit1 MIT2mit2* plants were sown on 0.5 × MS, stratified for 48 h, and grown for 3.5, 6, and 14 days. The same plant bearing three cotyledons was photographed. (**C**) Control wild type plants grown for 14 days. Bars = 100 µm in (**A**), 500 µm in (**B**), and 1000 µm in (**C**).

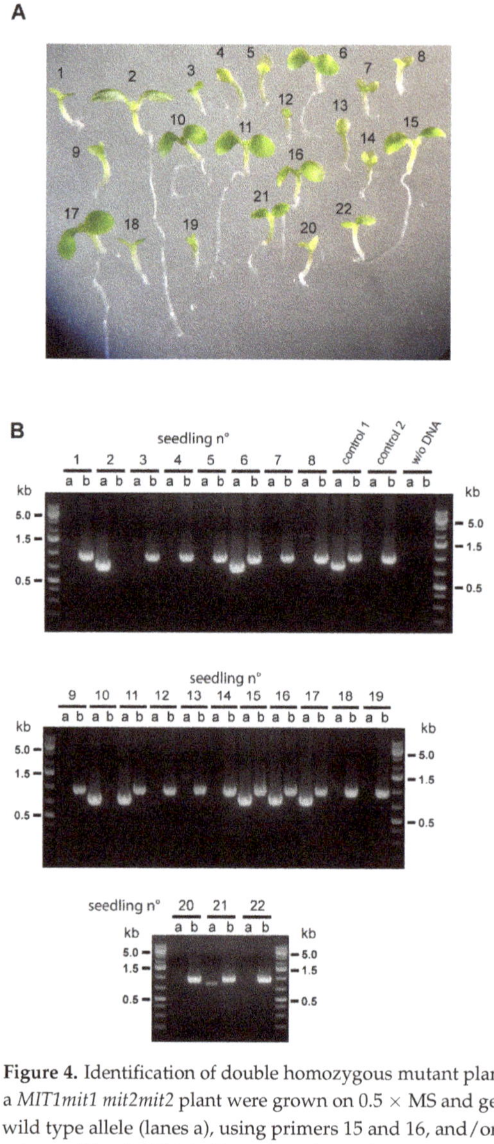

Figure 4. Identification of double homozygous mutant plants. (**A**) Twenty-two abnormal seeds from a *MIT1mit1 mit2mit2* plant were grown on 0.5 × MS and genotyped (**B**) for the presence of the *MIT1* wild type allele (lanes a), using primers 15 and 16, and/or the *mit1-1* mutant allele (lanes b), using primers 15 and LBb1.3. Two plants have the *MIT1MIT1* genotype, six plants have the *MIT1mit1* genotype, and fourteen plants have the *mit1mit1* genotype and are thus double homozygous mutant plants (*mit1mit1 mit2mit2*). Control 1 corresponds to DNA from a previously characterized double heterozygous plant (*MIT1mit1 MIT2mit2*), and control 2 to a homozygous *mit1-1mit1-1* plant. In the figures, n° correspond to number.

Five individual siliques from *MIT1mit1 mit2mit2* plants were used in two experiments to genotype all seedlings (established or not) grown from normal and abnormal seeds: 16.0 ± 3.5% of total seeds were double homozygous mutants (forty-three out of two hundred and sixty-four total seeds in the five siliques), and 27.7 ± 8.9% of the double homozygous mutants possessed three cotyledons (eleven plants out of forty-three). It is important to point out that all genotyped 3-cotyledon plants in this and other experiments were double homozygous mutants.

2.3. Growth of Double Homozygous Mutant Plants Is Severely Affected

Viable double homozygous *mit1-1mit1-1 mit2-1mit2-1* plants were easily identified by their severe phenotype. First, as already mentioned, germination and early post-germinative growth were slower than those of wild type plants and plants with three mutated alleles (Figure 4, Supplementary Figure S5). For instance, in one experiment, at 13 days, when all Col0 seedlings (90 out of 90) were at least at stage 1.02 on day 13 according to Boyes et al. (2001) [26], only one out of 59 (1.9%) of the double homozygous mutant seedlings attained this stage. Three weeks after germination, only 1.7% (1/59) and 10% (6/59) of these seedlings were at stages 1.06 and 1.04, respectively.

Growth of double homozygous mutant plants on soil was severely affected throughout the entire life cycle, and senescence was delayed by 1.5–2 months (Supplementary Figure S7). Drastic reduction of MIT expression has pleiotropic effects on double homozygous plants (Figure 5), including pinoid stems (Figure 5A,B) similar to those observed for mutants of auxin efflux carriers (PIN) (39 out of 46 plants, 85%), stems terminated at either cauline leaves (25 plants, 54%; Figure 5B), a unique flower (21 plants, 46%; Figure 5C), or multiple floral buds and cauline leaves (18 plants, 39%; Figure 5D). Furthermore, phyllotaxis in the appearance of cauline leaves was altered: there were a higher number of these leaves in some stems of 15 plants (33%, Figure 5E), their position was less regular (Figure 5D–F), and in some cases three cauline leaves were found at the same position (in seven plants, fifteen percent, Figure 5F). Additionally, some enlarged stems were found in five plants (eleven percent of the plants), as if two stems had been fused (Figure 5G, also visible in the plant shown in Figure 5F).

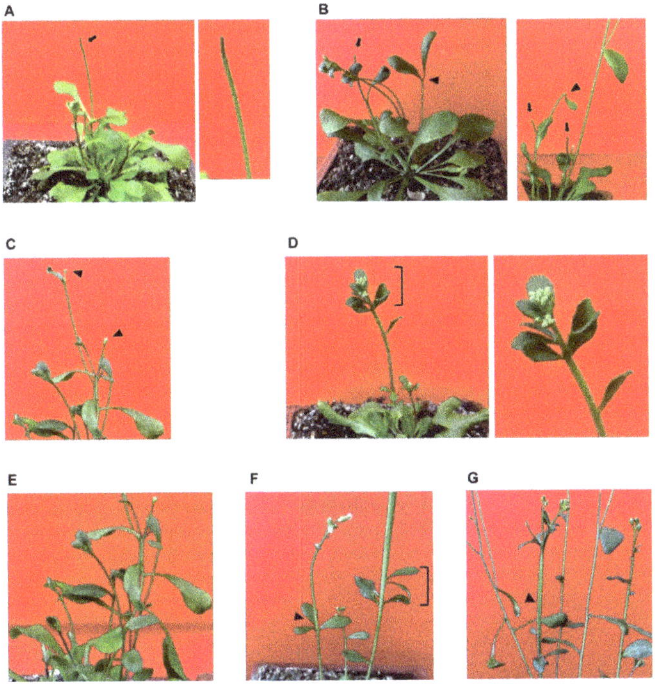

Figure 5. Phenotypic alterations observed in double homozygous plants with drastic reduction of MIT expression. (**A**) A pinoid stem (arrow) with the corresponding enlarged view. (**B**) Two plants with stems terminated at cauline leaves (arrowheads) and which also have pinoid stems (arrows). (**C**) A plant with stems terminated at unique flowers (arrowheads). (**D**) Multiple floral buds and cauline leaves at stem tip. (**E**) Higher number and irregular appearance of cauline leaves. (**F**) Three cauline leaves at the same position (arrowhead). (**G**) Enlarged stem (arrowhead).

All double mutant plants showed, alongside some normal flowers, flowers with all their structures (sepals, petals, anthers, and pistils) altered (Figure 6). This resulted in seventeen out of forty-six plants (thirty-seven percent) being unable to give seeds, and the remaining plants showed a reduced seed set (less than fifty seeds in one to six siliques for seventeen plants, between fifty and two hundred and fifty seeds in six to sixteen siliques for ten plants, and more than six hundred seeds for two plants with at least eighty siliques).

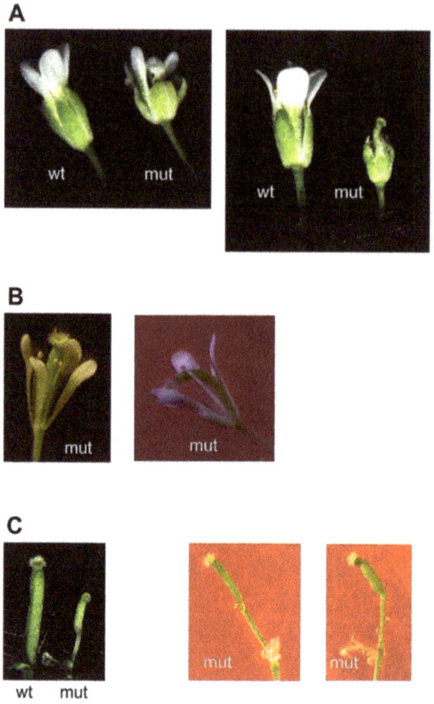

Figure 6. Altered flowers in double homozygous plants with drastic reduction of MIT expression. (**A**) Two different flowers from double homozygous *mit1-1 mit2-1* plants (mut) were compared with wild type (wt) flowers: smaller sepals, smaller petals, lack of anthers (right panel) were visible. (**B**) Two additional mutated flowers from which sepals have been excised: the four petals are heterogeneous in shape, abnormally positioned (right panel), only four and three anthers are present, pistil appears either normal (left panel) or curved (right panel). (**C**) Sepals, petals, and anthers have been excised from wild type and mutated flowers; pistils with a stem-like base are frequent in mutated flowers.

2.4. The Next Generation of Double Homozygous mit1-1 mit2-1 Mutant Plants Showed a Normal Phenotype

When seeds obtained from *mit1-1mit1-1 mit2-1mit2-1* plants were sown, almost all germinated (97.2 ± 5.0%), and plant establishment was variable (55.5 ± 17.9%). Most importantly, plant growth was similar to that of wild type plants; for instance, all established seedlings were at stage 1.0 at 7 days and at stage 1.02–1.03 at 2 weeks, and this similarity extends to vegetative and reproductive growth. From now on, these plants have been designated as "compensated" double homozygous mutant plants. Their genotype was verified by PCR to be *mit1-1mit1-1 mit2-1mit2-1*.

This intriguing result led us to analyze *MIT2* expression by RT-qPCR in these "compensated" double homozygous mutant plants and compare it with that observed in plants showing an affected phenotype (the first generation of double homozygous mutant plants, arising from seeds obtained from plants carrying three mutated alleles) (Figure 7). Interestingly, *MIT2* expression is significantly higher in compensated plants (60.9% that of wild type plants) compared with affected plants (11.0–18.5% that of wild type plants).

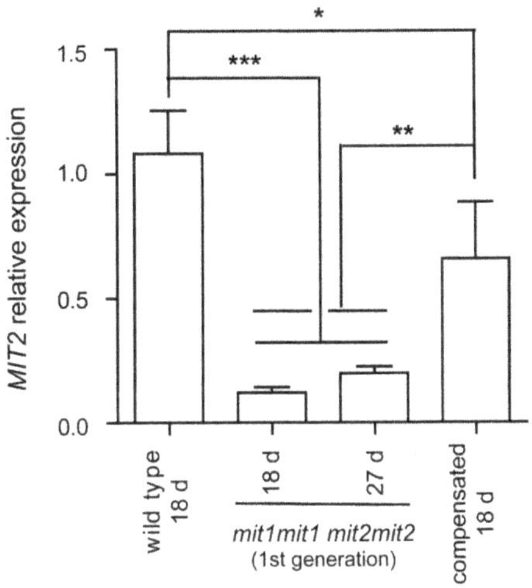

Figure 7. MIT2 expression in double homozygous mutant plants. *MIT2* mature transcript levels were determined by RT-qPCR and normalized to *TIP41-like* transcript levels. Means ± SD of four biological replicates is shown. RNAs were prepared from seedlings at stages 1.03–1.04 [26], i.e., 18 day old seedlings for wild type and "compensated" double homozygous mutant plants (*mit1-1mit1-1 mit2-1mit2-1*) and 27 day old seedlings for the first generation of the double homozygous mutant plants (same developmental stage), and also from 18 day old seedlings from these last plants (stage 1.0). The primers used for *MIT2* were primers 20 (encompassing exon junction) and 21 (Supplementary Table S1). Statistically significant differences were determined by one-way ANOVA followed by Tukey's multiple comparison test (***, $p < 0.001$; **, $p < 0.01$; *, $p < 0.05$).

These results are consistent with the view that increased splicing of the *MIT2* intron containing the T-DNA is responsible for phenotypic recovery of the double homozygous mutant plants and may be related to a relatively recently described phenomenon called "T-DNA suppression" (see Discussion).

2.5. Mitochondrial Stress Markers UPOX and MSM1 Are Upregulated in the First Generation of mit1-1 mit2-1 Double Mutant Plants

Marker genes for the mitochondrial response to stress have been identified [27], and they include the genes encoding the mitochondrial proteins alternative oxidase 1A and UPOX (upregulated by oxidative stress) [28,29]. On the other hand, Van Aken and Whelan (2012) [30] were able to identify marker genes that respond to mitochondrial and chloroplast dysfunction (e.g., *UPOX*) or are specific for mitochondrial dysfunction (e.g., *MSM1*, for Mitochondrial Stress Marker 1, also designated *At12cys-2*, [31]. We evaluated the expression of these three genes, *AOX1a*, *UPOX*, and *MSM1*, and found that *UPOX* and *MSM1* are significantly upregulated in the first generation of double homozygous mutant plants (*mit1-1mit1-1 mit2-1mit2-1*) but return to wild type levels in the "compensated" second generation plants (Figure 8). In contrast, *AOX1a* transcript levels were not significantly altered in any genotype. These results suggest some degree of mitochondrial perturbation in the first generation of double homozygous mutant plants with a drastic reduction of *MIT* expression.

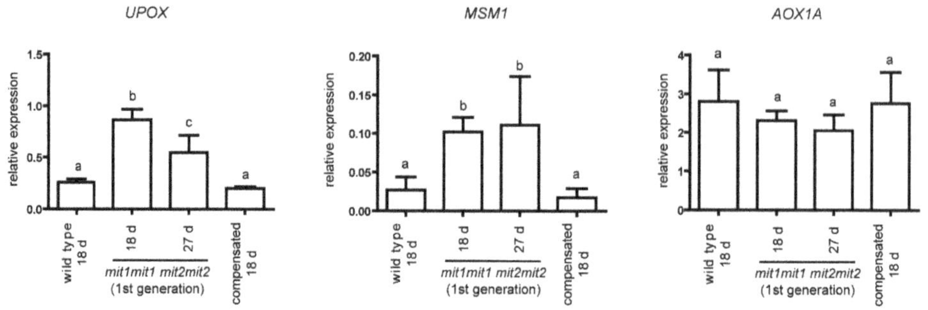

Figure 8. Expression of UPOX, MSM1, and AOX1A in double homozygous mutant plants. Transcript levels were determined by RT-qPCR and normalized to *TIP41-like* transcript level. Means ± SD of four biological replicates is shown. RNAs were prepared from seedlings at stages 1.03–1.04, i.e., 18 day old seedlings for wild type and "compensated" double homozygous mutant plants (*mit1-1mit1-1 mit2-1mit2-1*) and 27 day old seedlings for the first generation of the double homozygous mutant plants (same developmental stage), and also from 18 day old seedlings from these last plants (stage 1.0). Primer sequences are indicated in Supplementary Table S1. Statistically significant differences were determined by one-way ANOVA followed by Tukey's multiple comparison test (same letter indicates no significant differences). For *UPOX*: a-b differences, $p < 0.001$; b-c differences, $p < 0.01$; for MSM1: a-b differences, $p < 0.05$.

Unfortunately, we were unable to purify mitochondria from first-generation double homozygous mutant plants (*mit1-1mit1-1 mit2-1mit2-1*) with an affected phenotype. To do this, it would be necessary to grow plants with three mutated alleles, collect seeds, manually separate "abnormal" seeds, and grow plants from these seeds, thus, it was unfeasible to obtain enough biological material. Thus, mitochondria were purified from "compensated" double homozygous mutants (*mit1-1mit1-1 mit2-1mit2-1*) and wild type seedlings as described in Supplementary Methods.

Targeted proteomic analysis was performed on four biological replicates of both compensated and wild type mitochondria by Selective Reaction Monitoring (SRM) mass spectrometry. In this way, more than one hundred proteins (listed in Supplementary Table S3) were quantified, allowing a focused dissection of responses in the TCA cycle, electron transport chain, mitochondrial localized iron-related proteins, and MIT1/2 proteins (Supplementary Figure S8A). Significant differences in protein levels between compensated and wild type plants were found only for MIT1 and MIT2. The specific quantified peptide for MIT1 was found in wild type cells but was below the limit of detection in mitochondria from "compensated" double homozygous plants (as expected for a knockout mutation). The specific peptide for MIT2 in compensated plants was 30.9 ± 12.4 (SD)% the level found in wild type mitochondria, confirming that *mit2-1* is a knockdown allele.

Despite this large reduction in MIT abundance (absence of MIT1 and 30% of MIT2), no differences were observed in the oxygen consumption rate of isolated mitochondria (Supplementary Figure S8B). Furthermore, no differences were detected in the abundance or native size of respiratory complexes or in complex I activity when analyzed by BN gel electrophoresis (Supplementary Figure S8C). Complex I activity was assessed as it is the respiratory complex having the highest number of iron ions in its structure, present as iron-sulfur centers. These results show that, at least when plants are grown under standard conditions, mitochondria with a drastic reduction in MIT are not functionally impaired.

2.6. Double Homozygous Mutant Plants Misregulate Genes Involved in Iron Uptake, Root Development, and Stress-Related Response

Since RNA-seq analysis required less biological material, we were able to perform this analysis with the first generation of double homozygous mutant plants (*mit1-1mit1-1 mit2-1mit2-1*), which showed an affected phenotype, and we compare this transcriptome with that of "compensated" plants having the same genotype and that of wild type plants. Total RNA was prepared from three biological replicates of 18 day old wild type seedlings and compensated double homozygous mutant seedlings. For double homozygous plants of the first generation (thus presenting a severe phenotype), 27 day old seedlings were considered in order to compare plants at the same developmental stage, in this case, 1.04 of Boyes et al. (2001) [26]. Poly A-enriched RNA fractions were employed to construct libraries for Illumina sequencing (see Methods).

Differentially expressed genes between genotypes were identified (padj < 0.05, $\log_2$ [fold change] > 1 in any condition, LRT test) and grouped in two clusters by k-means (Figure 9). Interestingly, no significant differences (except for *MIT* expression) were found between wild type plants and compensated plants (Wald test, padj = 3.08×10^{-18}, $\log_2$ fold change = −4.2), further supporting the conclusion that partial expression (around 30%) of one of the two *MIT* genes is sufficient for normal plant growth and development. In cluster 1 (408 genes, listed in Supplementary Table S4), expression is downregulated in double homozygous mutant plants (1st generation) when compared with wild type and compensated plants. In cluster 2 (360 genes, listed in Supplementary Table S5), higher levels of transcripts are found in the double homozygous mutant plants (1st generation).

A

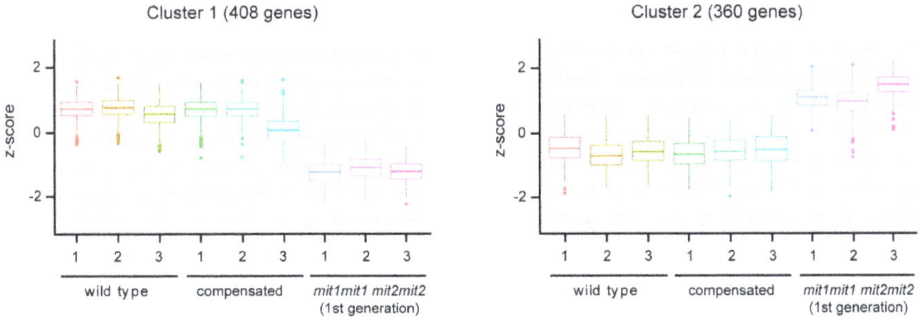

B

GO biological process	Fold Enrichment	FDR
cell-cell junction assembly (GO:0007043)	45.72	8.52E-04
coumarin metabolic process (GO:0009804)	38.10	1.38E-04
suberin biosynthetic process (GO:0010345)	24.94	1.27E-06
zinc ion transmembrane transport (GO:0071577)	14.91	3.94E-03
iron ion transport (GO:0006826)	14.69	8.55E-04
root hair elongation (GO:0048767)	7.12	2.82E-03
root hair cell differentiation (GO:0048765)	8.15	4.24E-08
trichoblast differentiation (GO:0010054)	9.08	2.65E-10
root epidermal cell differentiation (GO:0010053)	8.77	4.32E-11
root morphogenesis (GO:0010015)	5.85	6.10E-23
root system development (GO:0022622)	5.20	5.34E-34
plant-type cell wall modification (GO:0009827)	6.33	3.43E-02
cell wall organization or biogenesis (GO:0071554)	3.32	6.98E-10
unidimensional cell growth (GO:0009826)	3.04	9.83E-03

GO biological process	Fold Enrichment	FDR
production of siRNA involved in gene silencing by small RNA (GO:0070919)	24.35	4.37E-02
heterochromatin assembly (GO:0031507)	7.85	2.31E-02
regulation of jasmonic acid mediated signaling pathway (GO:2000022)	20.29	3.59E-08
jasmonic acid metabolic process (GO:0009694)	13.25	1.35E-04
plant organ formation (GO:1905393)	5.07	3.19E-02
anatomical structure formation involved in morphogenesis (GO:0048646)	4.12	5.04E-03
response to jasmonic acid (GO:0009753)	4.70	5.99E-11
flower development (GO:0009908)	3.46	4.42E-04
response to wounding (GO:0009611)	4.17	1.99E-11
glucosinolate metabolic process (GO:0019760)	3.95	2.26E-02
floral organ development (GO:0048437)	3.69	2.23E-02
response to water deprivation (GO:0009414)	2.95	4.28E-06
response to other organism (GO:0051707)	1.81	3.27E-03
defense response (GO:0006952)	1.68	3.28E-02

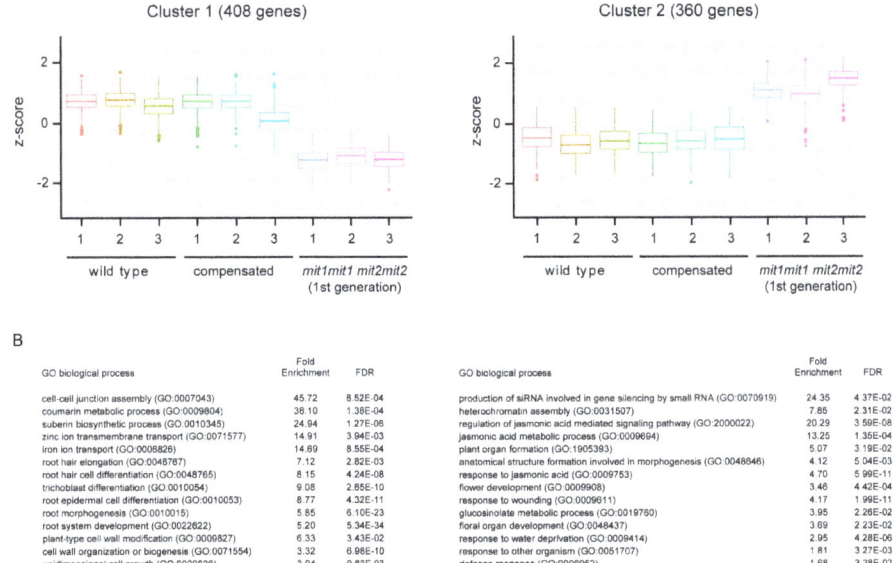

Figure 9. (**A**) K-means clustering of differentially expressed genes. RNAs were prepared from seedlings at the same developmental stage: wild type plants (18 days old), compensated double homozygous mutant plants (18 day old), and 1st generation of double homozygous mutant plants with an affected phenotype (27 day old). RNAseq was performed on three biological replicates (seedlings grown on different days) for each group. Differentially expressed genes (at least two-fold) were identified using DEseq2 software and clustered by k-means. (**B**) Some enriched GO terms (biological processes) in clusters 1 and 2. See Supplementary Tables S6 and S7 for more detailed information.

Enrichment of GO terms (biological processes) was analyzed in both clusters (Supplementary Tables S6 and S7).

Cluster 1 shows an overrepresentation of genes belonging to the Gene Ontology annotation categories of iron ion transport (GO:0006826; FDR 8.6×10^{-4}, 14.7 fold enrichment) and coumarin metabolic process (GO: 0009804; FDR 1.4×10^{-4}, 38.1 fold enrichment); coumarins being involved in iron chelation in the rhizosphere for incorporation into the plant [32–34]. For instance, *FRO2* (At1g01580, ferric reduction oxidase 2), *IRT1* (At4g19690, iron-regulated transporter 1), *IRT2* (At4g19680, iron-regulated transporter 2), *FIT1/FRU/bHLH29* (At2g28160, FER-like iron deficiency induced transcription factor), *IREG2/FPN2* (At5g03570, iron-regulated transporter 2, ferroportin 2), *BTSL2* (At1g74770, zinc finger BRUTUS-like protein 2), and *NAS2* (At5g56080, nicotianamine synthase 2), are in cluster 1. The downregulation of *FRO2* and *IRT1* was also independently verified by RT-qPCR (Figure 10). Other relevant overrepresented biological processes in cluster 1 are related to growth processes, including root hair elongation, plant-type cell wall modification, and unidimensional cell growth, which are consistent with the growth deficiency phenotypes presented in the double homozygous mutants (Supplementary Table S6).

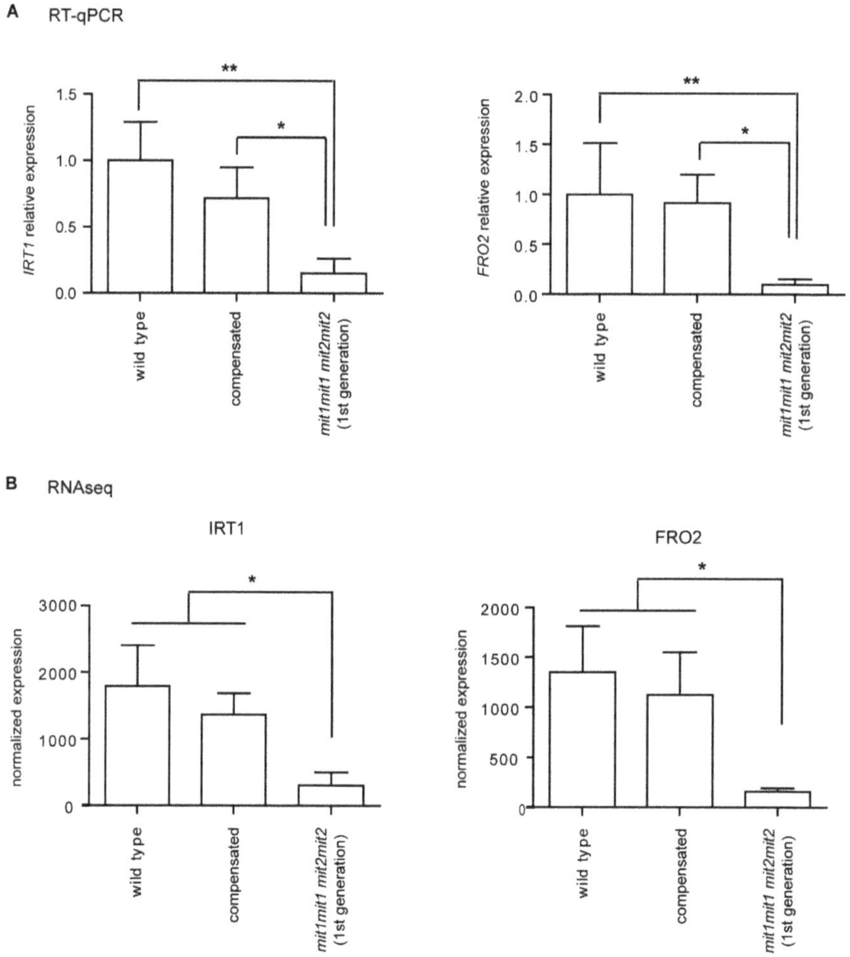

Figure 10. IRT1 and FRO2 expression in double homozygous mutant plants. (**A**) *IRT1* and *FRO2* mature transcript levels were determined by RT-qPCR and normalized to *TIP41-like* transcript levels. RNAs were prepared from seedlings at stages 1.03–1.04 (same developmental stage). Primers sequences are indicated in Supplementary Table S1. Means ± SD of three (*IRT1*) or four (*FRO2*) biological replicates are shown. Statistically significant differences were determined by one-way ANOVA followed by Tukey's multiple comparison test (**, $p < 0.01$; *, $p < 0.05$). (**B**) Normalized expression of *IRT1* and *FRO2* as determined by RNAseq. Means ± SD of the three biological replicates are shown; a one-way ANOVA was performed, followed by Tukey's multiple comparison test (*, $p < 0.05$).

Cluster 2, containing genes with increased expression relative to wild type and compensated plants, shows an overrepresentation of biological processes that can be related to plant defense, including different terms related to jasmonic acid (regulation of jasmonic acid-mediated signaling pathways, jasmonic acid metabolic process, response to jasmonic acid), regulation of defense response, response to other organisms, and glucosinolate metabolic process. As well, processes related to wounding and response to water deprivation were found in cluster 2. This suggests that double homozygous plants have a basal activation of stress-related responses, which might impact plant growth in these plants.

3. Discussion

Arabidopsis *MIT1* and *MIT2* encode proteins highly similar to rice MIT and with significant similarity to yeast mitochondrial iron transporters MRS3 and MRS4 (38% identity). MIT1 and MIT2, with their own targeting peptides (Supplementary Figure S1) or that of MRS3 [19], were able to complement the defect of the yeast *Δmrs3Δmrs4* mutant, demonstrating that they function as high-affinity iron transporters in yeast mitochondria and likely also in plant mitochondria.

The rice MIT function is essential, since the absence of its unique MIT encoding gene is embryo-lethal [18]. In contrast, in Arabidopsis, the two genes *MIT1* and *MIT2* appear to be redundant since individual mutants, including the knockout mutants *mit1-1* and *mit2-2*, were indistinguishable from wild type plants. Like in rice, MIT function is essential in Arabidopsis since, when crossing these two null mutants, no double homozygous plants could be obtained, and almost 25% of the seeds from plants with three mutated alleles (*MIT1mit1-1 mit2-2mit2-2*) aborted, as expected for a Mendelian segregation.

3.1. *Phenotypic Alterations of the mit1-1 mit2-1 Double Homozygous Mutants*

When crossing the knockout *mit1-1* mutant with the knockdown *mit2-1* mutant, we were able to obtain double homozygous mutant plants from the so-called "abnormal" seeds plated on MS x 0.5 (Figures 2 and 4). The 1st generation of these plants expressed only low levels of *MIT2* (10–20%, Figure 7 and Supplementary Figure S6) and showed striking phenotypes, highlighting the importance of MIT function. Pleiotropic defects include polycotyly (3-cotyledon embryos, Figure 3 and Table 2), retarded germination and early post-germinative growth with reduced establishment (Supplementary Figure S5, Figure 4, Table 2), delayed and altered reproductive development, including *pin* stems, abnormalities in phyllotaxy and in all organs of the flowers (Figures 5 and 6), and reduced seed set.

Some of these phenotypes are clearly indicative of defects in auxin signaling. For instance, it is well known that polar auxin transport is involved in cotyledon emergence [35,36], and both *pin* and *pid* mutants show polycotyly [37–40]. However, whereas the presence of one cotyledon is more frequent in *pin1* mutants (auxin efflux carrier, PIN-FORMED1), mutants in the ser/thr protein kinase PID (PINOID), which phosphorylates PIN1, show a higher frequency of 3-cotyledon embryos and thus resemble double homozygous *mit1-1 mit2-1* embryos. In all these cases, phenotype penetrance is incomplete. Furthermore, striking similarities are found between *pin*, *pid*, and *mit1-1mit2-1* mutants during reproductive development: inflorescence stems without flowers and cauline leaves (*pin* stems), alterations in cauline leaves' phyllotaxis, abnormal flowers, and reduced fertility (Figures 5 and 6) [37,41,42].

Although somewhat surprising at first sight, an interplay between mitochondrial function and auxin signaling is well documented [43–46]. The exact mechanisms linking, in each case, mitochondrial dysfunction and auxin signaling are not known, but several competing or complementary hypotheses have been proposed based on the deep connections between auxin synthesis, conjugation, and post-translational regulation of auxin signaling pathways [47]. These assembled examples and our results suggest that at least some of the phenotypes observed in the *mit1-1mit2-1* double homozygous mutants are mediated by defects in auxin homeostasis.

Alternatively, phenotypes such as reduced plant establishment, delayed germination, and slow early post-germinative growth are characteristic of mitochondrial deficiency. Mitochondria are expected to play a crucial role at these early stages, supplying energy and carbon skeletons for growth, and a role for respiratory complexes I, II, and IV has been described e.g., [48–53]. Unfortunately, as already mentioned, we were unable to obtain mitochondria from these seedlings to assess mitochondrial function, and only indirect evidence for mitochondrial dysfunction, i.e., higher transcript levels of *MSM1* and *UPOX*, was documented (Figure 8). However, Bashir et al. (2011) [18] characterized a knockdown mutant of the unique rice *MIT* gene (*mit-2*, T-DNA insertion in the gene promoter), which displays a less severe phenotype, allowing mitochondrial preparation from homozygous mutant plants. Those rice mitochondria contain less Fe and have less aconitase (an iron-sulfur protein) activity, thus supporting mitochondrial dysfunction due to iron deficiency, also in our Arabidopsis double mutant.

3.2. *Phenotypic Recovery of Double Homozygous mit1-1 mit2-1 in Next Generations*

When the few seeds obtained from the affected *mit1-1mit1-1 mit2-1mit2-1* plants were sown, plant growth was similar to that of wild type plants. In these "compensated" double homozygous mutant plants, *MIT2* expression is enhanced with respect to parent plants (Figure 7), and targeted proteomic analysis of purified mitochondria showed they differ significantly from wild type mitochondria only in MIT1 (undetectable) and MIT2 (30%) content (Supplementary Figure S8). Furthermore, no differences were observed in mitochondrial respiratory rate, respiratory complex size and abundance, or complex I activity. Thus, increased splicing of the *MIT2* intron containing the T-DNA is likely responsible for the phenotypic recovery of the double homozygous mutant plants. In the past years, a phenomenon called "T-DNA suppression" has been described, occurring when crossing two mutants with similar T-DNA insertions (e.g., two SALK lines). At least one of the T-DNA insertions must be intronic, and the "suppressed" phenotype is then caused by the T-DNA [54–57]. Although the mechanism is not well known, T-DNA hypermethylation and heterochromatinization are necessary, and the RdDM (RNA-dependent DNA methylation) pathway is involved [57]. In our experiments, the altered phenotypes observed in the first generation of double homozygous mutant plants are "suppressed" by an increase in *MIT2* intron splicing.

3.3. *Transcriptome of Double Homozygous mit1-1 mit2-1 Mutant Plants*

No significant differences (except for *MIT* expression) were found between wild type plants and "compensated" *mit1-1mit2-1* double homozygous mutant plants, supporting the conclusion that partial expression (around 30%) of MIT2 is sufficient for normal plant growth and development, at least under standard growth conditions. In contrast, in the first generation of *mit1-1mit2-1* double homozygous mutant plants, which expressed 10–20% of *MIT2* (at the transcript level), 408 genes are down-regulated (cluster 1) and 360 genes are upregulated (cluster 2) compared with wild type and compensated plants (Figure 9, Supplementary Tables S4 and S5).

Interestingly, the data suggest downregulation of the iron acquisition system. Besides *FRO2*, *IRT1* and *IRT2*, genes encoding either proteins involved in coumarin biosynthesis (F6′H1, feruloyl CoA ortho-hydroxylase 1; S8H, Scopoletin 8-hydroxylase; CYP82C4, fraxetin 5-hydroxylase) or coumarins export (ABCG37/PDR9) are included in cluster 1. These genes are part of the Fe deficiency response [58] and are regulated by the master transcription factor FIT1/FRU (Fe-deficiency-induced transcription factor), which is also found in cluster 1 and known to control expression of additional cluster 1 genes involved in Fe homeostasis: *IREG2*, iron-regulated transporter 2/ferroportin 2; *BTSL2*, E3 ubiquitin ligase BRUTUS-like protein 2; and *MTPA2*, metal tolerance protein A2. Therefore, the response to a deficiency in iron uptake by mitochondria may be opposite to the response observed under Fe deficiency reviewed in [34,58] and reminiscent of the root Fe exclusion strategy described in rice [59].

Other enriched GO categories in cluster 1 related to the root system may be indirectly relevant to iron acquisition (Supplementary Table S6). Given the role of the root epidermis, and in particular root hairs, in water and nutrient uptake [60–62], future work will be necessary to analyze root development in the mutant plants.

Relevant enriched GO terms in cluster 2 (upregulated in mutant plants) may be involved in the observed pleiotropic phenotypes (Supplementary Table S7), for instance plant organ formation, anatomical structure formation involved in morphogenesis, flower development, and floral organ development. Furthermore, the enriched GO categories "production of siRNA involved in gene silencing by small RNA", which has three RNA-dependent RNA polymerases (RDR1, RDR2, and RDR3), and "heterochromatin assembly", which contains the same three RDR, nucleolin 2, and two chromatin remodeling factors (chr31/SNF2 domain-containing protein CLASSY 3 and chr42/SNF2 domain-containing protein CLASSY 2), may be relevant to explain the T-DNA suppression phenomenon discussed above.

Although enriched GO terms related to auxin were not found, careful examination of the genes in both clusters (Supplementary Tables S4 and S5) highlighted a number of genes related to either auxin metabolism, transport, signaling, or response (9 in cluster 1 and 13 in cluster 2). For instance, PIN2 (At5g57090), ABCG37/PDR9 (At3g53480), ABCB11 (At1g02520), ERULUS (At5g61350), MYB93 (At1g34670), and YUCCA3 (At1g04610) are downregulated in mutant plants (cluster 1). Other genes were found to be upregulated (cluster 2), for example AIL7/PLT7 (At5g65510), LRP1 (At5g12330), SHI (At5g66350), SGR5 (At2g09140), SKP2A (At1g21410), ENP/MACCHI-BOU4/NPY1 (At4g31820), TCP18 (At3g18530), and GH3.5 (At4g27260). Whether these changes are related to the observed phenotypes similar to those of polar auxin transport mutants remains to be explored.

4. Materials and Methods

4.1. Plant Material and Growth Conditions

All *A. thaliana* plants used were from the Columbia (Col-0) region. Seeds were sown on half-concentrated MS agar medium and stratified for 48 h at 4 °C in the dark. After two weeks in a 16/8 h day/night cycle at 22 °C, seedlings were transferred to soil and grown under long-day conditions (16 h light/8 h dark).

Seeds from five T-DNA insertion mutants were obtained from the ABRC stock center: *mit1-1* (SALK_013388), *mit1-2* (SALK_208340C), *mit2-1* (SALK_096697), *mit2-2* (SAIL_653_B10, CS828300), and *mit2-3* (SALK_095187). For genotyping, DNA was extracted from either 15 day old seedlings or leaves of 4 week old plants and analyzed as described [63]. Primers used to amplify wild type and mutant alleles are described in Figure 1 and Supplementary Figures S2 and S3. To further characterize the T-DNA insertion in *mit2-1*, located in the intron spliced out in spite of its size, both T-DNA/*MIT2* junctions were amplified (Supplementary Figure S3). All amplified junctions were characterized by DNA sequencing.

Homozygous mutant *mit1-1* plants were crossed with either *mit2-1 or mit2-2* homozygous mutant plants. F1 seeds were germinated, and F1 plants were verified to be double heterozygous plants (genotype *MIT1mit1 MIT2mit2*). F2 seeds obtained from these selfed F1 plants were used to characterize the F2 generation and obtain the next generations (F3, F4, and so on).

4.2. Complementation of Mutant Δ*mrs3*Δ*mrs4* Yeast Cells

The Δ*mrs3*Δ*mrs4* strain (*MATa*, *ura3-52*, *leu2-3*, *112*, *trp1-1*, *his3-11*, *ade2-1*, *can1-100(oc)*, Δ*mrs3::kanMax* Δ*mrs4::kanMax*) was kindly provided by Liangtao Li and Diane Ward (Department of Pathology, School of Medicine, University of Utah) and is described in Li and Kaplan (2004) [64]. The constructs containing the *ADH1* promoter, *MIT1* (At2g30160) or *MIT2* (At1g07030) cDNAs, and the *ADH2* terminator were generated using in vivo assembly yeast recombinational cloning [65,66]. *ADH1* promoter and *ADH2* terminator were amplified by PCR using as templates a plasmid kindly provided by Dr. Luis Larrondo and Phusion High-Fidelity DNA polymerase (Thermo Scientific, https://www.thermofisher.com

(accessed on 27 February 2023)). For the promoter, either primers 1 and 2 (with an overlap in its 5′ end with the beginning of *MIT2* cds) or primers 1 and 3 (with an overlap in its 5′ end with the beginning of *MIT1* cds) were employed; for the terminator, either primers 8 (with an overlap in its 5′ end with the end of *MIT2* cds) and 9, or primers 10 (with an overlap in its 5′ end with the end of *MIT1* cds) and 9 were used. *MIT1* and *MIT2* cDNAs were obtained by RT-PCR. Total RNA was prepared from 15 day old seedlings with a Spectrum Plant Total RNA Kit (Sigma-Aldrich). cDNAs were synthesized with the Superscript First Strand synthesis system for RT-PCR (Invitrogen, Life Technologies), and PCR amplifications were performed with primers 4 and 5 for *MIT2* and primers 6 and 7 for *MIT1*. PCR products were co-transformed with the linear pRS426 plasmid into the BY4741 yeast strain (*MATa, his3Δ1, leu2Δ0, LYS2, met15Δ0, ura3Δ0*), and circular plasmids obtained from several ura^+ colonies were transferred to E. coli DH5α. Positive colonies were identified by PCR, plasmids were prepared (AxyPrep Plasmid Miniprep Kit), and construct integrity was verified by DNA sequencing. These plasmids were used to transform the Δ*mrs3*Δ*mrs4* strain, and transformants were selected by plating on synthetic defined (SD) medium without uracyl. Complementation assays were performed by growing yeast cells in liquid SD medium (without uracyl) to a DO_{600nm} of 1.5, concentrating five times and plating serial dilutions onto agarose-SD plates with or without 50 μM of the impermeable iron chelator bathophenanthroline disulfonate (BPDS).

4.3. Expression Analysis by RT-PCR and RT-qPCR

Total RNA was obtained from frozen seedlings with the TRIzol reagent, treated with DNase I (Promega, http://www.promega.com/ (accessed on 27 February 2023)), and quantified using a Nanodrop spectrophotometer (Thermo Scientific). cDNA synthesis was carried out on 1–2 μg of RNA, using oligodT as a primer and SuperScript II Reverse transcriptase (Thermo Scientific).

To analyze *MIT* expression by RT-PCR in individual mutants, total RNA was prepared from 15 day old seedlings. Then PCR reactions were performed with 1/10 of the cDNA and the following primer pairs (Supplementary Table S1): for *MIT1*, either primers 13 and 14 or 15 and 16; for *MIT2*, either primers 17 and 18 or 11 and 12.

For RT-qPCR, total RNA was obtained from seedlings at stages 1.02–1.04 (Boyes et al., 2001). RT-qPCR experiments were performed on 1/10 of the cDNA using the StepOne Plus Real-Time PCR System (Applied Biosystems, http://www.appliedbiosystems.com/ (accessed on 27 February 2023)) according to the manufacturer's instructions and the Brilliant III Ultra-fast SYBR GREEN QPCR reagents (Agilent). RNA levels were estimated considering the amplification efficiency of each primer pair and normalized relative to either the clathrin adaptor (At4g24550) or the TIP41L (At4g34270) transcripts as internal controls. The primer pairs (Supplementary Table S1) used for the clathrin adaptor were At4g24550F and At4g24550R, and for TIP41L, At4g34270F and At4g34270R. Those for *MIT2* are indicated in the figure legends. Primer sequences for *UPOX, MSM1, AOX1A, FRO2*, and *IRT1* are also shown in Supplementary Table S1.

4.4. Transcriptome Analysis by RNA Sequencing

Total RNA was prepared from seedlings (stage 1.04), using a Spectrum Plant Total RNA Mini Kit (Sigma-Aldrich). Following DNase I (Invitrogen, Life Technologies) treatment, RNA integrity was evaluated by capillary electrophoresis on a Qsep100 Biofragment analyzer (BiOptic, https://www.bioptic.com.tw (accessed on 27 February 2023)). Libraries for RNA-Seq were prepared with an Illumina TruSeq Stranded mRNA Kit, quantified by qPCR with a KAPA Library Quantification Kit (Universal) (Roche, https://sequencing.roche.com (accessed on 27 February 2023)), and sequenced using the NextSeq 500 System (Illumina, https://illumina.com (accessed on 27 February 2023)), considering 150 bp paired-end reads. Raw sequences were processed with Trimmomatic v.0.39 ([67]; http://www.usadellab.org (accessed on 27 February 2023)), using the following settings: ILLUMINACLIP:TruSeq3-PE.fa:2:30:10:2:keepBothReads LEADING:30 TRAIL-

ING:30 SLIDINGWINDOW:10:30 MINLEN:36. Filtered sequences were mapped to the Arabidopsis genome (The Arabidopsis Information Resource TAIR v.10, www.arabidopsis.org (accessed on 27 February 2023)) using HISAT2 v.2.1.0 with standard settings [68]; http://daehwankimlab.github.io/hisat2/ (accessed on 27 February 2023)). Count tables were generated using the featureCounts function from the Rsubread (v.2.10.4) library from R [69] and the Araport11 GTF gene annotation [70].

Differential gene expression between genotypes was analyzed with the DESeq2 package (v. 1.24; [71], using the likelihood ratio test (LRT). We considered differentially expressed genes those with $\log_2 > 1$ or $\log_2 < 1$, and adjusted p-value < 0.01. Clustering of differentially expressed genes was performed with the R library heatmap v. 1.0.12 (kmeans_k = 2).

Overrepresentation of Gene Ontology (GO) terms (biological processes) was analyzed for each cluster using the PANTHER Overrepresentation Test tool (http://geneontology.org (accessed on 27 February 2023)) [72].

Supplementary Materials: The following supporting information can be downloaded at: https://www.mdpi.com/article/10.3390/plants12051176/s1, Figure S1: Arabidopsis MIT1 and MIT2 complement the growth defect of yeast Δ*mrs3*Δ*mrs4* in Fe-deficient medium. Serial dilutions of the yeast Δ*mrs3*Δ*mrs4* strain transformed with a construct expressing either Arabidopsis MIT1 (AT2G30160) or MIT2 (AT1G07030) or with the pRS426 vector alone as a control were plated onto SD medium (solified with agarose to avoid Fe contamination and lacking uracyl) with or without the iron chelator BPDS; Figure S2: Characterization of mutant mit1 alleles. (**A**) Exon-intron structure of the MIT1 gene with exons represented by boxes. Horizontal arrows indicate the position of primers, and the T-DNA insertion sites are shown (T-DNA inserts are not drawn to scale). (**B**) Sequence of the MIT1/T-DNA junction in the mit1-1 mutant allele. PCR amplification was performed using primers 13 and LBb1.3, and the product was sequenced. T-DNA is inserted 595 bp downstream of the ATG initiation codon, interrupting codon 200. In the alignment, the first row corresponds to the left border T-DNA sequence (in blue), the third row to the wild type MIT1 allele sequence (in red), and the second row to the junction between MIT1 and T-DNA in the mutant allele (inserted nucleotides of unknown origin in green). (**C**) Sequence of the T-DNA/MIT1junction in the mit1-2 allele. PCR amplification was performed using primers LBb1.3 and 14. T-DNA is inserted 154 bp downstream of the ATG initiation codon, interrupting codon 53. In the alignment, the first row corresponds to the left border T-DNA sequence (in blue), the third row to the wild type MIT1 sequence (in red), and the second row to the junction between T-DNA and MIT1 in the mutant allele (inserted nucleotides of unknown origin in green); Figure S3: Characterization of mutant mit2 alleles. (**A**) Exon-intron structure of the MIT2 gene with exons represented by boxes. Horizontal arrows indicate the position of primers, and the T-DNA insertion sites are shown (T-DNA inserts are not drawn to scale). (**B**) Sequences of the T-DNA/MIT2 junctions in the mit2-1 mutant allele. PCR amplifications were performed with either primers LBb1.3 and 12 (3′ side of insertion) or primers 11 and RB (5′ side of insertion), and products were sequenced. T-DNA is inserted 419 bp downstream of the exon-intron junction; 18 nucleotides from the intron have been deleted (not shown), and 15 nucleotides of unknown origin are inserted (in green). In the alignments, the first rows correspond to the left or right border T-DNA sequences (in blue), the third row to the wild type MIT2 sequence (in red), and the second row to the junctions between MIT2 and T-DNA in mit2-1. (**C**) Sequence of the MIT2/T-DNA junction in the mit2-2 allele. PCR amplification was performed using primers 11 and LB1sail. T-DNA is inserted 712 bp downstream of the ATG initiation codon, interrupting codon 239. In the alignment, the first row corresponds to the left border T-DNA sequence (in blue), the third row to the wild type MIT2 sequence (in red), and the second row to the junction between MIT2 and T-DNA in the mutant allele. (**D**) Sequence of the MIT2/T-DNA junction in the mit2-3 allele. PCR amplification was performed using primers 11 and LBb1.3. T-DNA is inserted 500 bp downstream of the exon-intron junction. In the alignment, the first row corresponds to the left border T-DNA sequence (in blue), the third row to the wild type MIT2 sequence (in red), and the second row to the junction between MIT2 and T-DNA in the mutant allele; Figure S4: A significant proportion of seeds in plants with three mit mutated alleles have an abnormal phenotype. Seed set per silique was scored along the main inflorescence axis for four F3 MIT1mit1 mit2mit2 plants, four F3 *mit1mit1MIT2mit2* plants, and two wild type plants. Silique 5 (older) to Silique 16 (younger) were grouped in three groups of four siliques for each plant, and normal, abnormal, and aborted seeds were counted in each of the sixteen siliques of plants carrying three mutated alleles and the eight siliques of Col-0 plants. Error bars are SD; Figure S5: Abnormal seeds from plants with only one wild type MIT allele germinate slowly. Wild type seeds and both normal and abnormal seeds from *MIT1mit1-1 mit2-1mit2-1* plants were sown on 0.5 × MS, stratified for 48 h, and grown for 3 and 4 days before being photographed; Figure S6: MIT2 expression in double homozygous mutant plants. MIT2 expression levels were determined by RT-qPCR using primers 19 and 21 (Supplementary Table S1) and normalized to the clathrin adaptor transcript level. Means ± SD of three biological replicates is shown. RNAs were prepared from 6–12 seedlings at stage 1.02 (Boyes et al., 2001), i.e., 12 day old seedlings for the wild type and 19 day old seedlings for the first generation of the double homozygous mutant plants (*mit1-1mit1-1 mit2-1mit2-1*). Three asterisks indicate a value that was determined by the *t*-test to be significantly different from wild type ($p < 0.001$). To evaluate MIT2 gene expression, primers 19 and 21 were used; Figure S7: Growth of double homozygous mutant plants was retarded throughout the entire life cycle. Wild type Col-0 plants and double homozygous mutant plants (*mit1-1mit1-1 mit2-1mit2-1*, 1st generation) were grown as described in Materials and Methods, and three representative plants of each genotype were photographed at 2 months (**A**) and 3.5 months (**B**); Figure S8: Analysis of purified mitochondria from "compensated" double homozygous plants (*mit1-1mit1-1 mit2-1mit2-1*) and wild type plants. (**A**) Proteomic analysis of the purified mitochondria (four mitochondrial preparations for each genotype). Heat map showing the relative abundance in both genotypes of 274 peptides, corresponding to 108 polypeptides (Supplementary Table S3). Only the abundances of MIT1 (not detected in the mutant) and MIT2 were significantly different ($p < 0.001$, *t*-test) between wild type and compensated plants. (**B**) Oxygen consumption by purified mitochondria using pyruvate, malate, and ADP as substrates. Four mitochondrial preparations per genotype were tested, and means ± SD are shown. Difference between genotypes is not significant (*t*-test). (**C**) BN-PAGE separation of respiratory chain complexes and supercomplexes followed by staining for Complex I activity (NADH:UQ oxidoreductase; right panel) and Coomassie blue staining (left panel). Three replicates (each containing 0.3 mg mitochondrial protein) of purified mitochondria from wild type and "compensated" double homozygous seedlings were loaded. Mobility of respiratory complexes I, II, III, IV, and V and supercomplexes is indicated.

Author Contributions: Conceptualization, J.V., X.J., H.R.; methodology J.V., I.G., C.P.L., A.H.M., X.J., H.R.; software, J.V., E.A.V., C.P.L., A.H.M.; writing—original draft preparation, J.V., X.J., H.R.; writing—review and editing, J.V., E.A.V., C.P.L., A.H.M., X.J., H.R. All authors have read and agreed to the published version of the manuscript.

Funding: This work was supported by a grant from FONDECYT-Chile 1160334 to HR, Facultad de Ciencias Biológicas-PUC grant to HR, and CONICYT-Chile grant 21151344 to JVP, and the Australian Research Council to AHM (CE140100008, FL200100057).

Conflicts of Interest: The authors have no conflict of interest to declare.

Abbreviations

BN-PAGE: blue native polyacrylamide gel electrophoresis; bp, base pair; BPDS, bathophenanthroline disulfonate; nt, nucleotide; GO, gene ontology; MIT, mitochondrial iron transporter; qRT-PCR: quantitative reverse transcription-PCR; RNA-seq, RNA sequencing.

References

1. Papanikolaou, G.; Pantopoulos, K. Iron metabolism and toxicity. *Toxicol. Appl. Pharmacol.* **2005**, *202*, 199–211. [CrossRef]
2. Collins, J.; Anderson, G. *Physiology of the Gastrointestinal Tract*, 5th ed.; Elsevier: New York, NY, USA, 2012.
3. Curie, C.; Cassin, G.; Couch, D.; Divol, F.; Higuchi, K.; Le Jean, M.; Misson, J.; Schikora, A.; Czernic, P.; Mari, S. Metal movement within the plant: Contribution of nicotianamine and yellow stripe 1-like transporters. *Ann. Bot.* **2009**, *103*, 1–11. [CrossRef]
4. Santi, S.; Schmidt, W. Dissecting iron deficiency-induced proton extrusion in Arabidopsis roots. *New Phytol.* **2009**, *183*, 1072–1084. [CrossRef]
5. Robinson, N.J.; Procter, C.M.; Connolly, E.L.; Guerinot, M.L. A ferric-chelate reductase for iron uptake from soils. *Nature* **1999**, *397*, 694–697. [CrossRef]
6. Eide, D.; Broderius, M.; Fett, J.; Guerinot, M.L. A novel iron-regulated metal transporter from plants identified by functional expression in yeast. *Proc. Natl. Acad. Sci. USA* **1996**, *93*, 5624–5628. [CrossRef]

'. Vert, G.; Grotz, N.; Dedaldechamp, F.; Gaymard, F.; Guerinot, M.L.; Briat, J.F.; Curie, C. IRT1, an Arabidopsis transporter essential for iron uptake from the soil and for plant growth. *Plant Cell.* **2002**, *14*, 1223–1233. [CrossRef]
3. Guerinot, M.L.; Yi, Y. Iron: Nutritious, noxious, and not readily available. *Plant Physiol.* **1994**, *104*, 815–820. [CrossRef]
). Yi, Y.; Guerinot, M.L. Genetic evidence that induction of root Fe(III) chelate reductase activity is necessary foriron uptake under iron deficiency. *Plant J. Cell Mol. Biol.* **1996**, *10*, 835–844. [CrossRef] [PubMed]
.0. Briat, J.F.; Dubos, C.; Gaymard, F. Iron nutrition, biomass production, and plant product quality. *Trends Plant Sci.* **2015**, *20*, 33–40. [CrossRef] [PubMed]
.1. Schmidt, S.B.; Eisenhut, M.; Schneider, A. Chloroplast Transition Metal Regulation for Efficient Photosynthesis. *Trends Plant Sci.* **2020**, *25*, 817–828. [CrossRef] [PubMed]
.2. Duy, D.; Wanner, G.; Meda, A.R.; Wiren, N.; von Soll, J.; Philippar, K. PIC1, an Ancient Permease in Arabidopsis Chloroplasts, Mediates Iron Transport. *Plant Cell* **2007**, *19*, 986–1006. [CrossRef]
.3. Jeong, J.; Cohu, C.; Kerkeb, L.; Pilon, M.; Connolly, E.L.; Guerinot, M. Lou. Chloroplast Fe(III) chelate reductase activity is essential for seedling viability under iron limiting conditions. *Proc. Natl. Acad. Sci. USA* **2008**, *105*, 10619–10624.
.4. Divol, F.; Couch, D.; Conéjéro, G.; Roschzttardtz, H.; Mari, S.; Curie, C. The Arabidopsis YELLOW STRIPE LIKE4 and 6 transporters control iron release from the chloroplast. *Plant Cell.* **2013**, *25*, 1040–1055. [CrossRef] [PubMed]
15. Kim, L.J.; Tsuyuki, K.M.; Hu, F.; Park, E.Y.; Zhang, J.; Iraheta, J.G.; Chia, J.C.; Huang, R.; Tucker, A.E.; Clyne, M.; et al. Ferroportin 3 is a dual-targeted mitochondrial/chloroplast iron exporter necessary for iron homeostasis in Arabidopsis. *Plant J.* **2021**, *107*, 215–236. [CrossRef] [PubMed]
16. Tan, Y.F.; O'Toole, N.; Taylor, N.L.; Millar, A.H. Divalent metal ions in plant mitochondria and their role in interactions with proteins and oxidative stress-induced damage to respiratory function. *Plant Physiol.* **2010**, *152*, 747–761. [CrossRef]
17. Haferkamp, I.; Schmitz-Esser, S. The plant mitochondrial carrier family: Functional and evolutionary aspects. *Front Plant Sci.* **2012**, *3*, 2. [CrossRef] [PubMed]
18. Bashir, K.; Ishimaru, Y.; Shimo, H.; Nagasaka, S.; Fujimoto, M.; Takanashi, H.; Tsutsumi, N.; An, G.; Nishizawa, N.K. The rice mitochondrial iron transporter is essential for plant growth. *Nat. Commun* **2011**, *2*, 322. [CrossRef]
19. Jain, A.; Dashner, Z.S.; Connolly, E.L. Mitochondrial iron transporters (MIT1 and MIT2) are essential for iron homeostasis and embryogenesis in Arabidopsis thaliana. *Front. Plant Sci* **2019**, *10*, 1449. [CrossRef]
20. Walter, P.B.; Knutson, M.D.; Paler-Martinez, A.; Lee, S.; Xu, Y.; Viteri, F.E.; Ames, B.N. Iron deficiency and iron excess damage mitochondria and mitochondrial DNA in rats. *Proc. Natl. Acad. Sci. USA* **2002**, *99*, 2264–2269. [CrossRef]
21. Vigani, G.; Maffi, D.; Zocchi, G. Iron availability affects the function of mitochondria in cucumber roots. *New Phytol.* **2009**, *182*, 127–136. [CrossRef]
22. Vigani, G.; Solti, A.; Thomine, S.B.; Philippar, K. Essential and Detrimental-an Update on Intracellular Iron Trafficking and Homeostasis. *Plant Cell Physiol.* **2019**, *60*, 1420–1439. [CrossRef] [PubMed]
23. Vigani, G.; Bashir, K.; Ishimaru, Y.; Lehmann, M.; Casiraghi, F.M.; Nakanishi, H.; Seki, M.; Geigenberger, P.; Zocchi, G.; Nishizawa, N.K. Knocking down mitochondrial iron transporter (MIT) reprogramsprimary and secondary metabolism in rice plants. *J. Exp. Bot.* **2016**, *67*, 1357–1368. [CrossRef] [PubMed]
24. Foury, F.; Roganti, T. Deletion of the mitochondrial carrier genes MRS3 and MRS4 suppresses mitochondrial iron accumulation in a yeast frataxin-deficient strain. *J. Biol. Chem.* **2002**, *277*, 24475–24483. [CrossRef]
25. Mühlenhoff, U.; Stadler, J.A.; Richhardt, N.; Seubert, A.; Eickhorst, T.; Schweyen, R.J.; Lill, R.; Wiesenberger, G. A specific role of the yeast mitochondrial carriers Mrs3/4p in mitochondrial iron acquisition under iron-limiting conditions. *J. Biol. Chem.* **2003**, *278*, 40612–40620. [CrossRef] [PubMed]
26. Boyes, D.C.; Zayed, A.M.; Ascenzi, R.; McCaskill, A.J.; Hoffman, N.E.; Davis, K.R.; Gorlach, J. Growth stage-based phenotypic analysis of Arabidopsis: A model for hgh throughput functional genomics in plants. *Plant Cell* **2001**, *13*, 1499–1510. [CrossRef] [PubMed]
27. Van Aken, O.; Zhang, B.; Carrie, C.; Uggalla, V.; Paynter, E.; Giraud, E.; Whelan, J. Defining the mitochondrial stress response in Arabidopsis thaliana. *Mol. Plant* **2009**, *2*, 1310–1324. [CrossRef] [PubMed]
28. Clifton, R.; Lister, R.; Parker, K.L.; Sappl, P.G.; Elhafez, D.; Millar, A.H.; Day, D.A.; Whelan, J. Stress-induced co-expression of alternative respiratory chain components in Arabidopsis thaliana. *Plant Mol. Biol.* **2005**, *58*, 193–212. [CrossRef]
29. Ho, L.H.; Giraud, E.; Uggalla, V.; Lister, R.; Clifton, R.; Glen, A.; Thirkettle-Watts, D.; Van Aken, O.; Whelan, J. Identification of regulatory pathways controlling gene expression of stress-responsive mitochondrial proteins in Arabidopsis. *Plant Physiol.* **2008**, *147*, 1858–1873. [CrossRef]
30. Van Aken, O.; Whelan, J. Comparison of transcriptional changes to chloroplast and mitochondrial perturbations reveals common and specific responses in Arabidopsis. *Front. Plant Sci.* **2012**, *3*, 281. [CrossRef]
31. Wang, Y.; Lyu, W.; Berkowitz, O.; Radomiljac, J.D.; Law, S.R.; Murcha, M.W.; Carrie, C.; Teixera, P.F.; Kmiec, B.; Duncan, O.; et al. Inactiavtion of complex I induces the expression of a twin cysteine protein that targets and affects cytosolic, chloroplastidic and mitochondrial function. *Mol. Plant* **2016**, *9*, 696–710. [CrossRef]
32. Rajniak, J.; Giehl, R.F.H.; Chang, E.; Murgia, I.; von Wirén, N.; Sattely, E.S. Biosynthesis of redox-active metabolites in response to iron deficiency in plants. *Nat. Chem. Biol.* **2018**, *14*, 442–450. [CrossRef]
33. Tsai, H.H.; Rodríguez-Celma, J.; Lan, P.; Wu, Y.C.; Vélez-Bermúdez, I.C.; Schmidt, W. Scopoletin 8-hydroxylase-mediated fraxctin production is crucial for iron mobilization. *Plant Physiol.* **2018**, *177*, 194–207. [CrossRef] [PubMed]

34. Riaz, N.; Guerinot, M.L. All together now: Regulation of the iron deficiency response. *J. Exp. Bot.* **2021**, *72*, 2045–2055. [CrossRef]
35. Liu, C.; Xu, A.; Chua, N. Auxin polar transport is essential for the establishment of bilateral simmetry during early plant embryogenesis. *Plant Cell* **1993**, *5*, 621–630. [CrossRef] [PubMed]
36. Benková, E.; Michniewicz, M.; Sauer, M.; Teichmann, T.; Seifertová, D.; Jürgens, G.; Friml, J. Local, efflux-dependent auxin gradients as a common module for plant organ formation. *Cell* **2003**, *115*, 591–602. [CrossRef]
37. Bennett, S.R.M.; Alvarez, J.; Bossinger, G.; Smith, D.R. Morphogenesis in pinoid mutants of Arabidopsis thaliana. *Plant J.* **1995**, *8*, 505–520. [CrossRef]
38. Benjamins, R.; Quint, A.; Weijers, D.; Hooykaas, P.; Offringa, R. The PINOID protein kinase regulates organ development in Arabidopsis by enhancing polar auxin transport. *Development* **2001**, *128*, 4057–4067. [CrossRef] [PubMed]
39. Aida, M.; Vernoux, X.; Furutani, M.; Traas, J.; Tsaka, M. Roles of PIN-FORMED1 and MONOPTEROS in pattern formation of the apical region of the Arabidopsis embryo. *Development* **2002**, *129*, 3965–3974. [CrossRef]
40. Furutani, M.; Vernoux, T.; Traas, J.; Kato, T.; Tasaka, M.; Aida, M. PIN-FORMED1 and PINOID regulate boundary formation and cotyledon development in Arabidopsis embryogenesis. *Development* **2004**, *131*, 5021–5030. [CrossRef]
41. Okada, K.; Ueda, J.; Komaki, M.K.; Bell, C.J.; Shimura, Y. Requirement of the auxin polar transport system in early stages of Arabidopsis floral bud formation. *Plant Cell* **1991**, *3*, 677–684. [CrossRef]
42. Christensen, S.K.; Dagenais, N.; Chory, J.; Weigel, D. Regulation of auxin response by the protein kinase PINOID. *Cell* **2000**, *100*, 469–478. [CrossRef]
43. Kerchev, P.I.; De Clercq, I.; Denecker, J.; Mühlenbock, P.; Kumpf, R.; Nguyen, L.; Audenaert, D.; Dejonghe, W.; Van Breusegem, F. Mitochondrial perturbation negatively affects auxin signaling. *Mol. Plant* **2014**, *7*, 1138–1150. [CrossRef]
44. Ivanova, A.; Law, S.R.; Narsai, R.; Duncan, O.; Lee, J.-H.; Zhang, B.; Van Aken, O.; Radomiljac, J.D.; van der Merwe, M.; Yi, K.; et al. A functional relationship between auxin and mitochondrial retrograde signaling regulates Alternative Oxidase1a expression in Arabidopsis. *Plant Physiol.* **2014**, *165*, 1233–1254. [CrossRef]
45. Ohbayashi, I.; Huang, S.; Fukaki, H.; Song, X.; Sun, S.; Morita, M.T.; Tasaka, M.; Millar, A.H.; Furutani, M. Mitochondrial pyruvate dehydrogenase contributes to auxin-regulated organ development. *Plant Physiol.* **2019**, *180*, 896–909. [CrossRef] [PubMed]
46. Tivendale, N.D.; Belt, K.; Berkowitz, O.; Whelan, J.; Millar, A.H.; Huang, S. Knockdown of Succinate Dehydrogenase Assembly Factor 2 Induces Reactive Oxygen Species-Mediated Auxin Hypersensitivity Causing pH-dependent Root Elongation. *Plant Cell Physiol.* **2021**, *62*, 1185–1198. [CrossRef] [PubMed]
47. Tivendale, N.D.; Millar, A.H. How is auxin linked with cellular energy pathways to promote growth? *New Phytol.* **2022**, *233*, 2397–2404. [CrossRef] [PubMed]
48. De Longevialle, A.F.; Meyer, E.H.; Andrés, C.; Taylor, N.L.; Lurin, C.; Millar, A.H.; Small, I.D. The pentatricopeptide repeat gene OTP43 is required for trans-splicing of the mitochondrial nad1 intron 1 in Arabidopsis thaliana. *Plant Cell* **2007**, *19*, 3256–3265. [CrossRef]
49. Meyer, E.H.; Tomaz, T.; Carroll, A.J.; Estavillo, G.; Delannoy, E.; Tanz, S.K.; Small, I.D.; Pogson, B.J.; Millar, A.H. Remodeled respiration in ndufs4 with low phosphorylation efficiency suppresses Arabidopsis germination and growth and alters control of metabolism at night. *Plant Physiol.* **2009**, *151*, 603–619. [CrossRef]
50. Roschzttardtz, H.; Fuentes, I.; Vásquez, M.; Corvalán, C.; León, G.; Gómez, I.; Araya, A.; Holuigue, L.; Vicente-Carbajosa, J.; Jordana, X. A nuclear gene encoding the iron-sulfur subunit of mitochondrial complex II is regulated by B3 domain transcription factors during seed development in Arabidopsis. *Plant Physiol.* **2009**, *150*, 84–95. [CrossRef] [PubMed]
51. Kühn, K.; Obata, T.; Feher, K.; Bock, R.; Fernie, A.R.; Meyer, E.H. Complete mitochondrial complex I deficiency induces an up-regulation of respiratory fluxes that is abolished by traces of functional complex I. *Plant Physiol.* **2015**, *168*, 1537–1549. [CrossRef]
52. Restovic, F.; Espinoza-Corral, R.; Gómez, I.; Vicente-Carbajosa, J.; Jordana, X. An active mitochondrial complex II present in mature seeds contains an embryo-specific iron–sulfur subunit regulated by ABA and bZIP53 and is involved in germination and seedling establishment. *Front. Plant Sci.* **2017**, *8*, 277. [CrossRef]
53. Kolli, R.; Soll, J.; Carrie, C. OXA2b is crucial for proper membrane insertion of COX2 during biogenesis of Complex IV in Plant Mitochondria. *Plant Physiol.* **2019**, *179*, 601–615. [CrossRef]
54. Xue, W.; Ruprecht, C.; Street, N.; Hematy, K.; Chang, C.; Frommer, W.B.; Persson, S.; Niittylä, T. Paramutation-like Interaction of T-DNA loci in Arabidopsis. *PLoS ONE* **2012**, *7*, e51651. [CrossRef] [PubMed]
55. Gao, Y.; Zhao, Y. Epigenetic suppression of T-DNA insertion mutants in arabidopsis. *Mol. Plant.* **2013**, *6*, 539–545. [CrossRef] [PubMed]
56. Sandhu, K.S.; Koirala, P.S.; Neff, M.M. The ben1-1 brassinosteroid-catabolism mutation is unstable due to epigenetic modifications of the intronic T-DNA insertion. *G3 Genes Genomes Genet.* **2013**, *3*, 1587–1595. [CrossRef]
57. Osabe, K.; Harukawa, Y.; Miura, S.; Saze, H. Epigenetic Regulation of Intronic Transgenes in Arabidopsis. *Sci. Rep.* **2017**, *7*, 45166. [CrossRef]
58. Kobayashi, T.; Nishizawa, N.K. Iron uptake, translocation, and regulation in higher plants. *Annu. Rev. Plant Biol.* **2012**, *63*, 131–152. [CrossRef]
59. Aung, M.S.; Masuda, H. How does rice defend against excess iron?: Physiological and molecular mechanisms. *Front. Plant Sci.* **2020**, *11*, 1102. [CrossRef]

60. Gilroy, S.; Jones, D.L. Through form to function: Root hair development and nutrient uptake. *Tr. Plant Sci.* **2000**, *5*, 56–60. [CrossRef] [PubMed]
61. Müller, M.; Schmidt, W. Environmentally induced plasticity of root hair development in Arabidopsis. *Plant Physiol.* **2004**, *134*, 409–419. [CrossRef]
62. Grierson, C.; Nielsen, E.; Ketelaarc, T.; Schiefelbein, J. Root Hairs. *Arab. Book* **2014**, *12*, e0172. [CrossRef] [PubMed]
63. Leōn, G.; Holuigue, L.; Jordana, X. Mitochondrial complex II is essential for gametophyte development in Arabidopsis. *Plant Physiol.* **2007**, *143*, 1534–1546. [CrossRef] [PubMed]
64. Li, L.; Kaplan, J. A mitochondrial-vacuolar signaling pathway in yeast that affects iron and copper metabolism. *J. Biol. Chem.* **2004**, *279*, 33653–33661. [CrossRef] [PubMed]
65. Oldenburg, K.R.; Vo, K.T.; Michaelis, S.; Paddon, C. Recombination-mediated PCR-directed plasmid construction in vivo in yeast. *Nucleic Acids Res.* **1997**, *25*, 451–452. [CrossRef]
66. Gibson, D.G.; Benders, G.A.; Axelrod, K.C.; Zaveri, J.; Algire, M.A.; Moodie, M.; Montague, M.G.; Craig Venter, J.; Smith, H.O.; Hutchison, C.A., III. One-step assembly in yeast of 25 overlapping DNA fragments to form a complete synthetic Mycoplasma genitalium genome. *Proc. Nat. Acad. Sci. USA* **2008**, *105*, 20404–20409. [CrossRef]
67. Bolger, A.M.; Lohse, M.; Usadel, B. Trimmomatic: A flexible trimmer for Illumina sequence data. *Bioinformatics* **2014**, *30*, 2114–2120. [CrossRef]
68. Kim, D.; Paggi, J.M.; Park, C.; Bennett, C.; Salzberg, S.L. Graph-based genome alignment and genotyping with HISAT2 and HISAT-genotype. *Nat. Biotechnol.* **2019**, *37*, 907–915. [CrossRef]
69. Liao, Y.; Smyth, G.; Shi, W. The R package Rsubread is easier, faster, cheaper and better for alignment and quantification of RNA sequencing reads. *Nucleic Acids Res.* **2019**, *47*, 8. [CrossRef]
70. Cheng, C.Y.; Krishnakumar, V.; Chan, A.P.; Thibaud-Nissen, F.; Schobel, S.; Town, C.D. Araport11: A complete reannotation of the Arabidopsis thaliana reference genome. *Plant J.* **2017**, *89*, 789–804. [CrossRef]
71. Love, M.I.; Huber, W.; Anders, S. Moderated estimation of fold change and dispersion for RNA-seq data with DESeq2. *Genome Biol.* **2014**, *15*, 550. [CrossRef]
72. Mi, H.; Muruganujan, A.; Casagrande, J.T.; Thomas, P.D. Large-scale gene function analysis with the panther classification system. *Nat. Prot.* **2013**, *8*, 1551–1566. [CrossRef] [PubMed]

plants

MDPI

Article

Coated Hematite Nanoparticles Alleviate Iron Deficiency in Cucumber in Acidic Nutrient Solution and as Foliar Spray

Amarjeet Singh [1,2], Fruzsina Pankaczi [1,2], Deepali Rana [1,3], Zoltán May [4], Gyula Tolnai [5] and Ferenc Fodor [1,*]

1 Department of Plant Physiology and Molecular Plant Biology, ELTE Eötvös Loránd University, Pázmány Péter Lane 1/c, 1117 Budapest, Hungary; amarpc@student.elte.hu (A.S.); pankafru@gmail.com (F.P.); deepalirana944@gmail.com (D.R.)
2 Doctoral School of Biological Sciences, ELTE Eötvös Loránd University, Pázmány Péter Lane 1/c, 1117 Budapest, Hungary
3 Doctoral School of Environmental Sciences, ELTE Eötvös Loránd University, Pázmány Péter Lane 1/a, 1117 Budapest, Hungary
4 Institute of Materials and Environmental Chemistry, Research Centre for Natural Sciences, Eötvös Loránd Research Network, Magyar Tudósok Blvd. 2, 1117 Budapest, Hungary; may.zoltan@ttk.mta.hu
5 8/A Kondorosi Street, H-1116 Budapest, Hungary; gyula.tolnai@yahoo.com
* Correspondence: ferenc.fodor@ttk.elte.hu

Abstract: Micronutrient iron (Fe) deficiency poses a widespread agricultural challenge with global implications. Fe deficiency affects plant growth and immune function, leading to reduced yields and contributing to the global "hidden hunger." While conventional Fe-based fertilizers are available, their efficacy is limited under certain conditions. Most recently, nanofertilizers have been shown as promising alternatives to conventional fertilizers. In this study, three nanohematite/nanoferrihydrite preparations (NHs) with different coatings were applied through the roots and shoots to Fe-deficient cucumber plants. To enhance Fe mobilization to leaves during foliar treatment, the plants were pre-treated with various acids (citric acid, ascorbic acid, and glycine) at a concentration of 0.5 mM. Multiple physiological parameters were examined, revealing that both root and foliar treatments resulted in improved chlorophyll content, biomass, photosynthetic parameters, and reduced ferric chelate reductase activity. The plants also significantly accumulated Fe in their developing leaves and its distribution after NHs treatment, detected by X-ray fluorescence mapping, implied long-distance mobilization in their veins. These findings suggest that the applied NHs effectively mitigated Fe deficiency in cucumber plants through both modes of application, highlighting their potential as nanofertilizers on a larger scale.

Keywords: iron; micronutrient; nanofertilizers; nanoferrihydrite; nanohematite; XRF mapping

Citation: Singh, A.; Pankaczi, F.; Rana, D.; May, Z.; Tolnai, G.; Fodor, F. Coated Hematite Nanoparticles Alleviate Iron Deficiency in Cucumber in Acidic Nutrient Solution and as Foliar Spray. *Plants* **2023**, *12*, 3104. https://doi.org/10.3390/plants12173104

Academic Editor: Michael Moustakas

Received: 5 August 2023
Revised: 25 August 2023
Accepted: 27 August 2023
Published: 29 August 2023

1. Introduction

Iron (Fe) is an essential micronutrient for plants, playing a vital role in various metabolic processes, such as photosynthesis, respiration, and chlorophyll biosynthesis. It acts as an electron donor or acceptor in enzymes in the form of Fe–sulfur clusters, heme, and free Fe ion [1]. As a constituent of the photosynthetic electron transport chain, Fe plays a crucial role in plant growth and development [2].

Despite being the fourth most abundant element in the Earth's crust, plants continue to suffer from Fe deficiency [3]. Fe deficiency is a common problem in agriculture, particularly in alkaline or calcareous soils where higher pH levels prevent the uptake of Fe [4]. Fe deficiency manifests in plants as chlorosis or yellowing of the leaves, stunted growth, and reduced crop yields. The symptoms are more pronounced in young leaves and can result in poor root development, reduced yield, or even complete crop failure [5]. Fe deficiency can also reduce the production of proteins involved in the photosynthetic apparatus, decreasing photosynthetic efficiency. This includes alterations in the thylakoid membrane structures, photosynthetic electron transport chain, and reduced formation of iron–sulfur complexes [6]. Chloroplast biogenesis and differentiation are also negatively affected by Fe deficiency [7].

Plants have evolved two main strategies to acquire Fe from the soil. Dicots and non-graminaceous monocots use reduction-based strategy I. In this strategy, Fe is reduced from the ferric (Fe^{3+}) to ferrous (Fe^{2+}) form by a membrane-bound ferric reductase oxidase (FRO). The reduced Fe is then transported into the root epidermal cells via another membrane-localized iron-regulated transporter (IRT1). Phenolics, particularly coumarins and other secondary metabolites such as flavins, are also secreted by the roots and can directly reduce Fe^{3+} and chelate both Fe (III) and Fe (II), thus improving Fe mobilization and reduction [8]. Strategy II is based on chelation of Fe and is used by graminaceous plants. It is dependent upon the release of mugineic acid family phytosiderophores (MAs) in the rhizosphere that chelate ferric iron. In rice, transporter of mugineic acids (TOM1) is involved in the secretion of deoxy mugineic acid (DMA) from the root to the rhizosphere in Fe deficiency [9]. The chelated Fe-phytosiderophore complex is then transported into the root via yellow stripe (YS) and yellow stripe-like (YSL) transporters [10,11].

The traditional methods for alleviating Fe deficiency in agriculture include soil amendment with Fe-rich fertilizers, e.g., Fe sulfate, Fe chelate, or Fe oxide. In addition, use of Fe-rich organic matter, such as compost or green manure, is also considered. However, these have certain limitations. For example, Fe-rich fertilizers are usually ineffective in soils with high pH levels, as Fe is less soluble in alkaline conditions [12]. Moreover, even though they are cost-effective, higher amounts need to be applied in soil, which can prove to be toxic to plants and cause leaf damage or stunted growth. Fe fertilization with these may provide a temporary boost to Fe levels in plants, but the effects are not long-lasting and repeated applications may be necessary [13,14]. In addition, poor distribution of these fertilizers may result from leaching or fixation in soil, thus decreasing their effectiveness. On the other hand, chelated fertilizers, such as Fe-EDDHA, are effective even in alkaline conditions, but they are uneconomic and their application is limited to cash crops [15].

Due to the limitations of conventional fertilizers, there is increasing interest in exploring more environmentally friendly solutions, such as manufactured nanomaterials. Nanomaterials have unique properties, such as high surface area and reactivity, making them ideal for plant nutrient supplementation. The use of nanoscale Fe particles can provide a more efficient and sustainable solution for Fe supplementation, as they can be more easily absorbed by plants, reducing the amount of total fertilizer required [16]. Additionally, the use of nanoscale Fe particles can help to alleviate the issue of soil alkalinity, as they can be designed to release Fe at a specific pH level, improving the availability of Fe for plants [17]. In agriculture, Fe-based nanomaterials have been studied for their potential to alleviate Fe deficiency in crops. Recent studies have demonstrated that the application of nanofertilizers, both through the root system and as foliar treatment, can dramatically enhance plant growth and yield [18–20].

Fe exists as four main different forms in soils: (1) as Fe^{2+} released from primary minerals such as silicates, borates, and sulfates; (2) as ferric oxyhydroxides after oxidation and precipitation under aerobic conditions within the pH range of 5 to 8 (also known as pedogenic Fe); (3) as soluble and exchangeable Fe; and (4) as short-range ordered crystalline minerals, including ferrihydrite and schwertmannite, which are bound to organic matter [21]. Fe oxides that are most abundant in soil with low solubility include geothite (α-FeOOH) and hematite (α-Fe_2O_3) [16,22].

Fe oxides are known to have high affinity for water, with the ability to absorb up to one mole of excess H_2O per mole of Fe oxide. The affinity of these oxides for water is further increased by their small particle sizes, particularly at the nanoscale. In fact, nanoparticles of hematite can form naturally in soil through nucleation when groundwater becomes saturated with clusters of the same mineral phase. These nanoparticles are an important source of bioavailable Fe, which is essential for plants and microorganisms to thrive, grow, and diversify [23–25].

In previous work, we compared various aspects of nanoferrihydrite and nanohematite suspensions, including their utilization in providing Fe to plants [26]. It was found that nanohematites performed better than nanoferrihydrites, and surfactants such as polyethylene glycol polymers may increase the efficiency of nanohematites. In the present investigation, nanocolloid suspensions of hematites/ferrihydrites with different surfactant coatings were assessed for their effectiveness as supplements in nutrient solutions and as foliar sprays to address Fe deficiency in plants. By using these nanohematite/nanoferrihydrite (NH) preparations, this study sought to identify the most effective application method to enhance plant growth and alleviate Fe deficiency symptoms in cucumber model plants.

2. Results

2.1. Greening and Chlorophyll Concentration

In this study, three NH colloid suspensions differing mainly in the surfactant (PEG-1500, NH-S1; Emulsion 104D, NH-S2; and SOLUTOL HS 15, NH-S3) used for stabilization were used as amendments to nutrient solutions or as foliar fertilizers. For further characterization of the NHs, please see Section 4.2.

The roots of 2-week-old Fe-deficient (dFe) cucumber plants were supplied with a nominal concentration of 20 μM Fe in NH-S1-3 at two different pH values (unbuffered acidic pH 6.0 and alkaline pH buffered at 8.5). Plants at acidic pH showed a remarkable increase in chlorophyll (Chl) a+b content (Table 1). A significant 20–21-fold increase was observed for all three NHs in comparison with dFe plants. The Chl a/b ratio decreased when plants were supplied with NHs compared to dFe plants. Greening was also assessed using a SPAD instrument. Within 24 h of NH supplementation, plants showed an increase in the SPAD index. All three NH-treated plants exhibited a 13–14-fold increase in the SPAD values of the 1st leaf after 6 days of treatment compared with the corresponding dFe plants. For the 2nd leaves as well, in comparison to dFe plants, the SPAD values showed an 11-fold increase. In general, the relative SPAD values for dFe leaves continued to decline with time (Figure 1a,b). At pH 8.5, there was no greening at all and the Chl a/b ratio did not change significantly either. (SPAD values are not shown).

Table 1. The chlorophyll content of dFe plants treated with NH-S1-3 supplied in nutrient solutions at two different pH values. Data are shown as mean ± SD (n = 5). Significant differences between data are shown by different letters, $p < 0.05$.

Treatment	Chl a+b (μg/g FW)		Chl a/b	
	Acidic pH	Alkaline pH	Acidic pH	Alkaline pH
NH-S1	1924.8 ± 287.3 [a]	171.9 ± 28.99 [b]	3.0 ± 0.05 [b]	3.0 ± 0.22
NH-S2	2063.9 ± 76.3 [a]	173.8 ± 28.30 [b]	3.0 ± 0.09 [b]	3.0 ± 0.57
NH-S3	2092.9 ± 171.8 [a]	168.2 ± 39.80 [b]	2.9 ± 0.03 [b]	2.7 ± 0.28
sFe	2555.28 ± 191.76 [a]	2273.06 ± 246.92 [a]	2.4 ± 0.14 [b]	2.8 ± 0.03
dFe	92.2 ± 27.7 [b]	162.6 ± 70.97 [b]	4.9 ± 1.03 [a]	2.8 ± 0.54

In a preliminary experiment, it was found that NHs applied alone as foliar treatment did not induce Chl synthesis. For this reason, we applied citric acid (Cit), ascorbic acid (Asa), and glycine (Gly) as pre-treatments before the application of NHs. The leaves of plants with foliar application of NH-S1-3 demonstrated remarkable greening regardless of the pre-treatment received. The SPAD values of both the 1st and 2nd leaves increased significantly, ranging from 1.3- to 3-fold compared to their original values (Figure 1c,d) for all NHs. However, the greening of leaves was not uniform across the blade, with small, dark green spots corresponding to NH droplets clearly visible on the lamina (Supplementary Figure S1). Nevertheless, with time, the young 2nd leaves exhibited subsequent growth in surface area and increased overall greening.

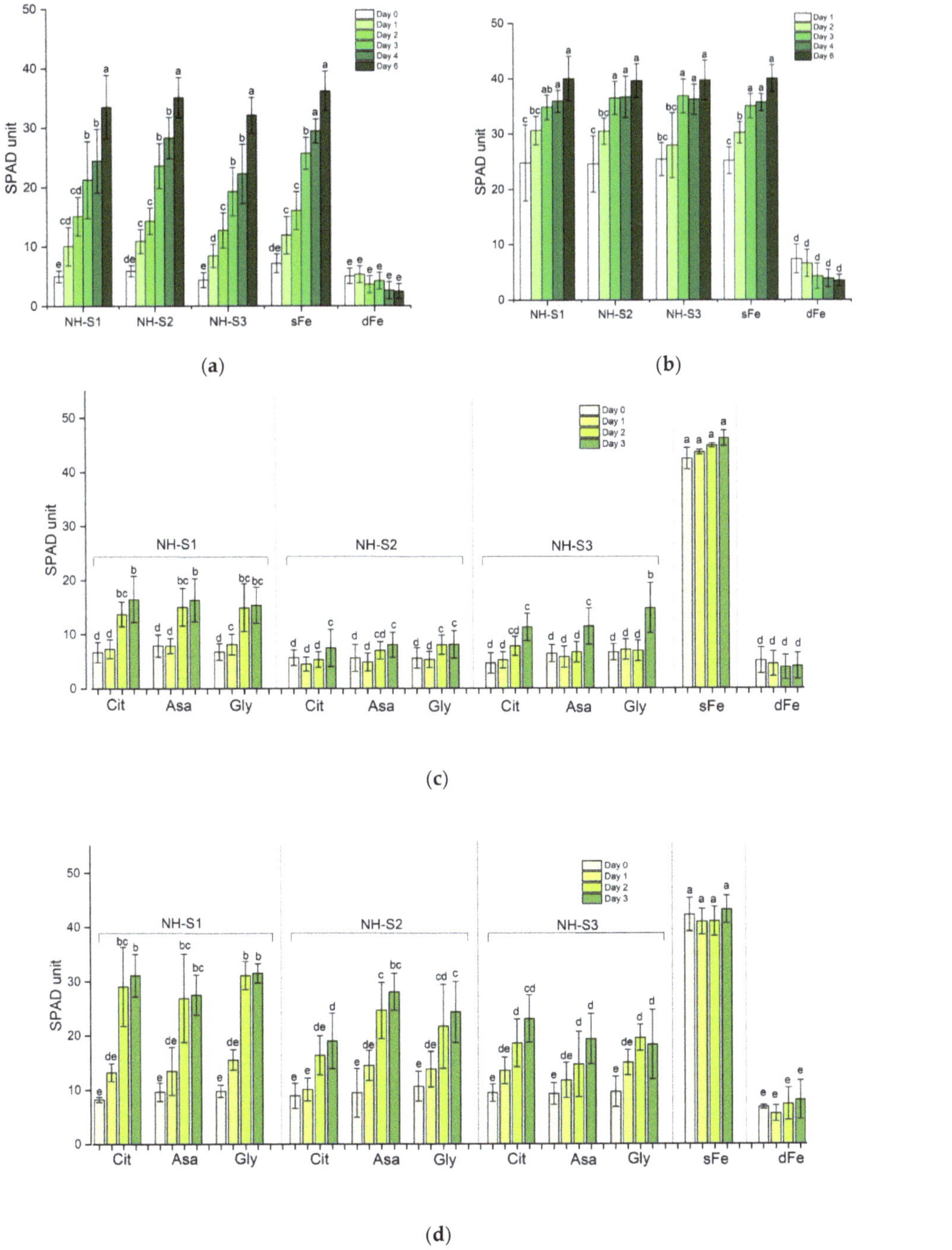

Figure 1. SPAD values of the 1st and 2nd leaves of dFe plants supplied with NH-S1-3 in unbuffered nutrient solution (**a**,**b**) and as foliar treatment combined with citric acid (Cit), ascorbic acid (Asa), or glycine (Gly) pre-treatment (**c**,**d**) (dFe and sFe plants served as controls). Data are shown as mean ± SD (n = 5). Significant differences between data are shown by different letters, $p < 0.05$.

2.2. Biomass and pH

Plants that were supplied with NHs in nutrient solutions at acidic pH showed an increase in biomass after 6 days of treatment (Figure 2). NH-S2-treated plants showed maximum gain in biomass, with a significant 2-fold increase in dry weight (DW) upon comparison with dFe plants. Importantly, the biomass recorded for this treatment was higher than that of Fe-citrate-treated control plants. The other two NHs also resulted in enhanced biomass compared to dFe plants, in the order NH-S2 > NH-S3 = NH-S1, but the change was not significant. For plants grown at alkaline pH, no significant change in biomass was recorded when compared with dFe plants (data not shown).

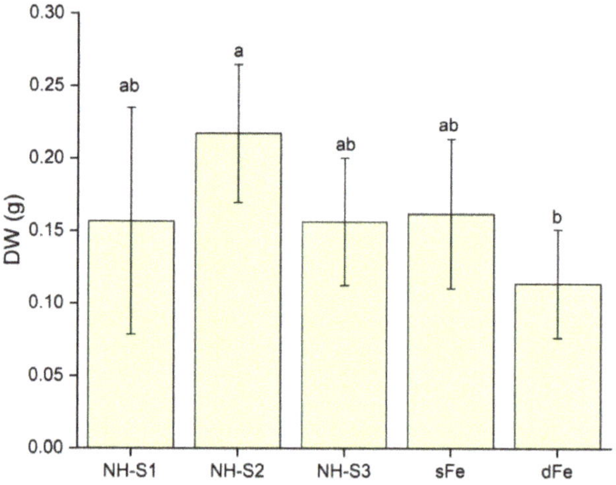

Figure 2. Total dry weight (DW) of plants grown in unbuffered nutrient solutions with 3 different NHs after 6 days. Data are presented as mean ± SD (n = 5) significant differences between data are shown by different letters, $p < 0.05$.

Regarding the pH of the unbuffered nutrient solution, there was smaller decrease in the pH values of plants treated with NH-S1-3 at the end of experiment than that of dFe plants (Table 2). As anticipated, the pH of the dFe plants decreased by approximately one pH unit. The final pH difference between nutrient solutions of the NH-treated and untreated dFe plants was significant whereas the 3 NHs were not different from each other. For plants at alkaline pH, the final pH of the nutrient solution did not deviate significantly from the original pH 8.5 as it was maintained by buffer (data not shown).

Table 2. The pH values of nutrient solutions 3 days after supplying NH-S1-3 to dFe plants in (unbuffered) nutrient solution or as foliar treatment. The initial pH values of the fresh nutrient solutions were 4.91 for dFe, 5.50 for sFe, 4.81 for NH-S1, 4.71 for NH-S2, and 4.66 for NH-S3. Data are presented as mean ± SD (n = 5). Significant differences between data are shown by different letters, $p < 0.05$, where statistical analysis was run independently for hydroponic and foliar treatments.

Treatment	Root Supply	Foliar Supply		
		Cit	Asa	Gly
NH-S1	4.27 ± 0.20^{B}	4.57 ± 0.03^{ab}	4.36 ± 0.19^{b}	4.04 ± 0.16^{b}
NH-S2	4.39 ± 0.25^{AB}	4.46 ± 0.17^{b}	4.13 ± 0.13^{b}	3.84 ± 0.15^{b}
NH-S3	4.37 ± 0.26^{AB}	4.60 ± 0.22^{ab}	4.89 ± 0.05^{ab}	4.18 ± 0.35^{b}
sFe	6.29 ± 0.08^{A}		5.97 ± 0.20^{a}	
dFe	3.88 ± 0.17^{C}		4.07 ± 0.36^{b}	

The pH of the nutrient solution of plants with foliar treatment, as an indicator of the downregulation of Fe-efficiency reactions, was slightly influenced by the type of pre-treatment they received. For all NHs, pre-treatments did not result in highly remarkable changes in the pH of the nutrient solution. Pre-treatment with Cit and in case of NH-S3 with Asa seemed to result in higher pH values than the other treatments. Surprisingly, the pH of plants pre-treated with Gly was remarkably low and for NH-S2, it was even lower than that of the dFe plants.

2.3. *Ferric Chelate Reductase Activity*

As the plants grown at alkaline pH did not show promising greening, further measurements were performed only for plants treated with nutrient solutions at unbuffered pH and with foliar sprays. Three days after treatment, the root ferric chelate reductase (FCR) activity was measured to evaluate the regeneration from Fe deficiency. The results shown in Figure 3a clearly demonstrate that the FCR activity of all NH-treated roots was significantly lower than that of dFe roots. The FCR activity of the roots of treated plants was in the range of that of sFe plants grown with Fe-citrate.

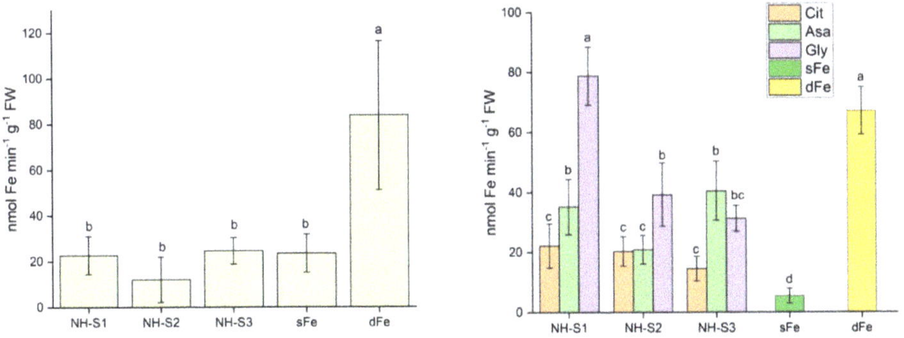

(a) (b)

Figure 3. Ferric chelate reductase activity of the root tips 3 days after supplying NH-S1-3 to dFe plants in (unbuffered) nutrient solution (**a**) or as foliar treatment (**b**). Data are presented as mean ± SD (n = 5). Significant differences between data are shown by different letters, $p < 0.05$.

The root FCR activity of the plants supplied with NHs as foliar spray after 3 days of treatment is shown in Figure 3b. All NH treatments resulted in significant suppression of FCR activity (except for NH-S1/Gly), suggesting that the plants could utilize Fe from the supplied NHs. Interestingly, the pre-treatment spraying solution had a significant impact modulating the effect of the NHs. The lowest FCR activity for all NHs was recorded in plants pre-treated with Cit, with significant 4.62-, 3.31-, and 3.02-fold decreases after NH-S3, NH-S2, and NH-S1 treatments, respectively, compared to the FCR activity of dFe roots. Similarly, Asa pre-treatment caused a significant suppression of root FCR activity for all three NHs, with 1.65-, 3.21-, and 1.88- fold decreases measured for NH-S3, NH-S2, and NH-S1, respectively. In contrast, Gly pre-treatment had a relatively lesser effect on FCR activity, with a 2-fold decrease only for NH-S3 and 1.7-fold decrease for NH-S2. For NH-S1, Gly pre-treated plants had comparable FCR activity to that of dFe plants. Therefore, these findings implied that pre-treatment of plants with Cit, Asa, and Gly had a significant repressive effect on root FCR activity for all three NHs, except for Gly pre-treatment in NH-S1-treated plants.

2.4. *Photosynthetic Efficiency*

Higher values of CO_2 assimilation rate (A_{net}) observed in Fe-citrate control and NH-treated plants indicated their greater ability to photosynthesize and convert CO_2 into organic carbon compounds (Table 3). However, a negative A_{net} value was recorded for dFe plants, indicating release instead of fixation of CO_2.

Table 3. Photosynthetic and Chl a fluorescence induction parameters of plants after 3 days of supplementation with NH-S1-3 (CO_2 assimilation, Anet; maximum quantum efficiency of PSII, Fv/Fm; actual quantum efficiency of PSII in light-adapted state, Fv′/Fm′; actual quantum efficiency of PSII, ΦPSII; quantum yield of CO2, ΦCO_2; non-photochemical quenching, NPQ). The NHs were applied either in nutrient solution or as foliar treatment after pre-treatment with Cit, Asa, or Gly. Untreated dFe and sFe plants served as controls. Data are presented as mean ± SD (n = 5). Statistical analysis was run independently for hydroponic and foliar treatments. Significant differences between data are shown by different letters, $p < 0.05$.

Root Supply	Assimilation (A_{net}) (μmol CO_2 m^{-2} s^{-1})	Fv/Fm	Fv′/Fm′	ΦPSII	ΦCO_2 (μmol CO_2 μmol^{-1} Photons)	NPQ
NH-S1	12.901 ± 2.526 ^{a}	0.798 ± 0.005 ^{a}	0.634 ± 0.034 ^{a}	0.478 ± 0.031 ^{a}	0.033 ± 0.006 ^{A}	0.914 ± 0.242 ^{a}
NH-S2	14.840 ± 1.426 ^{a}	0.798 ± 0.001 ^{a}	0.641 ± 0.025 ^{a}	0.488 ± 0.029 ^{a}	0.036 ± 0.003 ^{A}	0.832 ± 0.173 ^{a}
NH-S3	14.251 ± 0.560 ^{a}	0.799 ± 0.004 ^{a}	0.637 ± 0.001 ^{a}	0.487 ± 0.010 ^{a}	0.033 ± 0.002 ^{A}	0.921 ± 0.124 ^{a}
sFe	14.156 ± 3.421 ^{a}	0.806 ± 0.004 ^{a}	0.650 ± 0.027 ^{a}	0.491 ± 0.046 ^{a}	0.033 ± 0.007 ^{A}	0.845 ± 0.130 ^{a}
dFe	−0.506 ± 0.123 ^{b}	0.570 ± 0.047 ^{b}	0.389 ± 0.125 ^{b}	0.090 ± 0.009 ^{b}	0.003 ± 0.003 ^{b}	0.574 ± 0.130 ^{b}
Foliar supply						
NH-S1						
Cit	9.596 ± 2.374 ^{ab}	0.805 ± 0.002 ^{a}	0.568 ± 0.007 ^{a}	0.388 ± 0.015 ^{ab}	0.025 ± 0.004 ^{a}	1.351 ± 0.105 ^{a}
Asa	10.725 ± 3.810 ^{ab}	0.807 ± 0.006 ^{a}	0.587 ± 0.045 ^{a}	0.426 ± 0.041 ^{ab}	0.028 ± 0.006 ^{a}	1.309 ± 0.317 ^{a}
Gly	11.193 ±3.895 ^{ab}	0.809 ± 0.002 ^{a}	0.593 ± 0.015 ^{a}	0.426 ± 0.031 ^{ab}	0.029 ± 0.008 ^{a}	1.204 ± 0.128 ^{a}
NH-S2						
Cit	6.423 ± 0.356 ^{b}	0.772 ± 0.002 ^{a}	0.532 ± 0.042 ^{a}	0.324 ± 0.046 ^{b}	0.024 ± 0.004 ^{a}	1.111 ± 0.199 ^{a}
Asa	9.618 ± 2.917 ^{ab}	0.777 ± 0.008 ^{a}	0.586 ± 0.047 ^{a}	0.401 ± 0.089 ^{ab}	0.026 ± 0.004 ^{a}	1.196 ± 0.447 ^{a}
Gly	10.241 ± 1.368 ^{ab}	0.776 ± 0.008 ^{a}	0.611 ± 0.026 ^{a}	0.434 ± 0.017 ^{a}	0.035 ± 0.011 ^{a}	0.867 ± 0.079 ^{a}
NH-S3						
Cit	4.835 ± 1.478 ^{b}	0.790 ± 0.009 ^{a}	0.579 ± 0.038 ^{a}	0.286 ± 0.047 ^{b}	0.020 ± 0.005 ^{a}	1.082 ± 0.387 ^{a}
Asa	6.308 ± 1.163 ^{b}	0.777 ± 0.009 ^{a}	0.596 ± 0.062 ^{a}	0.319 ± 0.033 ^{b}	0.023 ± 0.003 ^{a}	0.941 ± 0.501 ^{a}
Gly	5.999 ± 1.689 ^{b}	0.776 ± 0.013 ^{a}	0.588 ± 0.036 ^{a}	0.322 ± 0.046 ^{b}	0.022 ± 0.005 ^{a}	0.972 ± 0.320 ^{a}
sFe	14.770 ± 3.799 ^{a}	0.797 ± 0.006 ^{a}	0.637 ± 0.031 ^{a}	0.463 ± 0.051 ^{a}	0.040 ± 0.006 ^{a}	0.821 ± 0.186 ^{a}
dFe	−1.950 ± 0.656 ^{c}	0.473 ± 0.051 ^{b}	0.313 ± 0.076 ^{b}	0.043 ± 0.006 ^{c}	0.004 ± 0.001 ^{b}	0.462 ± 0.187 ^{b}

The efficiency of the photosystem in fixing CO_2 was assessed by measuring ΦCO_2, which represents the quantum yield of CO_2 fixation during photosynthesis. The dFe plants exhibited significantly lower ΦCO_2 values compared to Fe-citrate control and NH-treated plants, indicating poor performance in fixing the incoming CO_2 (Table 3). This reduced CO_2 fixation was attributed to the observed lower biomass as well as the reduced plant growth and productivity in dFe plants. Foliar spray-treated plants also showed a significant increase in ΦCO_2 (0.0262 ± 0.0070 μmol μmol^{-1} photons) compared to dFe plants (0.00499 ± 0.00214 μmol μmol^{-1} photons).

The maximum quantum efficiency of photosystem II (Fv/Fm) of plants supplied with NHs in nutrient solution was in the healthy range compared to sFe plants (Table 3). Similarly, NH-S1 applied via the foliar method, regardless of pre-treatment, as well as NH-S2 with Cit pre-treatment caused full recovery. NH-S2 with Asa and Gly and NH-S3 with all pre-treatments also caused a significant increase in the Fv/Fm values of leaves compared to dFe plants; however, they did not reach the optimal range.

Similarly, the actual quantum efficiency of photosystem II (PSII) in a light-adapted state (Fv′/Fm′) showed recovery for all NH-treated dFe cucumbers, as the values reached the same levels as Fe-citrate control plants, with a 68% recorded increase compared to dFe plants (Table 3). Additionally, for all foliar treatments, a significant average increase of 88% in Fv′/Fm′ (0.582 ± 0.044) was observed compared to dFe plants, which approached that of Fe-citrate control plants.

The actual quantum efficiency of photosystem II (ΦPSII) for all plants (root-applied NHs) was in the same range as that of Fe-citrate control plants (0.47 ± 0.028) (Table 3). The ΦPSII value was recorded to be much lower for dFe plants, which was 80% less than that of Fe-citrate control plants. Additionally, for foliar spray-treated plants, the ΦPSII values increased significantly, but slightly lower values were recorded for NH-S3 than for the other two NHs.

In the presence of Fe deficiency stress, the non-photochemical quenching (NPQ) values demonstrated a marked reduction, indicating an impairment of both electron transport and proton pumping, which are critical for photosynthesis (Table 3). Fe-deficient plants, therefore, face a challenge in curbing the excess energy that may lead to photodamage and reduced photosynthetic efficiency. However, plants treated with NHs, both via foliar and root application, as well as Fe citrate control plants, exhibited NPQ values that were comparable in range. This suggested that they effectively dissipated excess energy and potentially experienced photoprotection under conditions of stress.

2.5. Element Analysis

The 2nd leaves of plants supplied with NHs in nutrient solution were analyzed for Fe, Mn, and Zn concentrations. The NH-treated plants successfully internalized the Fe originating from NHs, as shown in Figure 4. The Fe concentration doubled in NH-treated plants, but it did not reach that of sFe plants. This was accompanied by a visible but yet non-significant reduction in the uptake of Mn and Zn. The levels of Mn and Zn rose more than 2.3- and 2.6-fold in dFe plants in the absence of Fe in comparison with Fe-citrate control plants.

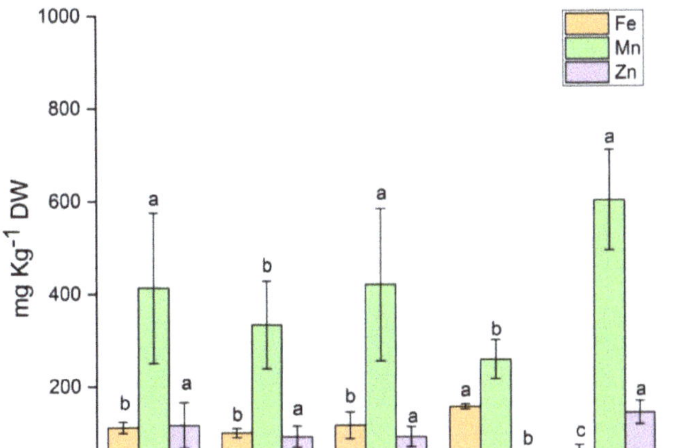

Figure 4. Element analysis of the 2nd leaves of plants treated with NH-S1-3 supplied in nutrient solution after 6 days. dFe and sFe plants served as controls. Data are presented as mean ± SD (n = 5). Statistical analysis was run independently for different elements. Significant differences between data are shown by different letters, $p < 0.05$.

2.6. X-ray Fluorescence Mapping of the Leaves

X-ray fluorescence mapping was performed on arbitrarily chosen sections of the 2nd leaves of plants to observe the distribution of Fe in comparison with well-detectable macroelements such as K and Ca (Figure 5). The macroelements showed well-characterized distributions in the leaves, with high intensity. In sFe plants, Ca showed accumulation especially in small-concentrated patches that were evenly distributed and in the veins. In dFe plants, the distribution was similar but the intensity was much lower, indicating lower concentrations. Potassium showed more uniform distribution in the interveinal sections, but its accumulation was confined to the major and minor veins in both sFe and dFe plants. This distribution was primarily observed in plants after application of foliar treatment. Iron accumulation was seen in sFe plants but not in untreated dFe plants. Patches of Fe were also seen on NH-S3/Cit and NH-S3/Gly leaves, obviously due to treatment solution dried on the surface, but not in other cases. A well-detectable Fe intensity was seen in NH-S1/Cit, NH-S1/Gly, NH-S2/Cit, and NH-S2/Asa, whereas in other treatments, the Fe intensity was much lower although the amount of Fe applied in the NH treatments was the same.

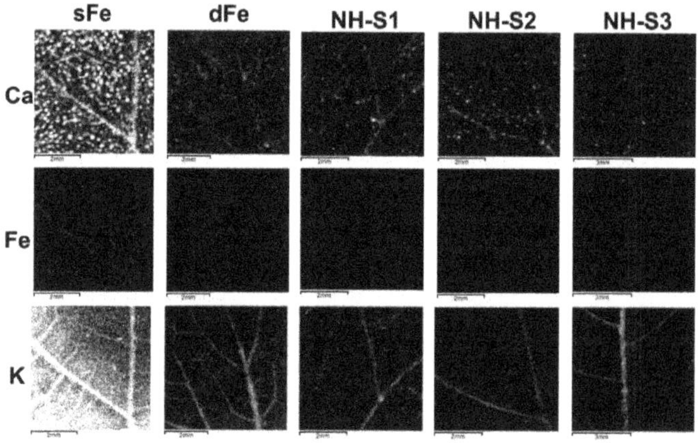

(**a**)

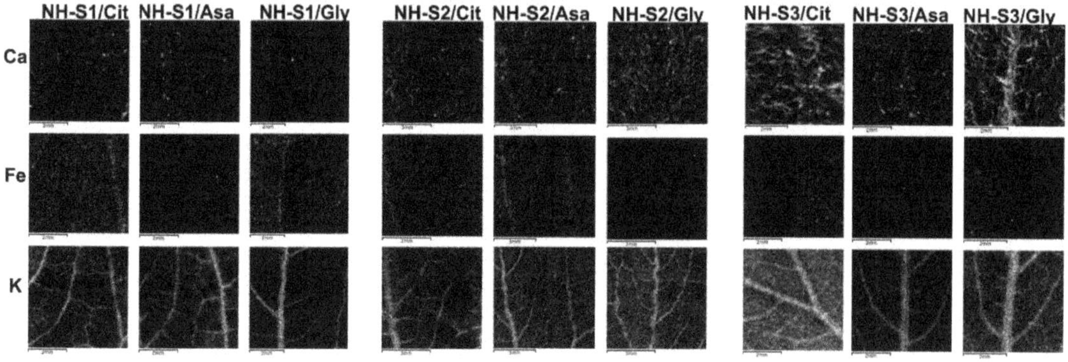

(**b**)

Figure 5. Representative X-ray fluorescence maps of arbitrarily selected leaf sections from the 2nd leaves of plants harvested 3 days following foliar treatment. Panel (**a**) demonstrates leaves of control plants (sFe and dFe) and plants treated with NH-S1-3 supplied in nutrient solution, while panel (**b**) exhibits leaves of plants treated with NH-S1-3 applied as foliar spray and pre-treated with Cit, Asa, or Gly. Data for optical images were collected for about 150 min at Kα emission for Ca, Fe, and K (peak emissions are 3.69, 6.40, and 3.31 keV, respectively).

3. Discussion

In this investigation, three NH colloid suspensions with different coatings, when supplemented in unbuffered nutrient solution at a nominal concentration of 20 μM, were found to ameliorate Fe deficiency symptoms in cucumber plants. The treated dFe plants showed enhancements in biomass, chlorophyll content, and improved photosynthetic efficiency. All of these indicate a metabolic shift towards Chl synthesis and assimilation of the photosynthetic apparatus. Furthermore, the acidification of the nutrient solution and FCR activity of the root tips decreased, which showed downregulation of the high-affinity Fe uptake system [27]. Element analysis of the 2nd leaves confirmed that the plants were able to efficiently take up and utilize Fe from all three NHs at unbuffered pH. On the contrary, when the NHs were supplied in nutrient solutions buffered to pH 8.5, no improvements were found in the measured parameters and the plants remained Fe deficient. Various studies have been carried out to understand the effect of hematite particles on plants. Boutchuen et al. [18] applied hematite NPs of ~16 nm at low (0.022 $g \cdot L^{-1}$ Fe) and high (1.1 $g \cdot L^{-1}$ Fe) concentrations to seeds of four different commercial crops, resulting in 230–280% increases in plant growth. The treatment also increased the survival span of plants, doubled fruit production per plant, accelerated fruit production by nearly two times, and produced healthy second-generation plants with slight species-specific variations. In a recent study by Ndou et al. [28], green-synthesized hematite NPs (average NP diameter of 18 nm) positively impacted sorghum bicolor growth and inhibited the oxidative damage of proteins, DNA, and lipids by enhancing nutrient uptake and osmoregulation under drought conditions. Rath et al. [29] also demonstrated enhanced seed germination for *Triticum aestivum*, fenugreek (*Trigonella foenum-graecum*), Bengal gram (*Cicer arietinum*), and broad bean (*Vicia faba*) in the presence of a low amount of hematite NPs (20 $mg \cdot L^{-1}$). Similarly, Pariona et al. [20] studied the effect of hematite NPs (100 nm, ovoid and rounded shape) on *Zea mays* and reported enhanced growth and chlorophyll content of maize seedlings. All of these results are in accordance with the positive effects of NHs found in our study.

We also studied the effectiveness of the three NHs as foliar fertilizers applied on dFe model plants. However, in a preliminary experiment, it was found that NHs applied on the leaves alone was not effective at all. Therefore, we examined the ability of three different pre-treatments (Cit, Asa, and Gly) to facilitate Fe uptake from the supplied NHs.

Various studies have explored the biostimulatory effects of organic and amino acids on plant growth and nutrient uptake enhancement under various biotic and abiotic stress conditions [30–32]. Rasp [33] found that foliar spraying of amino acids and $FeSO_4$ was more effective in promoting the absorption, translocation, and utilization of Fe in grapevine plants compared to $FeSO_4$ application alone. Additionally, studies by Tejada and Gonzalez [34] with asparagus (*Asparagus officinalis* L.) plants demonstrated the beneficial effects of amino acids, including increased root and shoot growth and chlorophyll content in leaves. Sánchez-Sánchez et al. [35,36] also observed that soil application of amino acids mixed with Fe-EDDHA improved Fe uptake and several fruit quality parameters in citrus and tomato plants. Most recently, Rajaie and Tavakoly [37] found that citric acid and sulfuric acid enhanced the leaf chlorophyll content of orange trees, indicating their role in the remobilization of already existing Fe in leaves. The improvement of plant Fe nutrition could be attributed to the ability of amino acids to act as natural chelators and efficient carriers of Fe into plants, as well as their effects on metabolism and plant physiology, such as the stimulation of H^+-ATPase and FCR activity and increased cell membrane permeability, as suggested by Sánchez-Sánchez et al. [35,36].

When we applied Cit, Asa, or Gly as a pre-treatment, it was found that Cit was the most effective regardless of the NH used. Plants treated with Cit had the lowest FCR activity and higher Fe accumulation compared to those treated with Gly or Asa. It is well established that Cit, succinic acid, and malic acid are involved in Fe metabolism [37]. Cit, specifically, serves as an intermediate in the Krebs cycle and forms a crucial Fe (III)–citrate complex that is highly mobile and facilitates the transport of Fe between various plant compartments over long distances [38,39]. The applied Cit may mobilize and complex ferric Fe from the applied NHs. Thus, Cit proved to be an effective pre-treatment solution for the NH foliar fertilizers, enhancing Fe utilization by the leaves and leading to the downregulation of root FCR activity in our model plants.

Another compound we applied as a pre-treatment, Asa, is a well-known acid for its ability to effectively scavenge oxidative stress and reduce the production of free radicals caused by abiotic stress factors [40]. Applying Asa to the leaves of plants can stimulate the plant's natural production of Asa, helping to mitigate stressful conditions [41,42]. Asa also plays a critical role in maintaining essential plant processes, such as photosynthesis, cell wall expansion, plant hormone production, regulation of antioxidant systems, and ion uptake, thereby increasing yield [43]. In vivo and in vitro studies have provided evidence that Asa can promote Fe solubility in mammals [44,45].

Several studies have investigated the use of Asa to promote plant growth. For instance, Shokr and Abdelhamid [46] found that foliar spraying of Asa resulted in improved vegetative growth and yield, and increased soluble solid substances and pigment constituents in pea (*Pisum sativum* L) plants compared to untreated controls (tap water). Similar results were observed in lettuce (*Lactuca sativa* L.) by Shafeek et al. [47] and in soybean (*Glycine max* L var. klark) by Mansour [48], where foliar spray of Asa led to significant increases in growth, yield, and nutritional content in the leaves. Ramírez et al. [49] reported that supplementing Arabidopsis seedlings with Asa had protective effects against Fe deficiency, maintaining chlorophyll content without increasing the internal Fe concentration. The authors suggested that this could be due to the antioxidant properties of Asa, as the protective effect was correlated with decreased levels of reactive oxygen species (ROS) and higher activity of ascorbate peroxidase, thereby maintaining cell redox homeostasis in Fe-deficient plants. In the present study, pre-treatment with Asa resulted in improved Fe utilization from the NHs by cucumber plants, similar to Cit pre-treatment, with lower root FCR activity (especially in case of NH-S2) compared to dFe plants.

Glycine is a commonly used amino acid in plant nutrition and is often utilized in the production of various amino-chelate fertilizers. External application of Gly can enhance the nitrogen status and concentration of mineral elements in plants [50]. Souri et al. [51] demonstrated that both foliar and soil application of Fe-glycine chelate resulted in a substantial improvement in the growth, yield, and quality characteristics of bean plants (*Phaseolus vulgaris* L.), even more so than Fe-EDDHA. Similarly, the application of Gly through Hoagland nutrient solution significantly increased leaf SPAD value and the fresh and dry weights of shoots and roots in coriander (*Coriandrum sativum*) plants, as reported in [52]. Noroozlo et al. [53] also showed that foliar application of Gly led to a significant increase in leaf Fe concentration compared to control Romain lettuce (*Lactuca sativa* subvar Sahara) plants. The latest study by Xu et al. [54] provided evidence using dual-isotope labeling tests that the presence of nitrogen in glycine can negatively influence Zn absorption by the leaves of waxy corn (*Zea mays* L. var. *ceratina* Kulesh), despite the fact that ZnGly facilitated the storage of Zn in the seeds, with improved Zn use efficiency. In the current study, pre-treatment with Gly improved the Fe utilization efficiency from the NHs in cucumber plants, as indicated by the increase in chlorophyll content and photosynthetic efficiency. However, the pH values of the nutrient solutions remained in the range of dFe plants for all NHs and the root FCR activity decreased only in case of NH-S2 and NH-S3, whereas it remained high for NH-S1.

The utilization of Fe in the mesophyllum tissues of treated leaves was also examined using XRF spectroscopy. The highest intensity was found in the case of macroelements such as Ca and K. Ca distribution in the leaf is determined by its accumulation in the cell walls and apoplastic spaces, especially around stomatal guard cells [55]. This was clearly seen in sFe plants and those treated with the three NHs in nutrient solutions. However, the Ca distribution in plants treated with NHs through the leaves resembled that of dFe plants. K distribution was mostly confined to the main veins, as it is delivered to leaves through the xylem in high concentrations. This provides a perfect orientation in leaf surface images, which is particularly useful in analyses of trace elements in low concentrations, such as Fe [56].

In the case of NHs applied in nutrient solution, a healthy delivery of Fe was found as compared to its distribution in sFe plants. Certainly, the intensity was much lower as the time of Fe supply was much shorter. In case of foliar application, NH-S1/Cit, NH-S1/Gly, NH-S2/Cit, and NH-S2/Asa also resulted in Fe distribution similar to that in sFe plants, as the veins were identified and the interveinal sections also had homogeneous Fe intensity. In the case of NH-S3, Fe intensity was much less confined to the veins in Cit and Asa pre-treatments. As the Fe in foliar treatment of dFe plants was not delivered from the nutrient solution, we suggest that Fe intensity confined to the veins may indicate its accumulation in sieve elements for distribution in the plant.

4. Materials and Methods

4.1. Plant Material

Cucumis sativus L. cv. Joker seeds were used. Seeds were first germinated on wet filter paper in darkness at 30 °C for 2 days, followed by 24-hour incubation with 0.5 mM $CaSO_4$ solution in darkness to stimulate elongation of the hypocotyl. The seedlings were then transferred to a modified quarter-strength unbuffered Hoagland solution containing 1.25 mM KNO_3, 1.25 mM $Ca(NO_3)_2$, 0.5 mM $MgSO_4$, 0.25 mM KH_2PO_4, 11.6 μM H_3BO_3, 4.5 μM $MnCl_2$, 0.19 μM $ZnSO_4$, 0.12 μM Na_2MoO_4, and 0.08 μM $CuSO_4$ (Fe-deficient plants). Fe-sufficient plants were grown in the same solution but with 10 μM Fe (III)-citrate. Another set of plants was grown in nutrient solution buffered to pH 8.5. In order to maintain the desired pH value, 1 mM $KHCO_3$ and 0.3 g $CaCO_3$ were added to each pot. In this case, Fe-sufficient plants were grown with Fe (III)-EDDHA. Each plant was grown in a single plastic pot with 400 mL nutrient solution, which was replaced every other day for 2 weeks. The plants were grown in a climate-controlled growth chamber with a 120 $\mu mol\ s^{-1} m^{-2}$ photosynthetic photon flux density and 14/10 h (light/dark) photoperiod, with temperatures of 20–26 °C and a relative humidity of approximately 70/75%. The experiments were conducted using 2-week-old Fe-deficient plants.

4.2. Treatment with Nanomaterials

Three NH colloid suspensions were applied in the experiments, which were prepared together in one synthesis process investigated and described in our previous study [26]. The different suspensions characterized by transmission electron microscopy and X-ray diffraction were originally ferrihydrite with a particle size of 4–7 nm (PEG-1500 coating, NH-S1), nanohematite with a particle size of 12–25 nm (Emulsion 104D coating, NH-S2), and nanoferrihydrite/nanohematite with a particle size of 4–8 nm (SOLUTOL HS 15 coating, NH-S3) [26]. With aging there was a slow conversion of the particles towards nanohematite components. The NHs were stabilized with different coating agents: PEG-1500 (NH-S1), Emulsion 104D (NH-S2), and SOLUTOL HS 15 (NH-S3). The nanomaterial treatments were applied to Fe-deficient plants in nutrient solution or as foliar fertilizer. In the nutrient solutions, the equivalent amount of NH suspension was added to give a 20 μM nominal concentration of Fe. In the foliar treatment, the NH suspensions were applied to the 2nd leaves of plants twice a day for 2 days in a nominal concentration of 2 mM Fe. Prior to foliar spray treatment with the NH suspensions, the leaves were sprayed with Cit, Asa, or Gly solution at a final concentration of 0.5 mM, prepared in 0.02% nonit as a surfactant.

4.3. Physiological Parameters

Chlorophyll content was estimated non-destructively using a SPAD-502+ Chlorophyll Meter device (Konica-Minolta, Osaka, Japan) for 4 days starting from the day of treatment in both cases.

Three days post-treatment, disks were cut from the 2nd leaves using a cork borer and the chlorophyll was extracted in 80% acetone (V/V) and 5 mM Tricin buffer (pH 7.5) using a mortar and pestle. After centrifugation at 10,000× *g* for 5 min, the chlorophyll concentration was measured using a UV2101 spectrophotometer (Shimadzu, Kyoto, Japan) with the extinction coefficients of Porra et al. [57].

Total dry weight of the plants was determined after drying at 80 °C for 2 days.

Gas exchange and photosynthetic activity were measured using an LI-6800F portable photosynthesis system (LICOR Biosciences, Lincoln, NE, USA) with a 2 cm^2 aperture for leaf samples. The applied parameter settings were as follows: CO_2 concentrations of 400 μmol mol^{-1} air, relative humidity of 60%, with a flow rate of approximately 600 μmol s^{-1}. To induce Chl *a* fluorescence, leaves were dark-adapted in the leaf chamber of the device until reaching a stable fluorescence signal in approx. 20 min. An actinic radiation of 600 μmol photons m^{-2} s^{-1} PPFD was applied. Measurements of different photosynthetic parameters (A_{net}, Fv/Fm, Fv′/Fm′, ΦPSII, $ΦCO_2$, and NPQ) were performed between 9 a.m. and 5 p.m. on the 2nd leaves 3–4 days post-treatment. The parameters used in this study were defined and calculated using the instrument's internal software: https://www.licor.com/documents/ajncmgt9xtonajwvs3n6hxxw5u9dlfai (accessed on 5 August 2023).

4.4. X-ray Fluorescence Imaging

Once the treatment was completed, the 2nd leaves were collected and dried at 60 °C under press, ensuring a smooth surface of the leaf blade. X-ray fluorescence (XRF) analysis was conducted using an XGT-7200 V analytical imaging instrument (Horiba, Osaka, Japan) equipped with a Rh X-ray tube and silicon drift detector (SDD). The sample was positioned in the measuring chamber at atmospheric pressure and room temperature. Acceleration voltage and current of 50 kV and 1 mA, respectively, and an X-ray guide tube of 100 μm were employed. The element distribution maps were prepared using the characteristic Kα photons emitted by the Fe (6.405 keV), Ca (3.692 keV), and K (8.637 keV) atoms and detected by the SDD. A mapping area of 15.36 × 15.36 mm was analyzed, with 1000 s survey time per frame, and data were collected 5 times per pixel.

4.5. Ferric Chelate Reductase Assay

The ferric chelate reductase assay was conducted according to the method used by Kovács et al. [58]. Spectrophotometric measurement of the absorbance of the $[Fe(II)\text{-bathophenanthroline disulfonate}_3]^{4-}$ complexes was performed at a wavelength of 535 nm using a Shimadzu UV-2101PC spectrophotometer (Shimadzu, Kyoto, Japan). The extinction coefficient of the complex used for calculations was 22.14 mM^{-1} cm^{-1}, as reported by Smith et al. [59].

4.6. Element Analysis

Element concentration estimation was performed after acid digestion of dried plant samples. The samples were dried for two days at 85 °C and then digested in ccH_2O_2 for 1 h, followed by the addition of $ccHNO_3$, and incubation for 15 min at 60 °C and 45 min at 120 °C. The resulting solution was filtered through MN 640 W filter paper (Macherey-Nagel, Düren, Germany) and thoroughly homogenized. The filtrate was then analyzed for elemental content using an inductively coupled plasma–optical emission spectrometer (ICP-OES) (Spectro Genesis, SPECTRO, Freital, Germany) with the help of a multi-element standard for 33 elements (Loba Chemie Product code: I166N, Loba Chemie PVT, Mumbai, India) used for calibration.

4.7. Statistical Treatment

The experiments were performed twice. To analyze differences, ANOVA was performed with the Tukey-Kramer multiple comparison post hoc test using InStat v. 3.00 software (GraphPad Software, Inc., San Diego, CA, USA). A significant difference was considered when the similarity of the samples was less than $p = 0.05$.

5. Conclusions

In our study, three NH colloid suspensions with different coatings were applied as nanofertilizers through nutrient solution or as foliar spray to young dFe cucumber plants. The results demonstrated that the NHs effectively improved various physiological parameters and overall growth when administered via both routes, underscoring their efficacy as nanofertilizers. However, it was noted that the NHs did not yield significant growth and physiological improvement when applied in nutrient solution at alkaline pH levels, as none of the coating materials were effective in maintaining continuous Fe supply. The introduction of three different pre-treatments (Cit, Asa, and Gly) prior to foliar spraying remarkably facilitated Fe uptake by the leaves, leading to enhanced photosynthetic parameters and reduced root ferric chelate reductase activity, indicating the downregulation of Fe efficiency reactions. The different coatings did not result in significantly different micronutrient efficiency; they performed equally well. Overall, the NHs assessed in this study hold promise as valuable sources of Fe, offering a potential solution to address Fe deficiency in agricultural settings.

Supplementary Materials: The following supporting information can be downloaded at: https://www.mdpi.com/article/10.3390/plants12173104/s1, Figure S1: Representative 2nd leaves post NH supply.

Author Contributions: Conceptualization, F.F. and A.S.; methodology, F.F.; formal analysis, A.S., F.P., D.R., Z.M. and G.T.; data curation, A.S. and F.F.; writing—original draft preparation, A.S.; writing—review and editing, F.F.; visualization, A.S. and F.F.; project administration, F.F.; funding acquisition, F.F. All authors have read and agreed to the published version of the manuscript.

Funding: This research was funded by the National Research, Development and Innovation Office of Hungary (NKFIH; grant nos. K-124159 and K-115913) and by the European Structural and Investment Fund (grant no. VEKOP-2.3.3-15-2016-00008).

Institutional Review Board Statement: Not applicable.

Informed Consent Statement: Not applicable.

Data Availability Statement: The data presented in this study are available on reasonable request from the corresponding author.

Acknowledgments: The authors would like to thank Györgyi Balogh for technical assistance.

Conflicts of Interest: The authors declare no conflict of interest.

References

1. Aung, M.S.; Kobayashi, T.; Masuda, H.; Nishizawa, N.K. Rice HRZ Ubiquitin Ligases Are Crucial for the Response to Excess Iron. *Physiol. Plant.* **2018**, *163*, 282–296. [CrossRef]
2. Lemanceau, P.; Bauer, P.; Kraemer, S.; Briat, J.-F. Iron Dynamics in the Rhizosphere as a Case Study for Analyzing Interactions between Soils, Plants and Microbes. *Plant Soil* **2009**, *321*, 513–535. [CrossRef]
3. Valentinuzzi, F.; Pii, Y.; Carlo, P.; Roberto, T.; Fontanella, M.C.; Beone, G.M.; Astolfi, S.; Mimmo, T.; Cesco, S. Root-Shoot-Root Fe Translocation in Cucumber Plants Grown in a Heterogeneous Fe Provision. *Plant Sci.* **2020**, *293*, 110431. [CrossRef]
4. Zuo, Y.; Zhang, F. Soil and Crop Management Strategies to Prevent Iron Deficiency in Crops. *Plant Soil* **2011**, *339*, 83–95. [CrossRef]
5. Abadía, J.; Vázquez, S.; Rellán-Álvarez, R.; El-Jendoubi, H.; Abadía, A.; Álvarez-Fernández, A.; López-Millán, A.F. Towards a knowledge-based correction of iron chlorosis. *Plant Physiol. Biochem.* **2011**, *49*, 471–482. [CrossRef] [PubMed]
6. Sárvári, É.; Mihailova, G.; Solti, Á.; Keresztes, Á.; Velitchkova, M.; Georgieva, K. Comparison of Thylakoid Structure and Organization in Sun and Shade Haberlea Rhodopensis Populations under Desiccation and Rehydration. *J. Plant Physiol.* **2014**, *171*, 1591–1600. [CrossRef]
7. Kroh, G.E.; Pilon, M. Regulation of Iron Homeostasis and Use in Chloroplasts. *Int. J. Mol. Sci.* **2020**, *21*, 3395. [CrossRef] [PubMed]
8. Hindt, M.N.; Akmakjian, G.Z.; Pivarski, K.L.; Punshon, T.; Baxter, I.; Salt, D.E.; Guerinot, M. Lou BRUTUS and Its Paralogs, BTS LIKE1 and BTS LIKE2, Encode Important Negative Regulators of the Iron Deficiency Response in Arabidopsis Thaliana. *Metallomics* **2017**, *9*, 876–890. [CrossRef]
9. Nozoye, T.; Nagasaka, S.; Kobayashi, T.; Takahashi, M.; Sato, Y.; Sato, Y.; Uozumi, N.; Nakanishi, H.; Nishizawa, N.K. Phytosiderophore Efflux Transporters Are Crucial for Iron Acquisition in Graminaceous Plants. *J. Biol. Chem.* **2011**, *286*, 5446–5454. [CrossRef]

10. Kurt, F.; Filiz, E. Genome-Wide and Comparative Analysis of *BHLH38, BHLH39, BHLH100* and *BHLH101* Genes in *Arabidopsis*, Tomato, Rice, Soybean and Maize: Insights into Iron (Fe) Homeostasis. *Biometals* **2018**, *31*, 489–504. [CrossRef]
11. Selote, D.; Samira, R.; Matthiadis, A.; Gillikin, J.W.; Long, T.A. Iron-Binding E3 Ligase Mediates Iron Response in Plants by Targeting Basic Helix-Loop-Helix Transcription Factors1. *Plant Physiol.* **2015**, *167*, 273–286. [CrossRef]
12. Rajaie, M.; Tavakoly, A.R. Iron and/or Acid Foliar Spray versus Soil Application of Fe-EDDHA for Prevention of Iron Deficiency in Valencia Orange Grown on a Calcareous Soil. *J. Plant Nutr.* **2018**, *41*, 150–158. [CrossRef]
13. Zuluaga, M.Y.A.; Cardarelli, M.; Rouphael, Y.; Cesco, S.; Pii, Y.; Colla, G. Iron Nutrition in Agriculture: From Synthetic Chelates to Biochelates. *Sci. Hortic.* **2023**, *312*, 111833. [CrossRef]
14. Hochmuth, G. *Iron (Fe) Nutrition of Plants Function of Fe in the Plant*; Department of Soil and Water Sciences, UF/IFAS Extension; Institute of Food and Agricultural Sciences, University of Florida: Gainesville, FL, USA, 2011; pp. 1–8.
15. Cieschi, M.T.; Polyakov, A.Y.; Lebedev, V.A.; Volkov, D.S.; Pankratov, D.A.; Veligzhanin, A.A.; Perminova, I.V.; Lucena, J.J. Eco-Friendly Iron-Humic Nanofertilizers Synthesis for the Prevention of Iron Chlorosis in Soybean (Glycine Max) Grown in Calcareous Soil. *Front. Plant Sci.* **2019**, *10*, 413. [CrossRef] [PubMed]
16. Raliya, R.; Saharan, V.; Dimkpa, C.; Biswas, P. Nanofertilizer for Precision and Sustainable Agriculture: Current State and Future Perspectives. *J. Agric. Food Chem.* **2018**, *66*, 6487–6503. [CrossRef] [PubMed]
17. Das, A.; Singh, J.; Yogalakshmi, K.N. Laccase Immobilized Magnetic Iron Nanoparticles: Fabrication and Its Performance Evaluation in Chlorpyrifos Degradation. *Int. Biodeterior. Biodegrad.* **2017**, *117*, 183–189. [CrossRef]
18. Boutchuen, A.; Zimmerman, D.; Aich, N.; Masud, A.M.; Arabshahi, A.; Palchoudhury, S. Increased Plant Growth with Hematite Nanoparticle Fertilizer Drop and Determining Nanoparticle Uptake in Plants Using Multimodal Approach. *J. Nanomater.* **2019**, *2019*, 6890572. [CrossRef]
19. Tombuloglu, H.; Slimani, Y.; Tombuloglu, G.; Almessiere, M.; Baykal, A. Uptake and Translocation of Magnetite (Fe_3O_4) Nanoparticles and Its Impact on Photosynthetic Genes in Barley (*Hordeum Vulgare* L.). *Chemosphere* **2019**, *226*, 110–122. [CrossRef]
20. Pariona, N.; Martinez, A.I.; Hdz-García, H.M.; Cruz, L.A.; Hernandez-Valdes, A. Effects of Hematite and Ferrihydrite Nanoparticles on Germination and Growth of Maize Seedlings. *Saudi J. Biol. Sci.* **2017**, *24*, 1547–1554. [CrossRef]
21. Claudio, C.; Iorio, E.D.; Liu, Q.; Jiang, Z.; Barrón, V. Iron oxide nanoparticles in soils: Environmental and agronomic importance. *J. Nanosci. Nanotechnol.* **2017**, *17*, 4449–4460. [CrossRef]
22. Kraemer, S.M. Iron Oxide Dissolution and Solubility in the Presence of Siderophores. *Aquat. Sci.* **2004**, *66*, 3–18. [CrossRef]
23. Shimizu, K.; Tschulik, K.; Compton, R.G. Exploring the Mineral-Water Interface: Reduction and Reaction Kinetics of Single Hematite (α-Fe_2O_3) Nanoparticles. *Chem. Sci.* **2016**, *7*, 1408–1414. [CrossRef]
24. Colombo, C.; Palumbo, G.; He, J.Z.; Pinton, R.; Cesco, S. Review on Iron Availability in Soil: Interaction of Fe Minerals, Plants, and Microbes. *J. Soils Sediments* **2014**, *14*, 538–548. [CrossRef]
25. Mazeina, L.; Navrotsky, A. Enthalpy of Water Adsorption and Surface Enthalpy of Goethite (α-FeOOH) and Hematite (α-Fe_2O_3). *Chem. Mater.* **2007**, *19*, 825–833. [CrossRef]
26. Gracheva, M.; Klencsár, Z.; Kis, V.K.; Béres, K.A.; May, Z.; Halasy, V.; Singh, A.; Fodor, F.; Solti, Á.; Kiss, L.F.; et al. Iron Nanoparticles for Plant Nutrition: Synthesis, Transformation, and Utilization by the Roots of Cucumis Sativus. *J. Mater. Res.* **2022**, *38*, 1035–1047. [CrossRef]
27. Martín-Barranco, A.; Spielmann, J.; Dubeaux, G.; Vert, G.; Zelazny, E. Dynamic Control of the High-Affinity Iron Uptake Complex in Root Epidermal Cells. *Plant Physiol.* **2020**, *184*, 1236–1250. [CrossRef] [PubMed]
28. Ndou, N.; Rakgotho, T.; Nkuna, M.; Doumbia, I.Z.; Mulaudzi, T.; Ajayi, R.F. Green Synthesis of Iron Oxide (Hematite) Nanoparticles and Their Influence on Sorghum Bicolor Growth under Drought Stress. *Plants* **2023**, *12*, 1425. [CrossRef] [PubMed]
29. Rath, K.; Ranganathan, P.; Vasappa, R.K.; Balasundaram, S.T. Superparamagnetic Hematite Nanoparticle: Cytogenetic Impact on Onion Roots and Seed Germination Response of Major Crop Plants. *IET Nanobiotechnology* **2020**, *14*, 133–141. [CrossRef] [PubMed]
30. Zargar Shooshtari, F.; Souri, M.K.; Hasandokht, M.R.; Jari, S.K. Glycine Mitigates Fertilizer Requirements of Agricultural Crops: Case Study with Cucumber as a High Fertilizer Demanding Crop. *Chem. Biol. Technol. Agric.* **2020**, *7*, 19. [CrossRef]
31. Mosa, W.F.A.; Ali, H.M.; Abdelsalam, N.R. The Utilization of Tryptophan and Glycine Amino Acids as Safe Alternatives to Chemical Fertilizers in Apple Orchards. *Environ. Sci. Pollut. Res.* **2021**, *28*, 1983–1991. [CrossRef]
32. Sh Sadak, M.; Abdelhamid, M.T.; Schmidhalter, U. Effect of Foliar Application of Amino acids On Plant Yield And Some Physiological Parameters In Bean Plants Irrigated With Seawater. *Acta Biológica Colomb.* **2015**, *20*, 141–152.
33. Rasp, H. Control of Grape Chlorosis through Nutrient Application on Leaves. In *Foliar Fertilization, Proceedings of the First International Symposium on Foliar Fertilization, Berlin, Germany, 14–16 March 1985*; Springer Science & Business Media: Dordrecht, The Netherlands, 1986; Volume 22, p. 242.
34. Tejada, M.; Gonzalez, J.L. Influence of Foliar Fertilization with Amino Acids and Humic Acids on Productivity and Quality of Asparagus. *Biol. Agric. Hortic.* **2003**, *21*, 277–291. [CrossRef]
35. Sánchez-Sánchez, A.; Sánchez-Andreu, J.; Juárez, M.; Jordá, J.; Bermúdez, D. Humic Substances and Amino Acids Improve Effectiveness of Chelate Fe-EDDHA in Lemon Trees. *J. Plant Nutr.* **2002**, *25*, 2433–2442. [CrossRef]
36. Sánchez, A.S.; Juárez, M.; Sánchez-Andreu, J.; Jordá, J.; Bermúdez, D. Use of Humic Substances and Amino Acids to Enhance Iron Availability for Tomato Plants from Applications of the Chelate FeEDDHA. *J. Plant Nutr.* **2005**, *28*, 1877–1886. [CrossRef]

37. Marschner, H.; Römheld, V.; Kissel, M. Different Strategies in Higher Plants in Mobilization and Uptake of Iron. *J. Plant Nutr.* **1986**, *9*, 695–713. [CrossRef]
38. Durrett, T.P.; Gassmann, W.; Rogers, E.E. The FRD3-Mediated Efflux of Citrate into the Root Vasculature Is Necessary for Efficient Iron Translocation. *Plant Physiol.* **2007**, *144*, 197–205. [CrossRef] [PubMed]
39. Rellán-Álvarez, R.; Andaluz, S.; Rodríguez-Celma, J.; Wohlgemuth, G.; Zocchi, G.; Álvarez-Fernández, A.; Fiehn, O.; López-Millán, A.F.; Abadía, J. Changes in the Proteomic and Metabolic Profiles of Beta Vulgaris Root Tips in Response to Iron Deficiency and Resupply. *BMC Plant Biol.* **2010**, *10*, 120. [CrossRef] [PubMed]
40. Das, K.; Roychoudhury, A. Reactive Oxygen Species (ROS) and Response of Antioxidants as ROS-Scavengers during Environmental Stress in Plants. *Front. Environ. Sci.* **2014**, *2*, 53. [CrossRef]
41. Hasanuzzaman, M.; Bhuyan, M.H.M.B.; Zulfiqar, F.; Raza, A.; Mohsin, S.M.; Al Mahmud, J.; Fujita, M.; Fotopoulos, V. Reactive Oxygen Species and Antioxidant Defense in Plants under Abiotic Stress: Revisiting the Crucial Role of a Universal Defense Regulator. *Antioxidants* **2020**, *9*, 681. [CrossRef]
42. Kapoor, D.; Singh, S.; Kumar, V.; Romero, R.; Prasad, R.; Singh, J. Antioxidant Enzymes Regulation in Plants in Reference to Reactive Oxygen Species (ROS) and Reactive Nitrogen Species (RNS). *Plant Gene* **2019**, *19*, 100182. [CrossRef]
43. Foyer, C.H.; Noctor, G. Ascorbate and Glutathione: The Heart of the Redox Hub. *Plant Physiol.* **2011**, *155*, 2–18. [CrossRef] [PubMed]
44. Jackson, C.S.; Kodanko, J.J. Iron-Binding and Mobilization from Ferritin by Polypyridyl Ligands. *Metallomics* **2010**, *2*, 407–411. [CrossRef]
45. Atanassova, B.D.; Tzatchev, K.N. Ascorbic acid-important for iron metabolism. *Folia Medica* **2008**, *50*, 11. [PubMed]
46. Shokr, M.M.B.; Abdelhamid, M.T. Using Some Antioxidant Substances for Enhancing Thermotolerance and Improving Productivity of Pea (*Pisum Sativum* L.) Plants Under Local Environment of Early Summer Season. *Agric. Res. J.* **2009**, *9*, 69–76.
47. Shafeek, M.; Helmy, Y.; Marzauk, N.M.; Magda, A.S.; Nadia, M.O. Effect of foliar application of some antioxidants on growth, yield and chemical composition of Lettuce plants (*Lactuca Sativa* L.) under plastic house condition. *Middle East J. Appl. Sci.* **2013**, *3*, 70–75.
48. Mansour, M.M. Response of soybean plants to exogenously applied with ascorbic acid, zinc sulphate and paclobutrazol. *Rep. Opin.* **2014**, *6*, 17–25.
49. Ramírez, L.; Bartoli, C.G.; Lamattina, L. Glutathione and Ascorbic Acid Protect Arabidopsis Plants against Detrimental Effects of Iron Deficiency. *J. Exp. Bot.* **2013**, *64*, 3169–3178. [CrossRef]
50. Khan, S.; Yu, H.; Li, Q.; Gao, Y.; Sallam, B.N.; Wang, H.; Liu, P.; Jiang, W. Exogenous Application of Amino Acids Improves the Growth and Yield of Lettuce by Enhancing Photosynthetic Assimilation and Nutrient Availability. *Agronomy* **2019**, *9*, 266. [CrossRef]
51. Souri, M.K.; Naiji, M.; Aslani, M. Effect of Fe-Glycine Aminochelate on Pod Quality and Iron Concentrations of Bean (*Phaseolus Vulgaris* L.) Under Lime Soil Conditions. *Commun. Soil Sci. Plant Anal.* **2018**, *49*, 215–224. [CrossRef]
52. Mohammadipour, N.; Souri, M.K. Beneficial Effects of Glycine on Growth and Leaf Nutrient Concentrations of Coriander (*Coriandrum Sativum*) Plants. *J. Plant Nutr.* **2019**, *42*, 1637–1644. [CrossRef]
53. Noroozlo, Y.A.; Souri, M.K.; Delshad, M. Stimulation Effects of Foliar Applied Glycine and Glutamine Amino Acids on Lettuce Growth. *Open Agric.* **2019**, *4*, 164–172. [CrossRef]
54. Xu, M.; Du, L.; Liu, M.; Zhou, J.; Pan, W.; Fu, H.; Zhang, X.; Ma, Q.; Wu, L. Glycine-Chelated Zinc Rather than Glycine-Mixed Zinc Has Lower Foliar Phytotoxicity than Zinc Sulfate and Enhances Zinc Biofortification in Waxy Corn. *Food Chem.* **2022**, *370*, 131031. [CrossRef] [PubMed]
55. Mijovilovich, A.; Morina, F.; Bokhari, S.N.; Wolff, T.; Küpper, H.; Küpper, H. Analysis of Trace Metal Distribution in Plants with Lab-Based Microscopic X-Ray Fluorescence Imaging. *Plant Methods* **2020**, *16*, 82. [CrossRef]
56. Terzano, R.; Alfeld, M.; Janssens, K.; Vekemans, B.; Schoonjans, T.; Vincze, L.; Tomasi, N.; Pinton, R.; Cesco, S. Spatially Resolved (Semi)Quantitative Determination of Iron (Fe) in Plants by Means of Synchrotron Micro X-Ray Fluorescence. *Anal. Bioanal. Chem.* **2013**, *405*, 3341–3350. [CrossRef]
57. Porra, R.J.; Thompson, W.A.; Kriedemann, P.E. Determination of Accurate Extinction Coefficients and Simultaneous Equations for Assaying Chlorophylls a and b Extracted with Four Different Solvents: Verification of the Concentration of Chlorophyll Standards by Atomic Absorption Spectroscopy. *Biochim. Biophys. Acta-Bioenerg.* **1989**, *975*, 384–394. [CrossRef]
58. Kovács, K.; Kuzmann, E.; Tatár, E.; Vértes, A.; Fodor, F. Investigation of Iron Pools in Cucumber Roots by Mössbauer Spectroscopy: Direct Evidence for the Strategy I Iron Uptake Mechanism. *Planta* **2009**, *229*, 271–278. [CrossRef] [PubMed]
59. Smith, G.F.; McCurdy, W.H.; Diehl, H. The Colorimetric Determination of Iron in Raw and Treated Municipal Water Supplies by Use of 4:7-Diphenyl-1:10-Phenanthroline. *Analyst* **1952**, *77*, 418–422. [CrossRef]

plants

Communication

Application of Salicylic Acid Derivative in Modifying the Iron Nutritional Value of Lettuce (*Lactuca sativa* L.)

Barbara Frąszczak [1], Renata Matysiak [2], Marcin Smiglak [3,4], Rafal Kukawka [3,4], Maciej Spychalski [3] and Tomasz Kleiber [2,*]

1 Department of Vegetable Crops, Faculty of Agronomy, Horticulture and Bioengineering, Poznan University of Life Sciences, Dąbrowskiego 159, 60-594 Poznan, Poland; barbara.fraszczak@up.poznan.pl
2 Department of Plant Physiology, Faculty of Agronomy, Horticulture and Bioengineering, Poznan University of Life Sciences, Wołyńska 35, 60-637 Poznan, Poland; renata.matysiak@up.poznan.pl
3 Poznan Science and Technology Park, Adam Mickiewicz University Foundation, Rubież 46, 61-612 Poznan, Poland; marcin.smiglak@ppnt.poznan.pl or marcin.smiglak@innosil.pl (M.S.); rafal.kukawka@ppnt.poznan.pl (R.K.); maciej.spychalski@ppnt.poznan.pl (M.S.)
4 Innosil Sp. z o.o., Rubież 46, 61-612 Poznan, Poland
* Correspondence: tomasz.kleiber@up.poznan.pl

Abstract: The present experiment addressed the effects of foliar sprays of different iron (Fe) concentrations (mg L^{-1}), i.e., 2.8 (Fe I), 4.2 (Fe II), and 5.6 (Fe III), as well as an ionic derivative of salicylic acid (iSal) in two doses (10 and 20 mg L^{-1}) on lettuce yield, chlorophyll and carotenoids content, and fluorescence parameters. Chemicals were used individually and in combinations two times, 23 and 30 days after the plants were transplanted. This experiment was carried out in a climate chamber. The Fe and iSal applications generally (except Fe I iSal, 10 mg L^{-1}; Fe I iSal, 20 mg L^{-1}; and Fe III iSal, 20 mg L^{-1}) did not influence the fresh and dry matter content. The concentration of chlorophylls and carotenoids was reduced for all treatments in comparison to the control (without spraying). The Fe content in leaves was promoted in the Fe-treated plants (+70% for Fe III + iSal, 10 mg L^{-1}, and Fe I). The iSal treatment promoted the Mn content. For most combinations, the Zn and Cu accumulations, as well as the fluorescence parameters, decreased after the foliar spray applications. Overall, our study revealed the effectiveness of Fe-DTPA chelate, but not iSal, in increasing the Fe content of lettuce grown in soilless cultivation systems.

Keywords: Fe chelate; biofortification; foliar spraying; exogenous salicylic acid

Citation: Frąszczak, B.; Matysiak, R.; Smiglak, M.; Kukawka, R.; Spychalski, M.; Kleiber, T. Application of Salicylic Acid Derivative in Modifying the Iron Nutritional Value of Lettuce (*Lactuca sativa* L.). *Plants* **2024**, *13*, 180. https://doi.org/10.3390/plants13020180

Academic Editor: Ferenc Fodor

Received: 13 December 2023
Revised: 3 January 2024
Accepted: 5 January 2024
Published: 9 January 2024

1. Introduction

Lettuce is one of the most widely consumed vegetables worldwide, but its nutritional value has been underestimated. Lettuce is low in calories, fat, and sodium. It is a good source of fibre, iron, folate, and vitamin C. Lettuce is also a good source of various other health-beneficial bioactive compounds, like phenolic compounds and carotenoids [1].

Iron (Fe) is a very important micronutrient for both plants and humans. It participates in processes such as photosynthesis, respiration, and oxygenation [2]. However, about two billion people suffer from anaemia, primarily due to a diet low in Fe [3]. In addition, its phytoavailable concentration (10^{-17} M) does not reach the optimal range for plant growth (10^{-9}–10^{-4} M) [4]. It should be further noted that Fe is poorly absorbed by the human body, with only about 14–18% of Fe available in food being bioavailable. [5]. Therefore, one way to solve the above problems may be to increase Fe in plant foods [6]. This condition can be achieved by biofortifying plants with a specific element [7]. A safe way to carry this out is through soilless cultivation, where it is possible to control water availability, pH, and nutrient concentration in the root zone [8]. Biofortification is carried out by increasing the level of a specific element in the nutrient solution. Additionally, in the case of Fe, its bioavailability can be increased by managing the pH in the nutrient solution [9]. Since, as mentioned above, Fe is not easily assimilated by plants and excess Fe can be harmful to plants (DNA and protein damage) and cause stress conditions for plant growth [10], using biostimulants or growth inducers that will alleviate the effects of stress and increase Fe assimilation is most appropriate.

According to the European Council Regulation (EC) No. 2019/1009, certain substances, mixtures, and products of microorganisms have been defined as plant biostimulants, and they can be used as fertilizer products according to the European Union (EU). Their task is to stimulate the nutritional processes of plants independently of the nutrient content of the product for the sole purpose of improving one or more of the following plant or plant rhizosphere characteristics: nutrient use efficiency, tolerance to abiotic factors and stress, quality characteristics, or the availability of limited nutrients in the soil or rhizosphere [11].

Several studies have shown that biostimulants of various origins can improve the uptake of nutrients, including minerals, in plants [12,13]. Among other things, biostimulants improve the plant uptake of Ca, Mg, and K [14]. Plant biostimulants also enhance the ion ratios of Na^{+}:Ca^{2+} and Na^{+}:Mg^{2+} in lettuce leaves [15].

One natural biostimulant is willow bark. The extract of this herbal raw material contains salicylic acid (SA). One of its characteristics is its strong fungicidal properties [16]. Another way in which salicylic acid and its ionic derivatives, such as choline salicylate, interact with plants is by influencing signalling pathways, leading to the induction of systemic acquired resistance (SAR) [17,18]. In our previous studies, we demonstrated that the derivatization of another well-known plant resistance inducer from the benzothiadiazole group leads to an increased expression of genes encoding pathogenesis-related (PR) proteins [19]. The effectiveness of SAR inductors has been confirmed in practice in various tested crops, where the activation of the immune system was observed by a reduction in the occurrence of diseases [20,21]. Moreover, it was found that the induction of resistance is correlated with the stimulation of plant metabolism, providing lasting beneficial effects to a diverse range of crop plants [22,23].

Some researchers suggest that SA plays an important role not only in protecting plants from disease but also in thermogenesis, abiotic stress and salinity tolerance, DNA damage/repair, seed germination, fruit yield, etc. [24]. According to recent studies, it can also be used for rooting woody (lavender) and semi-woody (chrysanthemum) cuttings [25] due to the presence of indole 3-butyric acid (IBA), as well as maize seedling production under salt stress conditions [26]. There was also determined to be a positive effect of salicylic acid contained in willow extract on some macronutrient uptake in lettuce [15].

The chemical composition of willow extract and the amount of salicylate compounds can vary depending on the age of the plant, the date of harvest of the herbaceous material, tissue, genotype, and species, as well as various environmental factors [27], so it was reasonable to create a synthetic SA with established and stable properties.

The conducted study aimed to evaluate the possibility of iSal (an ionic derivative of salicylic acid) application to modify the iron nutritional status of lettuce (*Lactuca sativa* L. cv. 'Zeralda').

2. Results and Discussion

2.1. The Yield

The spray applications of both Fe and iSal did not affect the fresh weight of lettuce (Table 1). The application of the highest doses of Fe and Fe I iSal (10 mg L^{-1}), Fe I iSal (20 mg L^{-1}), and Fe III iSal (20 mg L^{-1}) increased dry matter yield relative to the control. The other combinations resulted in similar dry matter yields and DM percentages.

In earlier studies, increasing Fe dosage (ranging from 0.9 to 25 mg L^{-1}) caused a decrease in the fresh weight of biofortified plants [28]. This was related to Fe toxicity above a certain level of doses used in biofortification [9]. Excess Fe levels in plants cause an increase in reactive oxygen species (ROS) production, oxidative stress responses, and physiological disorders [29]. In the current study, no symptoms of toxicity were observed, which means that the doses used in this experiment were safe for plants [30]. According to Filho et al. [28], for *Cichorium intybus* cultivated in an NFT, the optimal Fe range was from 2.7 to 8.3 mg L^{-1}. In the current experiment, the Fe doses applied resulted in a higher dry matter yield and % dry matter content. In particular, the highest Fe dose (Fe III, 5.6 mg L^{-1}) resulted in a greater dry mass. Interestingly, in another experiment, the dry matter content (%) in lettuce increased as the Fe dose increased, and in turn, the dry matter yield decreased [31]. It is worth noting that in this study, there was no effect of iSal on the lettuce biomass.

Table 1. The influence of the Fe and iSal applications on the yields of plants.

Treatment	Fresh Yield (g)	Dry Matter Yield (g)	% DM
Control	23.22 a *	1.32 a	5.70 a
Fe I	25.17 a	1.62 cd	6.46 abc
Fe II	24.89 a	1.58 bcd	6.35 abc
Fe III	25.19 a	1.64 cd	6.51 bc
iSal 10 mg L^{-1}	23.57 a	1.38 ab	5.86 ab
iSal 20 mg L^{-1}	23.94 a	1.50 abcd	6.25 abc
Fe I iSal 10 mg L^{-1}	24.36 a	1.55 bcd	6.38 abc
Fe II iSal 10 mg L^{-1}	24.08 a	1.44 abc	6.01 abc
Fe III iSal 10 mg L^{-1}	23.06 a	1.48 abcd	6.42 abc
Fe I iSal 20 mg L^{-1}	24.33 a	1.57 bcd	6.50 abc
Fe II iSal 20 mg L^{-1}	24.18 a	1.50 abcd	6.20 abc
Fe III iSal 20 mg L^{-1}	25.02 a	1.66 d	6.62 c

* Data followed by the same letter do not differ significantly at α = 0.05 for each parameter.

2.2. Microelement Content and Uptake

The applied sprays with iron as well as iSal modified the content and uptake of microelements by the lettuce leaves (Table 2). Each of the foliar spray treatments increased the Fe content in the lettuce leaves. The highest amount was obtained for the highest level of Fe with 10 mg L^{-1} iSal (70% more than in the control). However, it is worth noting that spraying only iSal or spraying iSal at a dose of 20 mg L^{-1} with added Fe had less effect on the content of this microelement in the leaves compared to the other foliar spray treatments. Similarly, spraying only Fe significantly increased the Fe uptake by the lettuce plants compared to the control combination, but the application of only iSal did not affect Fe uptake as strongly. Iron bioavailability may be affected by polyphenols due to the high affinity of these compounds for this mineral [32]. This may have been the reason for the reduction in the Fe content in the leaves when a higher dose of iSal was applied. The use of Fe biofortification also proved to be effective in increasing the Fe content in the cultivation of lettuce in some previous studies [7,31,33].

The well-known antagonistic effect of Fe on Mn absorption was not found in the conducted studies. In general, the treatments resulted in a higher Mn content in the leaves and a higher uptake by the leaves compared to the control combination. However, the Mn content of the lettuce leaves and the uptake by the leaves varied widely. The lowest content was obtained for the Fe I iSal 10 mg L^{-1} combination. Also, the treatments of Fe I and Fe II with 10 mg L^{-1} of iSal obtained a lower uptake, similar to that of the control combination. The lettuce treated Fe I and with a 20 mg L^{-1} iSal dose was characterised by the highest Mn content and uptake, significantly higher compared to treating plants with Fe I alone. We can conclude that the low doses of applied Fe with high doses of iSal reduced the Mn content and uptake compared to the control. The application of 10 mg L^{-1} of iSal alone also had a positive effect on the Mn content, as did the application of Fe II and Fe III. Previous studies have shown that foliar Fe application had much less or no effect on Mn uptake and content in soybean [34] and chickpea plants [35]. In chickpeas, foliar Fe and Mn additions increased the average Fe concentration and uptake in shoots. The antagonism of the two elements mainly occurs in the soil fertilization of plants through the negative effect of Fe on the translocation of Mn from the root to the shoot [35]. Our research resulted in a low Fe:Mn ratio. The Fe: Mn ratio varied from 1.6 to 1.0 in the leaves, which resulted from high levels of Mn in the leaves. At the molecular level, excess Mn may prevent the uptake and translocation of other essential elements, including Fe [36]. However, there was no negative effect of high Mn levels on the uptake of Fe and Fe content in the leaves. In another study, the application of two levels of Fe (1 and 2 mmol L^{-1} of Fe, in chelate form) in a nutrient solution also significantly increased the Mn as well as Zn content of lettuce leaves [37]. However, in these studies, the Fe:Mn ratio was much higher at 3.5 and 4.0 for the Fe-enriched nutrient solutions and 2.2 for the control, respectively. This may have been because the plants were fertilised with Fe-enriched nutrient solution all the time.

Table 2. The influence of the Fe and iSal applications on the content (mg kg^{-1} D.M.) of metallic microelements in lettuce leaves and the uptake (µg·$plant^{-1}$) of them by leaves.

Treatment	Fe	Mn	Zn	Cu
		Content (mg kg^{-1} D.M.)		
Control	142.79 a *	132.18 b	39.42 ef	9.26 bc
Fe I	202.75 bc	149.50 cd	37.45 de	9.80 c
Fe II	191.30 bc	163.70 def	30.80 ab	8.23 a
Fe III	197.30 bc	163.67 def	30.67 ab	8.10 a
iSal 10 mg L^{-1}	177.27 bc	165.40 ef	37.37 de	8.73 ab
iSal 20 mg L^{-1}	176.17 bc	149.60 cd	41.53 f	7.90 a
Fe I iSal 10 mg L^{-1}	192.20 bc	117.70 a	37.23 de	8.23 a
Fe II iSal 10 mg L^{-1}	193.00 bc	143.23 bc	37.87 de	8.67 ab
Fe III iSal 10 mg L^{-1}	204.70 c	151.60 cde	34.70 cd	8.55 ab
Fe I iSal 20 mg L^{-1}	172.70 b	173.15 f	29.05 a	8.30 a
Fe II iSal 20 mg L^{-1}	175.80 bc	158.55 de	31.40 ab	8.40 ab
Fe III iSal 20 mg L^{-1}	183.95 bc	162.15 def	32.45 bc	8.80 ab
		Uptake (µg·$plant^{-1}$)		
Control	210.34 a	194.71 ab	58.06 cde	13.63 bc
Fe I	323.26 d	237.90 cdef	59.69 de	15.63 d
Fe II	301.39 cd	258.36 efg	48.59 ab	13.00 ab
Fe III	322.25 d	268.06 fg	50.11 ab	13.24 abc
iSal 10 mg L^{-1}	245.20 ab	228.96 cde	51.73 abc	12.06 a
iSal 20 mg L^{-1}	263.16 bc	223.63 bcd	62.00 e	11.82 a
Fe I iSal 10 mg L^{-1}	298.56 cd	182.50 a	57.83 cde	12.75 ab
Fe II iSal 10 mg L^{-1}	279.06 bcd	206.66 abc	54.75 bcd	12.48 ab
Fe III iSal 10 mg L^{-1}	302.64 cd	224.76 bcd	51.38 abc	12.61 ab
Fe I iSal 20 mg L^{-1}	272.88 bcd	273.63 g	45.91 a	13.12 abc
Fe II iSal 20 mg L^{-1}	276.41 bcd	248.51 defg	49.19 ab	13.17 abc
Fe III iSal 20 mg L^{-1}	304.35 cd	268.87 fg	53.63 bcd	14.54 cd

* Data followed by the same letter do not differ significantly at $\alpha = 0.05$ for each parameter.

The Fe treatment significantly reduced the Zn content uptake in the leaves. The content of this microelement was positively affected by spraying iSal at a dose of 20 mg L^{-1}. In contrast, spraying Fe alone (without Fe I) or Fe with an iSal dose of 20 mg L^{-1} significantly reduced the Zn content in the leaves. The highest uptake was also noted for the iSal 20 mg L^{-1} combination. However, there were no statistically significant differences between this combination and the control and Fe I and Fe I iSal 10 mg L^{-1} combinations. All the applied sprays except Fe I significantly reduced the Cu content in the leaves compared to the control combination. Most of the treatments also had a negative effect on Cu uptake. The best results were obtained for the Fe I and Fe III iSal 20 mg L^{-1} combinations as well as the control.

One of the biggest problems associated with plant biofortification is the antagonism among some nutrients. The enrichment of plants with one ion reduces the uptake of others [38]. Such a plant response may be due, among other things, to the fact that Fe and other elements share the same membrane transporters, resulting in the competition of iron with other cations [39,40]. It is worth noting that in some studies, an increase in the level of Fe in a medium also contributed to an increase in the level of Zn in lettuce leaves [37]. The effect of Fe on a plant's Zn and Cu uptake and their content in the leaves also depends on the form of fertilizer used for biofortification. For example, iron-ammonium sulphate increased Zn uptake compared to untreated control plants, and chelates increased the Cu content in African marigolds [41].

In the current study, the biostimulant iSal greatly increased the uptake and content of Mn and Zn in the plants, in contrast to the uptake and content of Fe. This also supports the idea that salicylic acid applications work as biostimulants rather than fertilizers and contribute to the induction of different metabolic pathways beyond providing nutrients to the plant [26].

2.3. Chlorophyll and Carotenoid Content

There was a tendency for the chlorophyll (Chl) and carotenoid (Car) content to decrease after the sprays were applied (Table 3). This is especially evident with the Fe sprays combined with iSal at a dose of 20 mg L^{-1}. For these combinations, the level of carotenoids was about 25% lower than in the control, and the chlorophyll content was about 28–20% lower. Other studies have also shown that high SA concentrations (1–5 mM) reduce Chl contents in various plant species. The lowest concentration (10^{-5} M) of SA generated the highest values for Chl content for a 60 d-stage *Brassica juncea* [42]. However, the values decreased as the concentration of SA increased and reached below that of the control at the maximum concentration (10^{-3} M) [42]. Also, in wheat and moong seedlings, as the concentration of applied salicylic acid (SA) increased, the Chl content significantly decreased [43]. According to these authors, SA induces an increase in the hydrogen peroxide (H_2O_2) content in plants. The increase in oxidative stress can cause a decrease in the total Chl content, or a decrease in the total Chl content can induce oxidative stress with an increase in SA concentration. In addition, to protect the photosynthetic apparatus from oxidative stress, carotenoid levels may be increased. However, this was not observed in the current study.

Table 3. The influence of the Fe and iSal applications on chlorophyll and carotenoids content.

Treatment	Chlorophyll *a* [mg g^{-1} d.m.]	Chlorophyll *b* [mg g^{-1} d.m.]	Chlorophyll *a* + *b* [mg g^{-1} d.m.]	Carotenoids [mg g^{-1} d.m.]
Control	11.16 e *	3.38 c	14.54 e	14.55 e
Fe I	9.60 bcd	2.93 abc	12.53 bcd	12.54 bcd
Fe II	10.94 de	3.26 c	14.20 de	14.20 de
Fe III	10.44 de	3.30 c	13.74 de	13.74 de
iSal 10 mg L^{-1}	10.21 cde	3.00 bc	13.21 cde	13.21 cde
iSal 20 mg L^{-1}	10.27 cde	3.05 bc	13.32 cde	13.33 cde
Fe I iSal 10 mg L^{-1}	10.38 cde	2.93 abc	13.31 cde	13.31 cde
Fe II iSal 10 mg L^{-1}	10.23 cde	3.02 bc	13.26 cde	13.26 cde
Fe III iSal 10 mg L^{-1}	9.85 bcde	3.00 bc	12.85 bcde	12.86 bcde
Fe I iSal 20 mg L^{-1}	8.72 ab	2.59 ab	11.31 ab	11.31 ab
Fe II iSal 20 mg L^{-1}	8.10 a	2.47 a	10.57 a	10.57 a
Fe III iSal 20 mg L^{-1}	9.00 abc	2.65 ab	11.65 abc	11.65 abc

* Data followed by the same letter do not differ significantly at $\alpha = 0.05$ for each parameter.

Fe spraying also reduced the Chl and Car content compared to the control combination. In contrast, in other studies, Fe application (1.02 and 2.02 mmol L^{-1}) promoted the Chls and Car content in lettuce [37], and the application of Fe NPs (0, 5, 10, 20 mg L^{-1}) resulted in an enhancement of both pigments' content in Red Sails Lettuce [33]. The increase in the carotenoid content was probably linked to the high ROS-scavenging ability of this antioxidant [37]. The current study clearly showed the multidirectional influence of Fe applications on the Chls and Car content in lettuce.

2.4. Fluorescence Parameters

The values of the minimum chlorophyll fluorescence (Fo), the maximum chlorophyll fluorescence (Fm), and the variable fluorescence (Fv) varied widely (Table 4). However, for most combinations, these values were lower compared to the control. Fo is the minimum fluorescence level, assuming that all antenna pigment complexes associated with the photosystem are open (dark-adapted) [44]. An increase in Fo indicates any difficulty and degradation of photosystem II (D1 protein and another part of the PS) or any disruption of energy transfer to the reaction centre [45]. This suggests that the lettuce plants were partially subjected to photosynthetic stress under the applied treatments.

Table 4. The influence of the Fe and iSal sprays on chlorophyll fluorescence parameters.

Treatment	Fo	Fm	Fv	Fv/Fo	Fv/Fm
Control	6 525 ef *	48 587 e	42 062 e	6.45 a	0.87 a
Fe I	6 892 f	44 185 cde	38 925 cde	5.65 a	0.88 a
Fe II	5 240 ab	36 513 a	32 437 a	6.19 a	0.89 a
Fe III	5 541 abc	40 018 abc	34 477 abc	6.23 a	0.86 a
iSal 10 mg L^{-1}	6 018 cde	43 344 bcd	37 326 bcd	6.20 a	0.86 a
iSal 20 mg L^{-1}	5 517 abc	40 253 abc	34 736 abc	6.30 a	0.86 a
Fe I iSal 10 mg L^{-1}	5 294 ab	38 115 a	32 821 a	6.19 a	0.86 a
Fe II iSal 10 mg L^{-1}	6 209 e	44 226 cde	38 017 bcde	6.13 a	0.86 a
Fe III iSal 10 mg L^{-1}	5 359 ab	39 003 ab	33 644 ab	6.28 a	0.86 a
Fe I iSal 20 mg L^{-1}	6 161 de	46 068 de	39 908 de	6.48 a	0.87 a
Fe II iSal 20 mg L^{-1}	5 664 bcd	40 718 abc	35 055 abc	6.19 a	0.86 a
Fe III iSal 20 mg L^{-1}	5 035 a	36 929 a	31 894 a	6.32 a	0.86 a

* Data followed by the same letter do not differ significantly at $\alpha = 0.05$ for each parameter.

Spraying the applied chemicals mostly lowered the efficiency of the PSII quantum field (lower Fv), which may have resulted in a greater dissipation of energy in the form of heat [46]. However, the lettuce yield was not affected.

In the present study, there was no effect of the applied sprays on the parameters Fv/Fo and Fv/Fm. The lack of differences among the combinations confirms the low sensitivity of these parameters to changes in the photochemical properties of PSII [47].

Additional treatments applied during plant growth can be stress factors for plants, causing photoinhibition and/or damage to the photosynthetic apparatus. One such treatment may be the application of biostimulants or intensive biofortification. In the present study, foliar spray treatments were shown to affect the photochemical efficiency parameters of PSII in different ways (Table 5). The analysis of PSII function, assessed by PSII photochemical efficiency parameters, showed that both Fe biofortification and iSal sprays can lead to chloroplast dysfunction in lettuce leaves.

Table 5. The influence of the Fe and iSal sprays on chlorophyll fluorescence parameters.

Treatment	Pi_Abs	ABS/RC	TRo/RC	ETo/RC	DIo/RC
Control	10.12 abc*	1.57 d	1.36 c	0.95 c	0.21 c
Fe I	11.41 c	1.31 bc	1.13 ab	0.73 ab	0.17 a
Fe II	10.54 bc	1.20 ab	1.11 ab	0.65 a	0.17 a
Fe III	11.44 c	1.14 a	1.05 ab	0.69 ab	0.16 a
iSal 10 mg L^{-1}	9.04 ab	1.33 bc	1.15 ab	0.75 ab	0.18 ab
iSal 20 mg L^{-1}	9.41 abc	1.26 abc	1.09 ab	0.71 ab	0.17 a
Fe I iSal 10 mg L^{-1}	9.73 abc	1.17 ab	1.01 a	0.65 a	0.16 a
Fe II iSal 10 mg L^{-1}	8.23 a	1.39 c	1.19 b	0.77 b	0.20 bc
Fe III iSal 10 mg L^{-1}	10.44 bc	1.20 ab	1.03 ab	0.69 ab	0.16 a
Fe I iSal 20 mg L^{-1}	9.40 abc	1.33 bc	1.15 ab	0.76 b	0.18 ab
Fe II iSal 20 mg L^{-1}	9.48 abc	1.27 abc	1.09 ab	0.71 ab	0.18 ab
Fe III iSal 20 mg L^{-1}	9.89 abc	1.21 ab	1.04 ab	0.68 ab	0.17 a

* Data followed by the same letter do not differ significantly at $\alpha = 0.05$ for each parameter.

For most combinations, the applied sprays did not result in a deterioration of the PSII viability index, i.e., a reduction in the value of the PI_abs parameter compared to the control combination. Previous studies have shown that differences in the PI_abs values can be attributed to genetic differences, physiological traits, and environmental conditions [48,49].

One of the protective mechanisms of the photosynthetic apparatus, especially PSII, against stress-induced damage is the slowing down of electron transport from reaction centres to plastoquinones [50,51]. In the conducted study, a significant reduction in the rate of electron transport (ETo/RC) was found for all the combinations compared to the control. At the same time, there was no increase in energy dissipation at the expense of heat (DIo/RC).

The applied treatments caused a decrease in the flow of absorbed energy through one active reaction centre (ABS/RC). There was a similar tendency for changes in the energy uptake by one active reaction centre (TR0/RC)—it decreased significantly as the foliar sprays were used. The TR0/RC changes indicate a decrease in the conversion efficiency of the excitation energy.

3. Materials and Methods

3.1. Plant Material and Growth Conditions

The experiment was conducted on lettuce cultivation (*Lactuca sativa* L. cv. 'Zeralda') in a growth chamber. NEONICA LED 240 (Poland) modules were used as the light source. The photosynthetic photon flux density (PPFD) was 140 μmol m^{-2} s^{-1}, with the following share of individual colours: R (red): 111.7 (μmol m^{-2} s^{-1}), G (green): 9.7 (μmol m^{-2} s^{-1}), and B (blue): 18.6 (μmol m^{-2} s^{-1}). The plants were exposed to light for 16 h; the temperature was maintained at 18/17 °C (day/night); and the RH was approximately 60–75%.

The experiment was established in a randomized design with 5 replications (a replication was one single plant). Seedlings were prepared 30 days before the vegetation experiment. The seeds were sown individually on multiple plates filled with standard peat substrate, as recommended for seedling preparation. The seedlings (in the 3–4-leaf phase) were put in drainless pots filled with perlite (V 500 cm^3). During the whole experiment, the plants were watered to a stable weight.

The plants were fertigated with a nutrient solution (NS) of the following chemical composition (mg dm^{-3}): N-NH_4, <15; N-NO_3, 160; P-PO_4, 40; K, 250; Ca, 150; Mg, 50; Fe, 0.58; Mn, 0.33; Zn, 0.21; Cu, 0.08; B, 0.2. It had a pH of 5.50 and an EC of 1.9 mS·cm^{-1}. The following fertilisers for hydroponic cultivation were used to prepare the nutrient solution: potassium nitrate (13% N-NO_3, 38.2% K), calcium nitrate (14.7% N-NO_3, 18.5% Ca), mono potassium phosphate (22.3% P, 28.2% K), potassium sulphate (44.8% K, 17% S), magnesium sulphate (9.9% Mg, 13% S), manganese sulphate (32.3% Mn), copper sulphate (25.6% Cu), borax (11.3% B), and sodium molybdate (39.6% Mo).

3.2. Foliar Application of Fe and iSal

The studied factors were Fe (1. factor) and iSal application (2. factor). The source of iron was Librel FeDP7 DTPA chelate (7% Fe; Royal Brinkman, Poznan, Poland).

For the research, an active substance in the form of an ionic derivative of salicylic acid (iSal), developed by Poznan Science and Technology Park (PSTP) and the Innosil research team, was used. Currently, the active substance is the subject of patent application PCT/PL/2023/050110 [52]. To prepare a working solution for spraying, the ionic derivative of salicylic acid was weighed and dissolved in water in an amount to prepare solutions with concentrations of 10 and 20 mg L^{-1}. The published results are part of preliminary studies conducted to file a patent application.

A foliar spray treatment (5 mL per 1 plant) was applied two times, 23 and 30 days after the transplantation to a stable place (24–25 and 26–27 BBCH-scale, respectively). The plants were treated with different chemicals: control (without spraying), iSal (10 and 20 mg L^{-1}), and Fe (3 levels, in mg L^{-1}: 2.8, 4.2, and 5.6, described, respectively, as Fe-I, Fe-II, and Fe-III) and the mixture. Ten days after the second spraying, the experiment was finished.

3.3. Biometrical and Chlorophyll Fluorescence Measurements

On the day of harvest (40th day after transplanting to the stable place, 28–29 BBCH-scale), the following parameters were determined: the weight of the lettuce leaves (the whole head, g), the dry matter yield (after drying for 24 h at 105 °C), and the dry matter content (% DM). The chlorophyll *a* fluorescence was measured using a PAR-FluorPen FP 110D fluorometer (Photon Systems Instruments Company (PSI), Drásov, Czech Republic). All the plants in the experiment were measured. Leaf fragments were shaded with a special leaf clip for 30 min. Then, the OJIP test was conducted to measure the following chlorophyll fluorescence parameters: F_0—the initial fluorescence, F_M—the maximum fluorescence intensity, F_V—the maximum variable fluorescence, F_V/F_M—the maximum photochemical quantum PSII after dark adaptation, ABS/RC—the light energy absorbed by the PSII antenna photon flux per active reaction centre, TR_0/RC—the total energy used to reduce QA by the unit reaction centre of PSII per energy captured by a single active RC, ET_0/RC—the rate of electron transport through a single RC, DI_0/RC—non-photochemical quenching per reaction centre of PSII; the total dissipation of energy not captured by the RC in the form of heat, fluorescence, and transfer to other systems, PI_{Abs}—the performance index (potential) for energy conservation from excitation to the reduction in intersystem electron acceptors [53].

3.4. Chloroplast Pigments

On the day of harvest, the leaf samples from all the tested plants were collected and stored at −20 °C until the analyses. The total chlorophyll and carotenoids content was determined according to the method of Hiscox and Israelstam [54]. The leaf samples (100 mg) were cut into pieces, and pigments were extracted at 65 °C using 5 cm^3 of dimethyl sulfoxide (DMSO). The optical density of the extracts was measured at 480, 649, and 663 nm. The content of total chlorophyll and carotenoids was calculated following the modified Arnon equations [55] and expressed in mg/g d.m.

3.5. Chemical Analysis

All analyses were conducted on the aerial parts of the plants. The samples were dried for 48 h at 45–50 °C to a stable mass and then ground. Before mineralisation, the plant material was dried for 1 h at 105 °C. To analyse the total content of Fe, Mn, Zn, and Cu, the plant material (2.5 g) was dissolved in a mixture of concentrated nitric (ultrapure) and perchloric acids (analytically pure)in a 3:1 ratio (30 cm^3) [56] (pp. 25–83). After mineralisation, the following measurements were taken: Fe, Mn, Zn, and Cu. These were measured with flame atomic absorption spectroscopy (FAAS) using the Carl Zeiss Jena 5 apparatus (Carl Zeiss Jena, Thornwood, NY, USA). The accuracy of the methods used for the chemical analyses and the precision of the analytical measurements of nutrient levels were tested by analysing the reference material of branched flour (*Pseudevernia furfuracea*), certified by the IRMM (Institute for Reference Materials and Measurements) in Belgium. The procedure was also verified with the LGC7162 reference material (LGC standards), with an average nutrient recovery of 96% (N, P, K, Ca, Mg, Fe, Mn, Zn).

3.6. Statistical Analysis

The study was conducted as a one-factor experiment. The results are the averages of five replications. The differences between the means were estimated using Duncan's test at a significance level of $\alpha = 0.05$. The data were statistically analysed using Statistica 13.3 software (StatSoft Inc., Tulsa, OK, USA).

4. Conclusions

The iron and iSal treatments did not affect the linearly fresh and dry matter yields of lettuce, probably because the concentrations of both compounds were within appropriate ranges and had no toxic effects on the plants. The foliar spray of Fe improved the Fe content of the plants and had no negative effect on the Mn content. However, the higher doses of Fe negatively affected the Zn and Cu content when also in combination with iSal. It should be noted that the application of only iSal at a dose of 20 mg L^{-1} did not reduce the Zn content in the plants compared to the other treatments. The study showed that foliar-applied Fe chelate is effective in the biofortification of lettuce.

However, exogenous iSal applied foliarly did not specifically positively affect the uptake or content of micronutrients in the lettuce, except manganese. In addition, iSal at a dose of 20 mg L^{-1} combined with Fe negatively affected the chlorophyll and carotenoid content.

To sum up, we conclude that the foliar spraying of chelate Fe-DTPA may be an alternative for increasing the concentration of this element in lettuce. However, the need for additional applications of exogenous iSal has not been proven in this experiment. This may have been since the lettuce plants were cultivated under optimal growth conditions without any stress factors.

Author Contributions: Conceptualization, T.K., M.S. (Marcin Smiglak) and R.K.; methodology, T.K.; validation, T.K. and B.F.; formal analysis, T.K. and M.S. (Maciej Spychalski); investigation, T.K. and R.M.; resources, T.K., M.S. (Marcin Smiglak), R.K. and M.S. (Maciej Spychalski); data curation, T.K.; writing—original draft preparation, B.F.; writing—review and editing, B.F., T.K., M.S. (Maciej Spychalski), R.K. and M.S. (Marcin Smiglak); visualization, B.F.; supervision, T.K.; project administration, T.K. and R.K.; funding acquisition, T.K. All authors have read and agreed to the published version of the manuscript.

Funding: This research was supported by The National Centre for Research and Development (Poland), project LIDER (LIDER13/0211/2022)—"Growth and development stimulants with immunity-inducing effect as an innovative product for use in the cultivation of agricultural consumer plants".

Data Availability Statement: Data are contained within the article.

Conflicts of Interest: The authors declare no conflicts of interest.

References

1. Kim, M.J.; Moon, Y.; Tou, J.C.; Mou, B.; Waterland, N.L. Nutritional value, bioactive compounds and health benefits of lettuce (*Lactuca sativa* L.). *J. Food Compos. Anal.* **2016**, *49*, 19–34. [CrossRef]
2. Abbaspour, N.; Hurrell, R.; Kelishadi, R. Review on iron and its importance for human health. *J. Res. Med. Sci.* **2014**, *19*, 164–174. [PubMed]
3. FAO & WHO. *Human Vitamin and Mineral Requirements*; Report of a Joint FAO/WHO Expert Consultation, Bangkok, Thailand; FAO & WHO: Rome, Italy, 2002; 341p.
4. Sperotto, R.A.; Ricachenevsky, F.K.; de Abreu Waldow, V.; Palma Fett, J. Iron biofortification in rice: It's a long way to the top. *Plant Sci.* **2012**, *190*, 24–39. [CrossRef] [PubMed]
5. Pasricha, S.R.; Tye-Din, J.; Muckenthaler, M.U.; Swinkels, D.W. Iron deficiency. *Lancet* **2021**, *397*, 233–248. [CrossRef] [PubMed]
6. Frąszczak, B.; Kleiber, T. Microgreens biometric and fluorescence response to iron (Fe) biofortification. *Int. J. Mol. Sci.* **2022**, *23*, 14553. [CrossRef] [PubMed]
7. Buturi, C.V.; Mauro, R.P.; Fogliano, V.; Leonardi, C.; Giuffrida, F. Mineral biofortification of vegetables as a tool to improve human diet. *Foods* **2021**, *10*, 223. [CrossRef]
8. Savvas, D.; Gruda, N. Application of soilless culture technologies in the modern greenhouse industry—A review. *Eur. J. Hortic. Sci.* **2018**, *83*, 280–293. [CrossRef]
9. Kobayashi, T.; Nozoye, T.; Nishizawa, N.K. Iron transport and its regulation in plants. *Free Radic. Biol. Med.* **2019**, *133*, 11–20. [CrossRef]
10. Zahra, N.; Hafeez, M.B.; Shaukat, K.; Wahid, A.; Hasanuzzaman, M. Fe toxicity in plants: Impacts and remediation. *Physiol. Plant.* **2021**, *173*, 201–222. [CrossRef]
11. EU. Regulation of the European Parliament and of the Council Laying Down Rules on the Making Available on the Market of EU Fertilising Products and Amending Regulations (EC) No 1069/2009 and (EC) No 1107/2009 and Repealing Regulation (EC) No 2003/2003. 2019. Available online: https://eur-lex.europa.eu/eli/reg/2019/1009/oj (accessed on 3 September 2023).
12. De Pascale, S.; Rouphael, Y.; Colla, G. Plant biostimulants: Innovative tool for enhancing plant nutrition in organic farming. *Eur. J. Hortic. Sci.* **2017**, *82*, 277–285. [CrossRef]
13. Abou-Sreea, A.I.B.; Rady, M.M.; Roby, M.H.H.; Ahmed, S.M.A.; Majrashi, A.; Ali, E.F. Cattle manure and bio-nourishing royal jelly as alternatives to chemical fertilizers: Potential for sustainable production of organic *Hibiscus sabdariffa* L. *J. Appl. Res. Med. Aromat. Plants* **2021**, *25*, 100334. [CrossRef]
14. Kumar, U.; Gulati, I.J.; Rathiya, G.R.; Singh, M.P. Effect of saline water irrigation, humic acid and salicylic acid on soil properties, yield attributes and yield of tomato (*Lycopersicon esculentum*). *Environ. Ecol.* **2017**, *35*, 449–453.

15. Yaseen, A.A.; Takacs-Hajos, M. The effect of plant biostimulants on the macronutrient content and ion ratio of several lettuce (*Lactuca sativa* L.) cultivars grown in a plastic house. *S. Afr. J. Bot.* **2022**, *147*, 223–230. [CrossRef]
16. da Rocha Neto, A.C.; Maraschin, M.; Di Piero, R.M. Antifungal activity of salicylic acid against *Penicillium expansum* and its possible mechanisms of action. *Int. J. Food Microbiol.* **2015**, *215*, 64–70. [CrossRef] [PubMed]
17. Kukawka, R.; Czerwoniec, P.; Lewandowski, P.; Pospieszny, H.; Smiglak, M. New ionic liquids based on systemic acquired resistance inducers combined with the phytotoxicity reducing cholinium cation. *New J. Chem.* **2018**, *42*, 11984–11990. [CrossRef]
18. Kukawka, R.; Spychalski, M.; Stróżyk, E.; Byzia, E.; Zajac, A.; Kaczyński, P.; Łozowicka, B.; Pospieszny, H.; Smiglak, M. Synthesis, characterization and biological activity of bifunctional ionic liquids based on dodine ion. *Pest Manag. Sci.* **2022**, *78*, 446–455. [CrossRef]
19. Frąckowiak, P.; Pospieszny, H.; Smiglak, M.; Obrępalska-Stęplowska, A. Assessment of the efficacy and mode of action of benzo(1,2,3)-thiadiazole-7-carbothioic acid s-methyl ester (bth) and its derivatives in plant protection against viral disease. *Int. J. Mol. Sci.* **2019**, *20*, 1598. [CrossRef]
20. Spychalski, M.; Kukawka, R.; Prasad, R.; Borodynko-Filas, N.; Stępniewska-Jarosz, S.; Turczański, K.; Smiglak, M. A new benzothiadiazole derivative with systemic acquired resistance activity in the protection of zucchini (*Cucurbita pepo convar. giromontiina*) against viral and fungal pathogens. *Plants* **2022**, *12*, 43. [CrossRef]
21. Smiglak, M.; Lewandowski, P.; Kukawka, R.; Budziszewska, M.; Krawczyk, K.; Obrępalska-Stęplowska, A.; Pospieszny, H. Dual functional salts of benzo[1.2.3]thiadiazole-7-carboxylates as a highly efficient weapon against viral plant diseases. *ACS Sustain. Chem. Eng.* **2017**, *5*, 4197–4204. [CrossRef]
22. Spychalski, M.; Kukawka, R.; Krzesiński, W.; Spiżewski, T.; Michalecka, M.; Poniatowska, A.; Puławska, J.; Mieszczakowska-Frąc, M.; Panasiewicz, K.; Kocira, A.; et al. Use of New BTH derivative as supplement or substitute of standard fungicidal program in strawberry cultivation. *Agronomy* **2021**, *11*, 1031. [CrossRef]
23. Jarecka-Boncela, A.; Spychalski, M.; Ptaszek, M.; Włodarek, A.; Smiglak, M.; Kukawka, R. The Effect of a New Derivative of Benzothiadiazole on the Reduction of Fusariosis and Increase in Growth and Development of Tulips. *Agriculture* **2023**, *13*, 853. [CrossRef]
24. Dempsey, D.A.; Klessig, D.F. How does the multifaceted plant hormone salicylic acid combat disease in plants and are similar mechanisms utilized in humans? *BMC Biol.* **2017**, *15*, 23. [CrossRef] [PubMed]
25. Wise, K.; Gill, H.; Selby-Pham, J. Willow bark extract and the biostimulant complex Root Nectar® increase propagation efficiency in chrysanthemum and lavender cuttings. *Sci. Hortic.* **2020**, *263*, 109108. [CrossRef]
26. Mutlu-Durak, H.; Yildiz Kutman, B. Seed treatment with biostimulants extracted from weeping willow (*Salix babylonica*) enhances early maize growth. *Plants* **2021**, *10*, 1449. [CrossRef] [PubMed]
27. Sharma, A.; Sidhu, G.P.S.; Araniti, F.; Bali, A.S.; Shahzad, B.; Tripathi, D.K.; Landi, M. The role of salicylic acid in plants exposed to heavy metals. *Molecules* **2020**, *25*, 540. [CrossRef] [PubMed]
28. Filho, A.B.C.; Cortez, J.W.M.; de Sordi, D.; Urrestarazu, M. Common chicory performance as influenced by iron concentration in the nutrient solution. *J. Plant Nutr.* **2015**, *38*, 1489–1494. [CrossRef]
29. Ravet, K.; Touraine, B.; Boucherez, J.; Briat, J.F.; Gaymard, F.; Cellier, F. Ferritins control interaction between iron homeostasis and oxidative stress in Arabidopsis. *Plant J.* **2009**, *57*, 400–412. [CrossRef]
30. Preciado-Rangel, P.; Valenzuela-García, A.A.; Pérez-García, L.A.; González-Salas, U.; Ortiz-Díaz, S.A.; Buendía-García, A.; Rueda-Puente, E.O. La biofortificación foliar con hierro mejora la calidad nutracéutica y la capacidad antioxidante en lechuga. *Terra Latinoam.* **2022**, *40*, 1–7. [CrossRef]
31. Giordano, M.; El-Nakhel, C.; Pannico, A.; Kyriacou, M.C.; Stazi, S.R.; De Pascale, S.; Rouphael, Y. Iron biofortification of red and green pigmented lettuce in closed soilless cultivation impacts crop performance and modulates mineral and bioactive composition. *Agronomy* **2019**, *9*, 290. [CrossRef]
32. Rousseau, S.; Kyomugasho, C.; Celus, M.; Hendrickx, M.E.G.; Grauwet, T. Barriers impairing mineral bioaccessibility and bioavailability in plant-based foods and the perspectives for food processing. *Crit. Rev. Food Sci. Nutr.* **2019**, *60*, 826–843. [CrossRef]
33. Sameer, A.; Rabia, S.; Khan, A.A.A.; Hussain, A.; Ali, B.; Zaheer, M.S.; Ali, H.; Sheteiwy, M.S.; Ali, S. Combined application of zinc oxide and iron nanoparticles enhanced Red Sails Lettuce growth and antioxidants enzymes activities while reducing the chromium uptake by plants grown in a Cr-contaminated soil. *Res. Sq.* 2023; *posted.* [CrossRef]
34. Moosavi, A.A.; Ronaghi, A. Influence of foliar and soil applications of iron and manganese on soybean dry matter yield and iron-manganese relationship in a calcareous soil. *Aust. J. Crop Sci.* **2011**, *5*, 1550.
35. Ghasemi-Fasaei, R.; Ronaghi, A.; Maftoun, M.; Karimian, N.; Soltanpour, P.N. Iron-manganese interaction in chickpea as affected by foliar and soil application of iron in a calcareous soil. *Commun. Soil Sci. Plant Anal.* **2005**, *36*, 1–9. [CrossRef]
36. Lešková, A.; Giehl, R.F.H.; Hartmann, A.; Fargašová, A.; von Wirén, N. Heavy metals induce iron deficiency responses at different hierarchic and regulatory levels. *Plant Physiol.* **2017**, *174*, 1648–1668. [CrossRef]
37. Buturi, C.V.; Sabatino, L.; Mauro, R.P.; Navarro-León, E.; Blasco, B.; Leonardi, C.; Giuffrida, F. Iron biofortification of greenhouse soilless lettuce: An effective agronomic tool to improve the dietary mineral intake. *Agronomy* **2022**, *12*, 1793. [CrossRef]
38. Przybysz, A.; Wrochna, M.; Małecka-Przybysz, M.; Gawrońska, H.; Gawroński, S.W. Vegetable sprouts enriched with iron: Effects on yield, ROS generation and antioxidative system. *Sci. Hortic.* **2016**, *203*, 110–117. [CrossRef]

39. Lombj, E.; Tearall, K.L.; Howarth, J.R.; Zhao, F.J.; Hawkesford, M.J.; McGrath, S.P. Influence of iron status on cadmium and zinc uptake by different ecotypes of the hyperaccumulator *Thlaspi caerulescens*. *Plant Physiol.* **2002**, *128*, 1359–1367. [CrossRef]
40. Rietra, R.P.J.J.; Heinen, M.; Dimkpa, C.O.; Bindraban, P.S. Effects of nutrient antagonism and synergism on yield and fertilizer use efficiency. *Commun. Soil Sci. Plant Anal.* **2017**, *48*, 1895–1920. [CrossRef]
41. Broschat, T.K.; Moore, K.K. Phytotoxicity of several iron fertilizers and their effects on Fe, Mn, Zn, Cu, and P content of African marigolds and zonal geraniums. *HortScience* **2004**, *39*, 595–598. [CrossRef]
42. Fariduddin, Q.; Hayat, S.; Ahmad, A. Salicylic acid influences net photosynthetic rate, carboxylation efficiency, nitrate reductase activity, and seed yield in *Brassica juncea*. *Photosynthetica* **2003**, *41*, 281–284. [CrossRef]
43. Moharekar, S.T.; Lokhande, S.D.; Hara, T.; Tanaka, R.; Tanaka, A.; Chavan, P.D. Effects of salicylic acid on chlorophyll and carotenoid contents on wheat and moong seedlings. *Photosynthetica* **2003**, *41*, 315–317. [CrossRef]
44. Gorbe, E.; Calatayud, A. Applications of chlorophyll fluorescence imaging technique in horticultural research: A review. *Sci. Hortic.* **2012**, *138*, 24–35. [CrossRef]
45. Calatayud, A.; Roca, D.; Martínez, P.F. Spatial-temporal variations in rose leaves under water stress conditions studied by chlorophyll fluorescence imaging. *Plant Physiol. Biochem.* **2006**, *44*, 564–573. [CrossRef]
46. Ruban, A.V.; Horton, P. Regulation of non-photochemical quenching of chlorophyll fluorescence in plants. *Funct. Plant Biol.* **1995**, *22*, 221–230. [CrossRef]
47. Force, L.; Critchley, C.; van Rensen, J.J.S. New fluorescence parameters for monitoring photosynthesis in plants. *Photosynth. Res.* **2003**, *78*, 17–33. [CrossRef] [PubMed]
48. Streb, P.; Shang, W.; Feierabend, J.; Bligny, R. Divergent strategies of photoprotection in high-mountain plants. *Planta* **2012**, *236*, 399–412. [CrossRef]
49. Brestic, M.; Zivcak, M.; Kalaji, H.M.; Carpentier, R.; Allakhverdiev, S.I. Photosystem II thermostability in situ: Environmentally induced acclimation and genotype-specific reactions in *Triticum aestivum* L. *Plant Physiol. Biochem.* **2015**, *96*, 191–200. [CrossRef]
50. Baker, N.R.; Rosenqvist, E. Applications of chlorophyll fluorescence can improve crop production strategies: An examination of future possibilities. *J. Exp. Bot.* **2004**, *55*, 1607–1621. [CrossRef]
51. Stirbet, A.; Lazár, D.; Kromdijk, J. Chlorophyll a fluorescence induction: Can just a one-second measurement be used to quantify abiotic stress responses? *Photosynthetica* **2018**, *56*, 86–104. [CrossRef]
52. Smiglak, M.; Pospieszny, H.; Kukawka, R.; Spychalski, M. Ionic Derivatives of Aromatic Carboxylic Acid for Use as Plant Stimulants, a Method of Plant Stimulation, and the Application of These Derivatives in the Production of Compositions for Plant Stimulation. PCT/PL2023/050101, 29 December 2023.
53. Strasser, R.J.; Tsimilli-Michael, M.; Srivastava, A. Analysis of the chlorophyll a fluorescence transient. In *Chlorophyll a Fluorescence*; Papageorgiou, G.C., Govindjee, Eds.; Springer: Dordrecht, The Netherlands, 2004; pp. 321–362.
54. Hiscox, J.; Israelstam, G. A method for the extraction of chlorophyll from leaf tissue without maceration. *Can. J. Bot.* **1979**, *57*, 1332–1334. [CrossRef]
55. Wellburn, A.R. The spectral determination of chlorophyll a and chlorophyll b, as well as total carotenoids, using various solvents with spectrophotometers of different resolution. *J. Plant Physiol.* **1994**, *144*, 30. [CrossRef]
56. IUNG. *Analytical Methods in Agricultural Chemistry Stations Part II. Plant Analyses*; Institute of Soil Science and Plant Cultivation: Puławy, Poland, 1972; pp. 25–83.

plants MDPI

Article

MsYSL6, A Metal Transporter Gene of Alfalfa, Increases Iron Accumulation and Benefits Cadmium Resistance

Miao Zhang †, Meng-Han Chang †, Hong Li, Yong-Jun Shu, Yan Bai, Jing-Yun Gao, Jing-Xuan Zhu, Xiao-Yu Dong, Dong-Lin Guo * and Chang-Hong Guo *

Heilongjiang Provincial Key Laboratory of Molecular Cell Genetics and Genetic Breeding, College of Life Science and Technology, Harbin Normal University, Harbin 150025, China; zhangmiaokid669@163.com (M.Z.); cmh6650@163.com (M.-H.C.); lihong1232580@163.com (H.L.); syjun2003@126.com (Y.-J.S.); baiyan789@163.com (Y.B.); jingyun20000528@163.com (J.-Y.G.); zhujingxuan99@163.com (J.-X.Z.); dx20211006@163.com (X.-Y.D.)

* Correspondence: guodonglin@hrbnu.edu.cn (D.-L.G.); kaku3008@hrbnu.edu.cn (C.-H.G.); Tel.: +86-04-51880-60781 (D.-L.G.)

† These authors contributed equally to this work.

Abstract: Iron (Fe) is necessary for plant growth and development. The mechanism of uptake and translocation in Cadmium (Cd) is similar to iron, which shares iron transporters. Yellow stripe-like transporter (YSL) plays a pivotal role in transporting iron and other metal ions in plants. In this study, *MsYSL6* and its promoter were cloned from leguminous forage alfalfa. The transient expression of MsYSL6-GFP indicated that MsYSL6 was localized to the plasma membrane and cytoplasm. The expression of *MsYSL6* was induced in alfalfa by iron deficiency and Cd stress, which was further proved by GUS activity driven by the *MsYSL6* promoter. To further identify the function of *MsYSL6*, it was heterologously overexpressed in tobacco. *MsYSL6*-overexpressed tobacco showed better growth and less oxidative damage than WT under Cd stress. *MsYSL6* overexpression elevated Fe and Cd contents and induced a relatively high Fe translocation rate in tobacco under Cd stress. The results suggest that *MsYSL6* might have a dual function in the absorption of Fe and Cd, playing a role in the competitive absorption between Fe and Cd. *MsYSL6* might be a regulatory factor in plants to counter Cd stress. This study provides a novel gene for application in heavy metal enrichment or phytoremediation and new insights into plant tolerance to toxic metals.

Keywords: cadmium stress; transporter; gene expression; Fe translocation; antioxidant activity

Citation: Zhang, M.; Chang, M.-H.; Li, H.; Shu, Y.-J.; Bai, Y.; Gao, J.-Y.; Zhu, J.-X.; Dong, X.-Y.; Guo, D.-L.; Guo, C.-H. *MsYSL6*, A Metal Transporter Gene of Alfalfa, Increases Iron Accumulation and Benefits Cadmium Resistance. *Plants* **2023**, *12*, 3485. https://doi.org/10.3390/plants12193485

Academic Editor: Ferenc Fodor

Received: 6 September 2023
Revised: 1 October 2023
Accepted: 3 October 2023
Published: 5 October 2023

1. Introduction

Iron (Fe) is an essential micronutrient playing an active role in plant photosynthesis, antioxidant defense systems, electron respiration, and embryogenesis. Under iron deficiency, plants show typical chlorosis symptoms and hampered crop yield and productivity [1]. Plants develop efficient absorption strategies to support the requirement of sufficient iron. Iron abundance in plants is tightly regulated by iron uptake, translocation, and recycling. In addition to playing an important role in the uptake process, iron transporters are responsible for the transport and distribution of iron in plants [2]. Cadmium (Cd) is one of the most phytotoxic elements that negatively affects plant metabolism, growth, and development and indirectly contributes to ROS formation by altering the antioxidant system in plants, reducing crop failure [3–6]. Since there is no Cd-specific transporter in plants, Cd is supposed to be transported in plants via cation transport systems. Being closely similar to the essential metal element Fe, Cd is often taken up by Fe transporters. For example, NRAMPs in *Arabidopsis thaliana*, *Oryza sativa* [7–9], and *Sedum alfredii* [10], IRTs in *Oryza sativa* [11], and OPT and ZIP in *Arabidopsis thaliana* [12,13] transport both Fe and Cd. Cd distribution throughout a plant is an intricate process controlled by root uptake, root-to-shoot translocation through the xylem, and the redistribution of Cd from the leaves to sink tissues [4,14]. The contribution of cation transporters to the management of stress is beneficial in plant improvement programs [15]. Metal transporter genes can be potentially used for engineering genotypes for phytoremediation or minimizing Cd in crops. The overexpression of *AtHMA3* [16], *OsNramp1* [8], *NtPIC1* [17], and *Miscanthus sacchariflorus MsYSL1* [18,19] enhances plant tolerance to Cd. The sodium/calcium exchanger-like gene *TaNCL2-A* plays a role in alleviating the toxic effects of Cd in conjunction with salinity and osmotic stress in Arabidopsis [20].

Cd competes with essential macro- and microelements at their absorption sites, disrupting the homeostasis of crucial microelements [6,21]. It has been postulated that Cd toxicity can be attributed, at least in part, to the perturbation of the metabolism of Fe [22]. The chlorosis of plant leaves under Cd stress is similar to the typical symptoms of iron deficiency [23]. Cd induces moderate-to-strong iron deficiency in leaves, which strongly affects photosynthesis [24]. Previous studies revealed that Cd significantly decreased the Fe content in shoots, but increased the Fe content in roots [22,25]. Fe significantly accumulated in root apoplasts and root cell walls under Cd stress [10]; Fe played a significant role in alleviating the damage caused by Cd toxicity through reducing Cd accumulation and increasing Cd detention, photosynthesis protection, antioxidant defense capacity enhancement, electrolyte leakage amelioration, and other ion status regulation [26–29]. Limiting Fe uptake through the downregulation of Fe acquisition mechanisms confers a Si-mediated alleviation of Cd toxicity [30]. The Fe regulation transcript factors (*FITs*) *AtbHLH38*, *AtbHLH39*, *bHLH104*, and *BTS* are involved in Cd tolerance [22,31,32]. Although the Cd-induced increase in root Fe contents has been described, and several possible explanations for such observations including impaired Fe uptake and translocation have been proposed, the underlying mechanism of Cd-induced changes in Fe homeostasis within plants is still not fully understood.

The yellow stripe-like transporter (YSL) family has attracted attention in recent years for its function in iron uptake from soil or iron translocation throughout whole plants [19]. During the last two decades, YSLs have been revealed to transport iron complexes with specific Fe chelators and phytosiderophores. YSLs participate in Fe homeostasis in graminaceous and nongraminaceous plants with distinguished functions in each plant type. In graminaceous plants, most of the YSLs take up the Fe^{3+}-mugineic acid (MA) complex [33,34]. Some YSLs transport Fe^{2+}-Niacinamide (NA) or Fe^{3+}-2′-deoxymugineic acid (DMA) over long distances and enable the distribution of Fe within the plant leaves, phloem, and reproductive organs [35–37]. In nongraminaceous plants, YSLs transport the Fe–NA complex over long distances and intracellularly distribute Fe in plants [38–40]. Beyond the essential functions in Fe transport, evidence supports the notion that YSLs are mediated in the processes of transporting Cu, Mn, Ni, and Cd in plants [19,41–43]. *YSL1* and *YSL3* in *Arabidopsis thaliana* [40], *SnYSL3* in *Solanum nigrum* [44], *BjYSL7* in *Brassica junce* [45], *YSL3* in peanut [46], *MsYSL1* in *Miscanthus sacchariflorus* [18], and *ZmYS1* in maize [47] were reported to be involved in the uptake, distribution, and translocation of Cd.

Leguminous plants can grow in poor and degraded soils. Meanwhile, perennial grasses occupy diverse soils worldwide, including many sites contaminated with heavy metals [26], whereas studies on Fe and Cd interaction have minimal information reported in legumes [48]. Alfalfa (*Medicago sativa* L.) is an essential forage of biological nitrogen fixation, biofuel, and animal feed. Investigators have supported the idea that alfalfa reaches a new steady state to acclimate under chronic Cd stress by adequately adjusting its metabolic composition [30,49]. An increased abundance of xylogalacturonan of *Medicago sativa* after long-term exposure to Cd might hinder Cd binding in the cell wall and is an important factor during tolerance acquisition [50]. At the same time, reducing Cd bioaccumulation and controlling Cd distribution in alfalfa deserve more attention. The overlap of Fe and Cd transport in previous reports reflects the importance of the transporter's function. The overexpression or elimination of transporter reduces Cd uptake in plants [51].

Moreover, several *YSL* genes have been found to enhance Cd resistance ability in transgenic plants [18,45]. It is beneficial to explore the potential of *YSL* in metal transport for genetic engineering. Several *YSL* genes can assist plants in participating in the process of Cd uptake, transport, and accumulation [52]. There are also some YSLs that can help the zinc–iron biofortification of wheat [53]. It was found that the rice YSL family is closely related to the evolution and function of other plants [54]. Whether YSL is involved in Cd uptake and transport has not been investigated in alfalfa yet. Therefore, we isolated iron transport genes *MsYSL6* from alfalfa, focusing on the Fe and Cd contents and the Cd-tolerant ability in *MsYSL6* transgenic plants. This research will provide new insight into plant metal transporters in heavy metal enrichment or phytoremediation.

2. Results

2.1. MsYSL6 Belongs to the YSL Transporter Family

The open reading frame (ORF) of *MsYSL6* (LOC_MG673951.1) was cloned from alfalfa by PCR (Supplementary Figure S1A). The full length of *MsYSL6* ORF contains 2 028 bp, encoding 675 amino acids. MsYSL6 protein harbors 12 transmembrane domains (Supplementary Figure S2). The phylogenetic tree was constructed using data of MsYSL6 and other 23 YSL proteins from eight plant species (Figure 1). MsYSL6 belonged to a branch with AtYSL6, OsYSL6, MtYSL6, GmYSL6, GsYSL6, VaYSL6, CcYSL6, and CaYSL6, indicating a close relationship in these YSLs. These findings indicated that MsYSL6 belonged to the YSL transporter family.

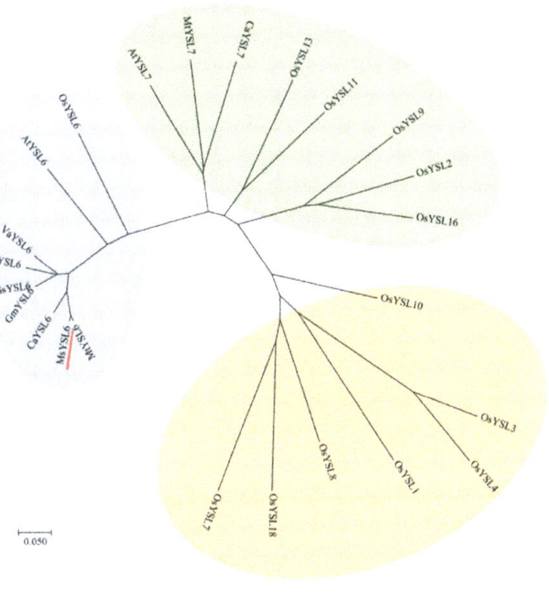

Figure 1. The phylogenetic analysis and domain characteristics of MsYSL6. The phylogenetic tree was constructed using data of MsYSL6 and other 23 YSL proteins from nine plant species, *OsYSL6* (NP_001406189.1), *AtYSL6* (NP_566806.1), *CcYSL6* (LOC109787669), *VaYSL6* (LOC108329245), *GsYSL6* (LOC114398115), *GmYSL6* (LOC100799897), *CaYSL6* (LOC101506832), *MtYSL6* (LOC11438343), *AtYSL7* (NC_003070.9), *MtYSL7* (LOC11415075), *CaYSL7* (LOC11415075), *OsYSL13* (LOC4336441), *OsYSL11* (LOC4336445), *OsYSL9* (LOC4336545), *OsYSL2* (LOC4330161), *OsYSL16* (LOC4336546), *OsYSL10* (LOC4337382), *OsYSL3* (LOC4338223), *OsYSL4* (LOC4338224), *OsYSL1* (LOC4326360), *OsYSL8* (LOC4328078), *OsYSL18* (LOC4327424), and *OsYSL7* (LOC4328077), using methods of neighbor-joining by MEGA 7.0. Red underline highlights the MsYSL6 protein. Ms (*Medicago sativa*); Mt (*Medicago truncatula*); Ca (*Cicer arietinum*); At (*Arabidopsis thaliana*); Os (*Oryza sativa*); Gm (*Glycine max*); Va (*Vigna angularis*); Gs (*Glycine soja*); Cc (*Cajanus cajan*).

2.2. The Subcellular Localization of MsYSL6

The subcellular localization of MsYSL6 was investigated using a fusion protein with GFP (Figure 2). The pBWA(V)HS-*MsYSL6*-Glosgfp was transformed into tobacco leaf cells. Apparent green fluorescence signals of MsYSL6-GFP were observed in the cell membrane and protoplast. The tobacco leaf cells expressing GFP alone exhibited weak and dispersive fluorescence inside the cells (Figure 3). Overlay images were created and showed that MsYSL6-GFP-mediated fluorescence does not co-localize with the chlorophyll-mediated fluorescence.

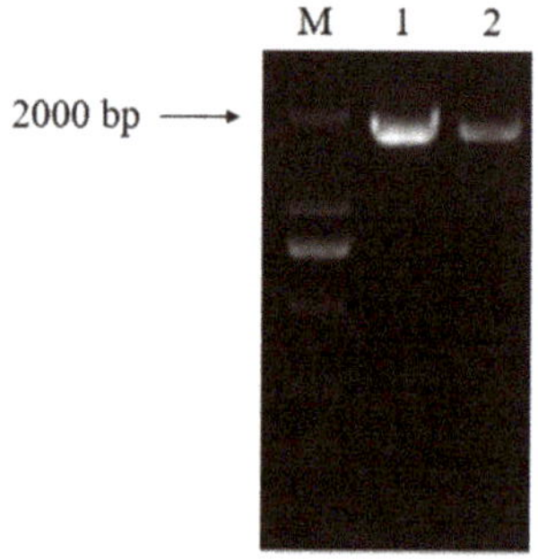

Figure 2. PCR amplification of *MsYSL6* promoter fragment. M, Marker DL 2000; 1–2, *MsYSL6* promoter.

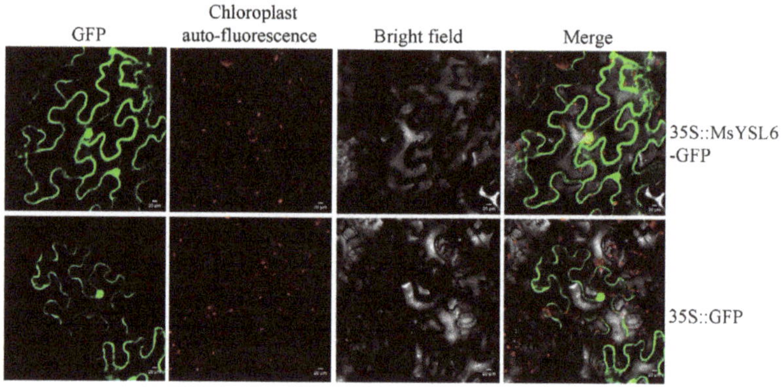

Figure 3. The subcellular localization of MsYSL6. The subcellular localization of MsYSL6 was investigated using a pBWA(V)HS-*MsYSL6*-Glosgfp vector. MsYSL6-GFP fusion protein transiently expressed in tobacco. Left to right: green fluorescence of GFP, red fluorescence of chloroplast spontaneous fluorescence, bright field, and merged microscope images (scale bars: 20 μm).

2.3. The Expression of MsYSL6 in Alfalfa

To gather insight into the expression of the *MsYSL6* gene in alfalfa, the transcript profiles of *MsYSL6* were monitored in alfalfa under iron deficiency or Cd treatment, with nonstress treatment as the control. *MsYSL6* showed tissue-specific expression in alfalfa in the control. The expression level was higher in the leaves than that in the roots and the stems. In alfalfa roots, the expression of *MsYSL6* was significantly upregulated by iron deficiency and Cd stress at 0.05 and 0.01 levels, respectively. In particular, *MsYSL6* expression increased nearly 60 folds in alfalfa roots under Cd treatment. In alfalfa stems, the expression of *MsYSL6* was significantly upregulated by iron deficiency and Cd stress at 0.05 level. Although the expression was higher than that in the roots and the stems, the expression of *MsYSL6* showed no significant differences in leaves under these three conditions (Figure 4A).

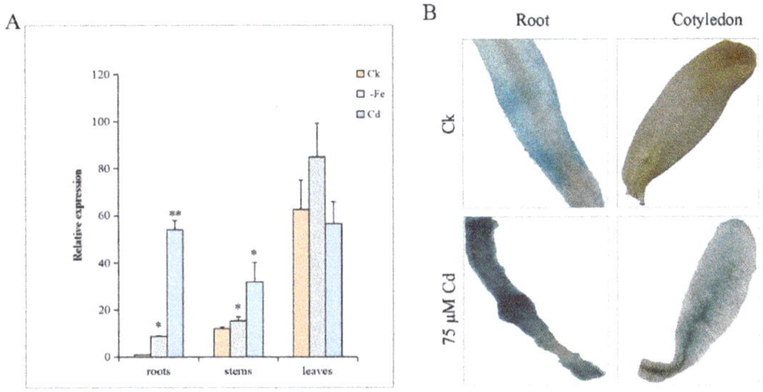

Figure 4. The expression of *MsYSL6* in response to iron deficiency and Cd. The 30-day-old hydroponic young alfalfa seedlings were treated with no Fe supply (–Fe), 0.5 mM $CdCl_2$ (Cd), and nonstress as the control (Ck) for 4 days. (**A**) The relative expression of *MsYSL6* in the roots, the stems, and the leaves of alfalfa. Data are means ± s.d. of three experimental replications. Statistical comparison was performed by ANOVA followed by an LSD test. Asterisks indicate significant differences between the treatment and the control (* $p < 0.05$; ** $p < 0.01$). (**B**) The GUS activity induced by *MsYSL6*pro in alfalfa hairy roots. The alfalfa hairy roots were treated with 75 μM $CdCl_2$ (Cd) and no stress as the control (Ck) for 15 days.

2.4. The GUS Activity

In order to determine the specific expression of *MsYSL6*, we fused the *MsYSL6* promoter in-frame to β-glucuronidase (GUS) reporter (Supplementary Figures S3 and S4). The pBI121-*MsYSL6*pro::GUS construct was transformed in alfalfa and induced hairy roots (Supplementary Figure S5). The alfalfa with hairy roots were grown in a medium containing 75 μM $CdCl_2$, with no Cd treatment as the control. The alfalfa hairy roots exhibited blue when the medium contained no Cd, and exhibited strong blue when the medium contained Cd. The alfalfa cotyledons exhibited no staining when the medium contained no Cd, and exhibited blue when the medium contained Cd (Figure 4B). The GUS activity driven by *MsYSL6*pro was enhanced by Cd indicating that *MsYSL6* responded to Cd stress, which is consistent with the upregulated expression of *MsYSL6* detected by qRT-PCR.

2.5. The Growth of MsYSL6 Transgenic Tobacco

To screen the effect of *MsYSL6* on Cd tolerance, the T_3 seeds of three *MsYSL6* overexpress lines (*MsYSL6*OE), namely, L5, L8, L9, and WT, were grown in the medium containing 50 μM $CdCl_2$, 75 μM $CdCl_2$, and 100 μM $CdCl_2$, with no Cd as the control. A preliminary observation showed that *MsYSL6*OE germination was similar to WT plants when grown in no Cd medium. The germination of *MsYSL6*OE and WT decreased following the increase of Cd concentration, and the germination of *MsYSL6*OE lines was significantly less affected than that of WT (Supplementary Figure S6, Table S1). The WT plant grown on 75 μM $CdCl_2$ medium exhibited shorter and smaller roots, while *MsYSL6*OE exhibited a normal root phenotype with minimal stunted growth. The *MsYSL6*OE plants showed significantly higher fresh weight and longer roots than the WT (Figure 5).

2.6. The Chlorophyll Content

The chlorophyll content decreased in *MsYSL6*OE and WT plants under Cd treatment more than that of the control. The chlorophyll content of *MsYSL6*OE leaves was significantly higher than that of WT in the control at 0.05 level. The chlorophyll content of L5 and L8 leaves was significantly higher than that of WT under Cd treatment at 0.05 level. The higher chlorophyll content showed that the *MsYSL6*OE transgenic tobacco was more tolerant to Cd stress (Figure 6A).

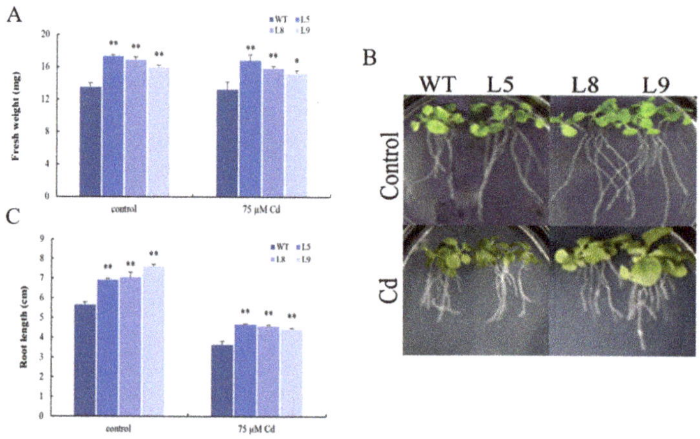

Figure 5. The growth of *MsYSL6*OE tobacco under Cd stress. The 4-week-old tobacco seedlings of three *MsYSL6OE* tobacco lines (L5, L8, and L9) and WT plants were treated with 75 μM $CdCl_2$ (Cd) and no stress as the control (Ck) for 10 d. The fresh weight and root length were detected. (**A**) The fresh weight of in *MsYSL6OE* tobacco lines and WT. (**B**,**C**) The root length of in *MsYSL6OE* tobacco lines and WT. The data presented are the means ± s.d. of three experimental replications. Statistical comparison was performed by ANOVA followed by an LSD test. Asterisks indicate significant differences (* $p < 0.05$; ** $p < 0.01$).

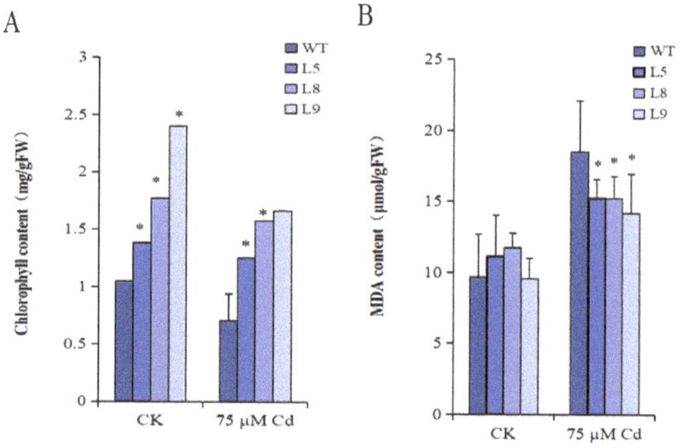

Figure 6. The chlorophyll and MDA content of *MsYSL6*OE tobacco under Cd stress. The 4-week-old seedlings of three *MsYSL6*OE tobacco lines (L5, L8, and L9) and WT plants were treated with 75 μM $CdCl_2$ (Cd) and no stress as the control (Ck) for 10 d. (**A**) The chlorophyll content in *MsYSL6OE* tobacco lines and WT. (**B**) The MDA content in *MsYSL6OE* tobacco lines and WT. The data presented are the means ± s.d. of three experimental replications. Statistical comparison was performed by ANOVA followed by an LSD test. Asterisks indicate significant differences (* $p < 0.05$).

2.7. The MDA Content

The malondialdehyde (MDA) content showed no difference in *MsYSL6*OE tobacco and WT in the control. Cd treatment increased the MDA content in the *MsYSL6*OE tobacco and WT plants. The MDA content in L5, L8, and L9 was significantly lower than that of WT under Cd treatment at 0.05 level (Figure 6B). The results showed that overexpress of *MsYSL6* protects plants from damage and leakage in the membrane caused by Cd stress.

2.8. The DAB and NBT Staining

Under Cd treatment, DAB staining showed that the leaf of WT was deeper brown while the leaves of L5, L8, and L9 were shallow brown. The staining was scattered at the leaves of L5, L8, and L9, and the brown color was less than that of WT. NBT staining showed that the leaf of WT was deeper blue, while the leaves of L5, L8, and L9 were shallow blue (Figure 7). The results represent that *MsYSL6*OE tobacco has less ROS production than WT under Cd stress.

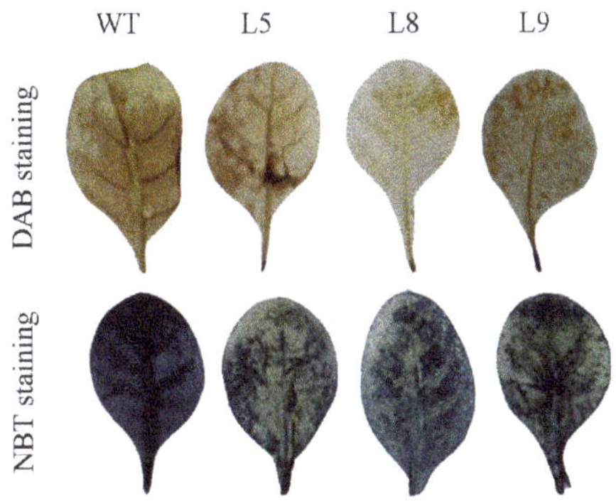

Figure 7. The DAB and NBT staining of *MsYSL6*OE tobacco under Cd stress. The 4-week-old tobacco seedlings of *MsYSL6OE* tobacco lines (L5, L8, and L9) and WT plants were treated with 75 µM $CdCl_2$ (Cd) and no stress as the control (Ck) for 10 d. The leaf was stained with nitro blue tetrazolium chloride (NBT) and diaminobenzidine (DAB). The leaf of WT showed deeper staining. The results represent that *MsYSL6OE* tobacco has less ROS than WT under Cd stress.

2.9. The Cd Content and Translocation Rate

The Cd accumulation in *MsYSL6*OE plants under 75 µM Cd treatment with a significant increase was observed compared with the WT plants (Figure 8A). The Cd content in L5, L8, and L9 roots was significantly higher than that of the WT at 0.05 level. The Cd content in shoots of L5 and L8 was significantly higher than that of the WT at 0.05 level, which showed no difference in L9 from that of the WT (Figure 8A). The Cd translocation rate was 31%, 34%, 38%, and 33% in L5, L8, L9, and WT, respectively.

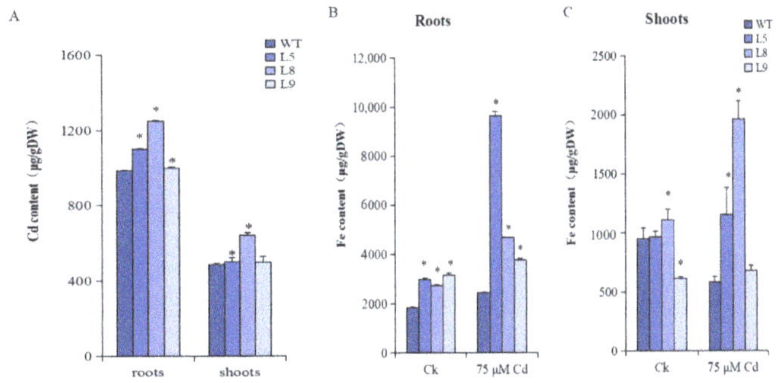

Figure 8. The Cd and Fe contents in *MsYSL6OE* tobacco and WT. The 4-week-old tobacco seedlings of three *MsYSL6OE* tobacco lines (L5, L8, and L9) and WT plants were treated with 75 µM $CdCl_2$ (Cd) and no Cd treatment as the control (Ck) for 10 d. (**A**) The Cd content in the roots and shoots of *MsYSL6OE* tobacco lines and WT. (**B**) The Fe content in the roots of *MsYSL6OE* tobacco lines and WT. (**C**) The Fe content in the shoots of *MsYSL6OE* tobacco lines and WT. Data are means ± s.d. of three experimental replications. Statistical comparison was performed by ANOVA followed by an LSD test. Asterisks indicate significant differences (* $p < 0.05$).

2.10. The Fe Content and Translocation Rate

In the roots, the Fe content of *MsYSL6*OE lines was significantly higher than that of WT under Cd or non-Cd treatment at 0.05 level. Notably, the Fe content in roots was increased in all lines by Cd treatment, and the increase of *MsYSL6*OE plants was greater than that of WT (Figure 8B). In the shoots, the Fe content was significantly higher in L8 and significantly lower in L9 than that in WT under non-Cd treatment; the Fe content was significantly higher in L5 and L8 than that in WT under Cd treatment at 0.05 level; the Fe content increased in *MsYSL6*OE shoots and decreased in WT shoots by Cd treatment (Figure 8C). Under non-Cd treatment, the Fe translocation rate accounted for 15%, 24%, and 13% in L5, L8, and L9, respectively, both lower than 29% in WT. Under Cd treatment, the Fe translocation rate accounted for 24%, 29%, and 16% in L5, L8, and L9, respectively, both lower than 34% in WT.

2.11. The Cu, Mn, and Zn Content

The Cu, Mn, and Zn content were also screened in *MsYSL6*OE plants and the WT plants. In plant roots, Cu contents were significantly higher and Mn contents were significantly lower in *MsYSL6*OE lines than that of WT plants under no-Cd treatment at 0.05 level; Cu contents and Mn contents in L5 and L8 were significantly higher and Mn content in L9 was significantly lower than that of the WT plants under Cd treatment at 0.05 level; Zn contents of *MsYSL6*OE lines showed no difference with the WT plants under all treatments (Figure 9A–C). In shoots, the Cu and Zn contents in L8 and L9 were significantly higher and Mn content in L5 shoots was significantly lower than that of the WT plants under no-Cd treatment at 0.05 level; the Mn contents in L5, L8, and L9 shoots were significantly lower than that of the WT plants at 0.05 level (Figure 9D–F). In general, Cu, Mn, and Zn contents in the roots and Cu contents in the shoots decreased, while Mn and Zn contents in the shoots of *MsYSL6*OE plants and the WT plants increased by Cd treatment.

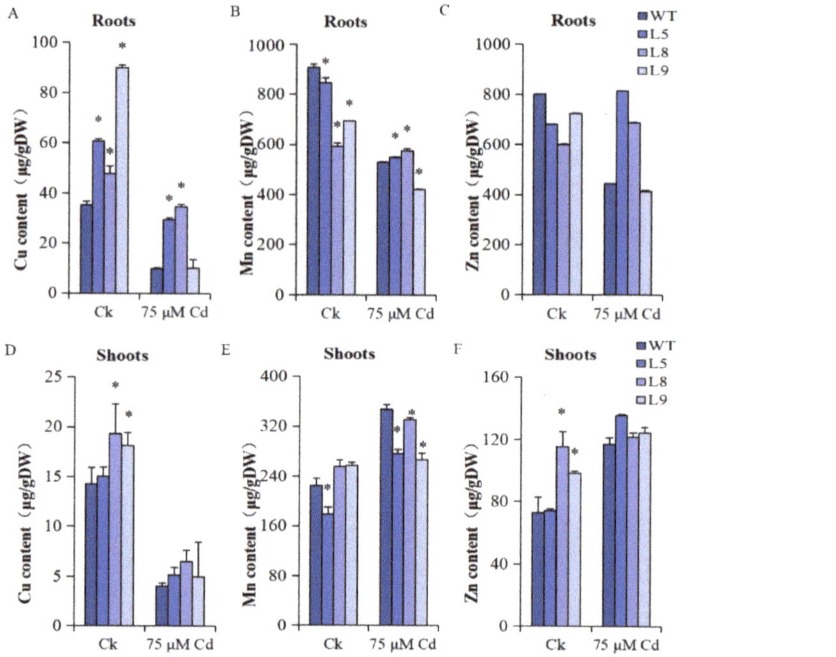

Figure 9. The Cu, Mn, and Zn contents in *MsYSL6*OE tobacco and WT. The 4-week-old tobacco seedlings of three *MsYSL6OE* tobacco lines (L5, L8, and L9) and WT plants were treated with 75 μM

CdCl2 (Cd) and no stress as the control (Ck) for 10 d. (**A**) The Cu content in the roots of *MsYSL6OE* tobacco lines and WT. (**B**) The Mn content in the roots of *MsYSL6OE* tobacco lines and WT. (**C**) The Zn content in the roots of *MsYSL6OE* tobacco lines and WT. (**D**) The Cu content in the shoots of *MsYSL6OE* tobacco lines and WT. (**E**) The Mn content in the shoots of *MsYSL6OE* tobacco lines and WT. (**F**) The Zn content in the shoots of *MsYSL6OE* tobacco lines and WT. The Cu, Mn, and Zn contents in the roots and shoots were assessed by ICP-OES. Data are means ± s.d. of three experimental replications. Statistical comparison was performed by ANOVA followed by an LSD test. Asterisks indicate significant differences (* $p < 0.05$).

3. Discussion

Although not essential for plant growth, Cd is readily taken up by roots and accumulated in plant tissues to high levels [26]. Cd uptake and transport in different plant species is linked to Fe due to their similar chemical characteristics. It has been reported that the iron transporter *OsNRAMP5* contributes to the uptake of Cd from the soil, and the total Cd content in the *OsNRAMP5*i plants decreases [9]. The upregulation of *IRT1* facilitated Cd uptake by roots, thus increasing the Cd concentration in plant tissues [32]. Transcript levels of *IRT1* are very low and transcript levels of *HMA2* are strongly elevated in *Arabidopsis halleri* from the most highly heavy metal-contaminated soil, which can account for its altered Cd handling [55]. *ABC, NRAMP,* and *ZIP* genes might play important roles in different levels of Cd accumulation in sunflower cultivars [56]. *MsYSL1 SnYSL3* and *BjYSL7* may be essential transporters for diverse metal–NAs to participate in the Cd detoxification by mediating the reallocation of other metal ions [18,40,45]. Studies have shown that most YSLs may function in the intracellular transport, translocation, and distribution of Fe and other metals in *Arabidopsis thaliana* and *Oryza sativa*. In this study, we obtained a novel YSL transporter gene, *MsYSL6,* from alfalfa. MsYSL6 belonged to a branch with AtYSL6, OsYSL6, and several YSL6s in leguminous plants, indicating a close relationship in these YSLs. A fundamental role has been demonstrated for Arabidopsis YSL6 in managing chloroplastic iron [38]. OsYSL6 was suggested to be a Mn–NA transporter [42].

Transporters play specific biological roles in mineral transport within the tissues where they are expressed. Data indicated that YSL expression was detected in three main territories: vascular tissues throughout the plant, pollen grains, and seeds [35,39,57,58]. *MsYSL6* was expressed in all vegetative tissues of alfalfa and was most expressed in the leaf. However, *MsYSL6* was induced by iron deficiency and Cd in alfalfa roots, consistent with *MsYSL1, SnYSL3,* and *BjYSL7,* suggesting an important role for *MsYSL6* in response to Cd stress, possibly functioning in Cd uptake [18,43,44]. The altered patterns of *MsYSL6* expression might be a response to Cd stress.

In plants, Cd toxicity disrupts photosynthesis, causes the production of reactive oxygen species (ROS), and increases antioxidant enzyme activity [6,21,27,59–61]. The present study also found that *MsYSL6* overexpression reduced ROS accumulation and mitigated the adverse effects on the growth of tobacco under Cd stress, in accordance with previous studies. The specific mechanism of *MsYSL6* conferring tolerance to Cd stress into tobacco might be involved in metal distribution.

The Fe-biofortified lentil genotype exhibited Cd tolerance by inciting an efficient antioxidative response to Cd toxicity [19]. Fe significantly reduced Cd transfer towards rice grains, which might be attributed to a sharp decrease in the proportion of bioavailable Cd in leaves [62]. The significantly high Fe content in *MsYSL6* transgenic tobacco indicated the primary function of *MsYSL6* in Fe uptake from the medium. Increasing supply of Fe remarkably reduced Cd accumulation in rice shoots, mainly because of inhibited translocation of Cd from rice roots to shoots [63]. Fe deficiency increased Cd uptake and accumulation in peanuts but decreased Cd translocation from roots to shoots [64]. In rice, low Fe or excess Fe facilitated the uptake of Cd in rice roots, as low Fe upregulated the expression of Cd-transport-related genes, and excess Fe enhanced Cd enrichment on the root by iron plaque [63]. Thus, the accumulation of Cd in *MsYSL6* transgenic tobacco might be due to the high expression level of *MsYSL6* in this study.

Many studies have found that Cd stress affects Fe concentration in plants. Cd-induced Fe accumulation in roots is mediated by upregulating Fe transporter genes such as *IRT1* [65]. The Cd and Fe accumulation in plants could be partially due to the crucial ligands for metal chelation [66]. Like *AhOPT3/6/7* and *AhYSL1/3* in peanuts, *MsYSL6* might be involved in the transport of Fe and/or Cd and Fe/Cd interactions [67]. Cd damages the photosynthetic apparatus, as well as the "Fe-deficiency-like" symptoms, and the supply of Fe nutrients saves the photosynthesis symptoms caused by Cd [68,69]. Under Cd stress, high Fe concentrations in *MsYSL6* confer the plant tolerance to Cd toxicity. We infer that *MsYSL6* might have a dual function in the absorption of Fe and, at the same time, Cd. It plays a role as a regulatory factor in the competition absorption between iron and Cd. *MsYSL6* increased the ratio of iron to Cd in the total amount and transferred more Fe to the above-ground part of the plant, thereby reducing the oxidative damage of Cd to the plant and resulting in the improvement of tolerance to Cd toxicity in *MsYSL6* transgenic tobacco lines.

The metal translocation rate represents the metal translocate capacity in plants. The Cd translocation ratio showed little difference in all tested plant lines. From the results, we hypothesize that *MsYSL6* might possibly contribute to the Cd uptake while not likely involved in Cd transport from roots to shoots. In this study, overexpression of *MsYSL6* did not increase the Fe transport rate, suggesting that *MsYSL6* might not be involved in the translocation of Fe from roots to shoots, while Cd stress increased the Fe transport rate in *MsYSL6*OE lines, which is similar to the idea that Cd exposure increases the root-to-shoot translocation ratios of Fe in *MsYSL1* and *SnYSL3* overexpressing *Arabidopsis thaliana*. The relationship between Cd stress and Fe uptake ability deserves further investigation [66]. Additionally, the absorption and transport of Mn, Zn, and Cu in *MsYSL6* transgenic tobacco were affected by Cd stress. Specialized cation transport mechanisms have been developed in plants, which can maintain a balance between the deficiency and toxicity of these ions [15]. The state of these metals could also have an effect on plant growth and is worth further study.

The phytosiderophores chelate iron were suggested to prevent graminaceous plants from Cd toxicity. In maize, phytosiderophore provides an advantage under Cd stress relative to Fe acquisition via ferrous Fe [47]. The nicotinamide synthase (NAS) synthesizes nicotinamide (NA), which plays an essential role together with YSLs in iron homeostasis in nongraminaceous plants. It has also been reported that Cd induces the expression of the *NAS* gene in *Medicago truncatula* [70]. *MsYSL6* may also achieve the effect of alleviating plant Cd toxicity by co-acting with *NAS*, despite reports claiming that YSL6 in *Arabidopsis thaliana* has an essential role in chloroplast development, likely in the transport of Fe required for chloroplast development [36]. In another study, AtYSL6 was localized to vacuole membranes and to internal membranes [71]. In this study, MsYSL6 was found to have no chloroplast localization properties, so it is unlikely that MsYSL6 is involved in Cd detoxification through this pathway. Our results support the idea that MsYSL6 may function as a Cd/Fe transport gene and may be useful for genetic engineering in cultivating Fe-fortified or Cd-tolerant crops.

4. Materials and Methods

4.1. Plant Growth Conditions

Alfalfa (*Medicago sativa* L. cv. Zhaodong) seeds were soaked in 70% ethanol and 10% NaClO for 10 min, then washed with H_2O three times. Seeds were placed onto sterile Whatman paper and germinated for 3 d in the dark. After germination, seedlings were transferred into 1/2 Hoagland's solution. The solution was renewed every 2 d. The seedlings were grown at 25 °C with 16 h light/8 h darkness. The wild-type tobacco (*Nicotiana tabacum*) seeds were grown on a sterile MS medium at 25 °C with 16 h light/8 h darkness. For detection of gene expression, 30-day-old hydroponic alfalfa seedlings were treated with iron deficiency (no Fe supply, -Fe), 0.5 mM $CdCl_2$ (Cd), and the control (no stress, Ck) for 4 days. For detection of the tolerance to Cd stress, T_3 of *MsYSL6*OE and WT tobacco seedlings were germinated and cultivated in MS medium for four weeks, and then transferred onto MS medium containing 75 μM $CdCl_2$ (Cd) for 10 d, with no $CdCl_2$ MS medium as the control (Ck).

4.2. RNA and DNA Extraction

The total RNA of plant tissues was isolated using the RNA prep Pure Plant Kit 175 (TianGen Biotech, Beijing, China). RNA derived from the leaf was reverse transcribed into cDNA by using Easy Script One-Step DNA Removal and cDNA Synthesis SuperMix (TransGen 177 Biotech, Beijing, China). The total RNA quality was performed by visualization on a 1 × TBE gel stained with ethidium bromide. Total DNA was extracted using a Plant Kit (OMEGA, New York, NY, USA, D3485-01).

4.3. Cloning of *MsYSL6* and *MsYSL6pro*

Alfalfa cDNA was used as the template to amplify the open reading frames (ORFs) of the *MsYSL6* gene. The *MsYSL6* gene amplified primers were designed according to alfalfa iron deficiency transcriptome data (Table S2). The PCR system was as follows: template 1 μL, each primer 0.4 μL, ExTaq DNA polymerase (5 U/μL) 0.05 μL, dNTP mix (2.5 mM) 0.8 μL, 10 × PCR Buffer 1 μL, add ddH_2O to 10 μL. PCR was performed using the following program: 94 °C for 5 min, 94 °C for 30 s, 55 °C for 30 s, 72 °C for 2.5 min, 95 °C for 15 s, for 30 cycles; 16 °C for 1 h. The PCR products were purified with a Gel DNA Purification Kit (OMGAE), and then cloned into pMD18T plasmids for sequencing. The amino acid sequence of MsYSL6 was identified using the SMART website (http://smart.embl-heidelberg.de/ accessed on 11 June 2022). To elucidate the evolutionary relationship, a BLASTP search was performed in the NCBI nr (nonredundant protein sequences) database. Amino acid sequences were aligned with MEGA7.0 software. An unrooted neighbor-joining tree was constructed. The transmembrane domains were predicted using the TMHMM Server v.2.0 website (http://www.cbs.dtu.dk/services/TMHMM/ accessed on 11 June 2022). Conserve protein motifs were predicted through the online program MEME (http://memesuite.org/tools/meme accessed on 11 June 2022).

The amplification primers of the *MsYSL6* promoter (*MsYSL6*pro) fragment were designed according to alfalfa genomic data (Table S2). The PCR system was as follows: template DNA 1 μL, each primer 0.4 μL, ExTaq DNA polymerase (5 U/μL) 0.05 μL, dNTP mix (2.5 mM) 0.8 μL, 10 × PCR Buffer 1 μL, add ddH_2O to 10 μL. The PCR was performed as follows: 94 °C for 5 min; 30 cycles of 94 °C for 30 s, 58.1 °C for 30 s, and 72 °C for 2 min; and a final extension at 72 °C for 10 min. The PCR products were separated and purified with a Gel DNA Purification Kit (OMEGA). Purified products were then cloned into pMD18T plasmids for sequencing. The cis-elements in *MsYSL6*pro were predicted online by PlantCARE.

4.4. Subcellular Localization of *MsYSL6*

The *MsYSL6* ORF without stop codes was cloned into the pBWA(V)HS-GLosgfp vector to construct a fusion plasmid. Then, the fusion plasmid was introduced into *Agrobacterium tumefaciens* GV3101 by freeze–thaw method and injected into tobacco leaf through the epidermis with a syringe. After cultivation for 2 d, fluorescence signals were recorded using a confocal laser scanning microscope (Nikon C2-ER Laser Scanning Confocal Microscope (Tokyo, Japan)). The green fluorescent protein was excited at 488 nm, and emissions were collected at 510 nm. The chlorophyll autofluorescence was excited at 640 nm, emissions were collected at 675 nm, and overlay images were created.

4.5. qRT-PCR

The qRT-PCR was performed by a Real-Time PCR System (Applied Biosystems 7300 Real-Time PCR System, Foster City, CA, USA), using specific primers (Table S2). *MsActin* and *NtGAPDH* were used as internal controls. The qRT-PCR system was as follows: cDNA 2 μL, primer Mix 1 μL, qPCR Mix 10 μL, 50 × ROX 0.4 μL, ddH_2O 6.6 μL, total 20 μL. The qRT-PCR was performed as follows: 95 °C for 10 min, 95 °C for 15 s, 55 °C for 30 s, 72 °C for 30 s, 95 °C for 15 s, 60 °C for 30 s, for 40 cycles; 95 °C for 15 s. The qRT-PCR results were quantified using the comparative $2^{-\Delta\Delta Ct}$ method. All of the qRT-PCR experiments were performed with three independent RNA samples.

4.6. GUS Assay

The *MsYSL6*pro was cloned to create a C-terminal translational fusion to the reporter gene GUS. The pBI121-*MsYSL6*pro::GUS was introduced into the *Agrobacterium rhizogenes* K599 and then infected on alfalfa roots to induce hairy roots. *MsYSL6*pro::GUS transformed explants were cultured with MS_1 medium with 50 μM kanamycin for 4 d and then cultured with MS_3 medium. After hairy root emergence, the transformants were transferred to an MS_3 medium supplemented with 75 μM $CdCl_2$ (Cd) and no stress (Ck) for 15 d. The alfalfa hairy roots were stained for the exam the GUS activity.

4.7. Plant Transformation and Selection

MsYSL6 was digested with XbaI and SmaI and cloned into a plant expression vector pBI121. The *Agrobacterium tumefaciens* GV3101 harboring the pBI121-35S:: *MsYSL6* was transformed into tobacco by leaf disc method. Subsequently, the transgenic T_0 plants were consecutively self-crossed and detected by PCR and qRT-PCR analysis with *MsYSL6* gene-specific primers. T_3 seeds were used for further analyses.

4.8. Phenotypic and Physiological Analysis

The 2-week-old tobacco seedlings of three *MsYSL6*OE tobacco lines (L5, L8, and L9) and WT were treated with 75 μM $CdCl_2$ (Cd) and no stress (Ck) for 15 d. The germinated seeds were recorded and used to calculate the germination rate. The fresh weight and root length of seedlings were measured and photographed. The chlorophyll was extracted by acetone according to Lichtenthaler [72] et al., and the absorption value was measured at 649 nm and 665 nm by an ultraviolet–visible spectrophotometer (Metash Instruments CO. UV-5100). The MDA content was measured at 450 nm according to Reshmi et al.[73].

4.9. Oxidative Analysis

The leaves from tobacco were infiltrated with NBT and DAB dyeing solution following published procedures. The leaf was infiltrated with 1 mg/mL nitro blue tetrazolium (NBT) (pH 7.8) for 40 to 60 min in the dark to detect O_2^- accumulation. The leaf was infiltrated with 1 mg/mL diaminobenzidine (DAB) (pH 7.0) for 8 h in the dark to detect H_2O_2 accumulation; the leaf was then decolorized by boiling in ethanol: glycerin (3:1) solution for 10 min, photographed after cooling.

4.10. Metal Content Determination

The 30-day-old tobacco seedlings of *MsYSL6* transgenic tobacco (*MsYSL6*OE) lines and wild-type plants (WT) were treated with 75 μM $CdCl_2$ for 10 d [74]. The shoots and roots of *MsYSL6*OE and WT were collected, and all samples were dried to measure weight. The dried and ground samples were placed into the digestive tube with 4 mL 65% HNO_3 and heated at 120 °C for 90 min. After cooling, 1 mL 30% H_2O_2 was added and it was heated at 120 °C for 90 min. After cooling, 750 μL 30% H_2O_2 was added and it was left to stand for 4 h, constant to 50 mL with ddH_2O. All standard solutions were analyzed with the ICP-OES method using ICP-OES8000 (Perkin Elmer, Waltham, MA, USA). All of the measurements were performed with three independent samples.

4.11. Statistical Analysis

All of the experiments were repeated three times independently. Data were analyzed by one-way analysis of variance (ANOVA) with LSD Test (least significant difference). The significance of the data was annotated with asterisks (*) based on the following criteria: *, $p < 0.05$, **, $p < 0.01$.

5. Conclusions

Understanding how Fe/Cd accumulates in plants provides guidance for crop cultivation. For the first time, the role in Fe/Cd transport of *MsYSL6* was revealed. This study was a preliminary characterization of the *MsYSL6* gene. Studies are needed to determine the subcellular localization of MsYSL6 protein using markers specific for chloroplasts and various organelles. Although studies showing that Fe/Cd uptake and accumulation mechanisms in plants are similar in some respects, increasing numbers of researchers have found that these mechanisms are distinct between Fe/Cd uptake and accumulation. *MsYSL6* might cooperate with NAS to exercise their biological functions, but it remains unclear. Further study is needed to further clarify these mechanisms.

Supplementary Materials: The following supporting information can be downloaded at: https://www.mdpi.com/article/10.3390/plants12193485/s1, Figure S1: The cloning of *MsYSL6* from alfalfa and the transformation into tobacco; Figure S2: The predicted transmembrane domains of MsYSL6 protein; Figure S3: Construction of pBI121-*MsYSL6*pro::GUS; Figure S4: Identification of pBI121-*MsYSL6*pro::GUS; Figure S5: The pBI121-*MsYSL6*pro::GUS induced alfalfa hairy roots; Figure S6: The germination of *MsYSL6*OE tobacco under Cd stress. Table S1: The germination rates of *MsYSL6* overexpressed tobacco lines and WT under Cd stress; Table S2: The primer sequence.

Author Contributions: Conceived this study: D.-L.G. and C.-H.G.; contributed to compiling and analyzing the data and wrote the manuscript: D.-L.G., M.Z., and M.-H.C.; performed the experiments: M.Z., M.-H.C., H.L., J.-Y.G., J.-X.Z., and X.-Y.D.; participated in the data analysis and supervised the writing of the manuscript: D.-L.G., Y.-J.S., and Y.B.; participated in photo production: D.-L.G., M.Z., and H.L. All authors have read and agreed to the published version of the manuscript.

Funding: This study was supported by the Key Research and Development Project of Heilongjiang Province (No. 2022ZX02B05), the National Natural Science Foundation of China (No. U21a20182, 31972507), Open Fund of the State Key Laboratory of Forest Genetics and Breeding (Northeast Forestry University No. 2015X), and Doctoral innovation project of Harbin Normal University (No. HSDBSCX2020-09).

Institutional Review Board Statement: Not applicable.

Informed Consent Statement: Not applicable.

Data Availability Statement: Not applicable.

Acknowledgments: We are grateful to the Animal Husbandry Institute of Heilongjiang Academy of Agricultural Sciences for providing alfalfa (*Medicago sativa* L. cv. Zhaodong) seeds.

Conflicts of Interest: The authors declare no conflict of interest.

References

1. Merry, R.; Espina, M.J.; Lorenz, A.J.; Stupar, R.M. Development of a controlled-environment assay to induce iron deficiency chlorosis in soybean by adjusting calcium carbonates, pH, and nodulation. *Plant Methods* **2022**, *18*, 36. [CrossRef]
2. Morrissey, J.; Guerinot, M.L. Iron uptake and transport in plants: The good, the bad, and the ionome. *Chem. Rev.* **2009**, *109*, 4553–4567. [CrossRef] [PubMed]
3. Hasan, S.A.; Fari, D.Q.; Ali, B.; Hayat, S.; Ahmad, A. Cd: Toxicity and tolerance in plants. *J. Environ. Biol.* **2009**, *30*, 165–174. [CrossRef]
4. De Maria, S.; Pus, C.R.M.; Rivelli, A.R. Cd accumulation and physiological response of sunflower plants to Cd during the vegetative growing cycle. *Plant Soil. Environ.* **2013**, *59*, 254–261. [CrossRef]
5. Han, Z.; Wei, X.; Wan, D.; He, W.; Wang, X.; Xiong, Y. Effect of Molybdenum on Plant Physiology and Cd Uptake and Translocation in Rape (*Brassica napus* L.) under Different Levels of Cd Stress. *Int. J. Environ. Res. Public Health* **2020**, *17*, 2355. [CrossRef]
6. Rodríguez, S.M.; Romero, P.M.C.; Paz, M.D.M.; Testi, L.P.S.; Risue, N.M.C.; Del, R.L.A.; Sanda, L.L.M. Cellular response of pea plants to Cd toxicity: Cross talk between reactive oxygen species, nitric oxide, and calcium. *Plant Physiol.* **2009**, *150*, 229–243. [CrossRef] [PubMed]
7. Th, M.S.; Wang, R.; Ward, J.M.; Crawford, N.M.; Schroeder, J.I. Cd and iron transport by members of a plant metal transporter family in Arabidopsis with homology to Nramp genes. *Proc. Natl. Acad. Sci. USA* **2000**, *97*, 4991–4996. [CrossRef]
8. Takahashi, R.; Ishimaru, Y.; Seno, U.T.; Shi, M.H.; Ishikawa, S.; Arao, T.; Naka, N.H.; Nishizawa, N.K. The OsNRAMP1 iron transporter is involved in Cd accumulation in rice. *J. Exp. Bot.* **2011**, *62*, 4843–4850. [CrossRef]

9. Ishi, M.Y.; Taki, H.R.; Bashir, K.; Shi, M.H.; Seno, U.T.; Sugi, M.K.; Ono, K.; Yano, M.; Ishikawa, S.; Arao, T.; et al. Characterizing the role of rice NRAMP5 in Manganese, Iron and Cd Transport. *Sci. Rep.* **2012**, *2*, 286. [CrossRef]
10. Chen, S.; Han, X.; Fang, J.; Lu, Z.; Qiu, W.; Liu, M.; Sang, J.; Jiang, J.; Zhuo, R. Sedum alfredii SaNramp6 Metal Transporter Contributes to Cd Accumulation in Transgenic Arabidopsis thaliana. *Sci. Rep.* **2017**, *7*, 13318. [CrossRef]
11. Naka, N.H.; Oga, W.I.; Ishi, M.Y.; Mori, S.; Nishi, Z.N.K. Iron deficiency enhances Cd uptake and translocation mediated by the Fe^{2+} transporters OsIRT1 and OsIRT2 in rice. *Soil Sci. Plant Nutr.* **2006**, *52*, 464–469. [CrossRef]
12. Mendoza, C.D.G.; Xie, Q.; Akma, J.G.Z.; Jobe, T.O.; Patel, A.; Stacey, M.G.; Song, L.; De, M.D.W.; Juri, S.S.S.; Stacey, G.; et al. OPT3 is a component of the iron-signaling network between leaves and roots and misregulation of OPT3 leads to an over-accumulation of Cd in seeds. *Mol. Plant* **2014**, *7*, 1455–1469. [CrossRef] [PubMed]
13. Spielmann, J.; Ahmadi, H.; Scheepers, M.; Weber, M.; Nitsche, S.; Carnol, M.; Bosman, B.; Kroy, M.J.; Motte, P.; Clemens, S.; et al. The two copies of the zinc and Cd ZIP6 transporter of Arabidopsis halleri have distinct effects on Cd tolerance. *Plant Cell Environ.* **2020**, *43*, 2143–2157. [CrossRef] [PubMed]
14. Zhao, Y.; Zhang, S.; Wen, N.; Zhang, C.; Wang, J.; Liu, Z. Modeling uptake of Cd from solution outside of root to cell wall of shoot in rice seedling. *Plant Growth Regul.* **2017**, *82*, 11–20. [CrossRef]
15. Sharma, A.; Sharma, H.; Upadhyay, S.K. *Cation Transporters in Plants*; Elsevier: Rio de Janeiro, Brazil, 2021.
16. Morel, M.; Crouzet, J.; Gra, V.A.; Au, R.P.; Leon, H.N.; Vavasseur, A.; Rich, A.P. AtHMA3, a P1B-ATPase allowing Cd/Zn/Co/Pb vacuolar storage in Arabidopsis. *Plant Physiol.* **2009**, *149*, 894–904. [CrossRef]
17. Gong, X.; Yin, L.; Chen, J.; Guo, C. Overexpression of the iron transporter NtPIC1 in tobacco mediates tolerance to Cd. *Plant Cell Rep.* **2015**, *34*, 1963–1973. [CrossRef]
18. Chen, H.; Zhang, C.; Guo, H.; Hu, Y.; He, Y.; Jiang, D. Overexpression of a *Miscanthus sacchariflorus* yellow stripe-like transporter MsYSL1 enhances resistance of Arabidopsis to Cd by mediating metal ion reallocation. *Plant Growth Regul.* **2018**, *85*, 101–111. [CrossRef]
19. Zhang, C.; Chen, H.; Guo, H.; Dean, J. MsYSL1 of Miscanthus sacchariflorus enhances resistance of Arabidopsis to cadmium by mediating metal ion reallocation. In Proceedings of the 7th Yangtze River Delta Plant Science Seminar and Youth Academic Report, Shanghai, China, 12 October 2018.
20. Shu, M.A.; Tyagi, S.; Sharma, Y.; Ma, D.H.; Sharma, A.; Pandey, A.; Sin, G.K.; Upadhyay, S.K. Expression of TaNCL2-A ameliorates cadmium toxicity by increasing calcium and enzymatic antioxidants activities in Arabidopsis. *Chemosphere* **2023**, *329*, 138636. [CrossRef]
21. Ban, S.R.; Priya, S.; Dikshit, H.K.; Jacob, S.R.; Rao, M.; Bana, R.S.; Ku, M.J.; Tripathi, K.; Ku, M.A.; Kumar, S.H.M.; et al. Growth and Antioxidant Responses in Iron-Biofortified Lentil under Cd Stress. *Toxics* **2021**, *9*, 182. [CrossRef]
22. Wu, H.; Chen, C.; Du, J.; Liu, H.; Cui, Y.; Zhang, Y.; He, Y.; Wang, Y.; Chu, C.; Feng, Z.; et al. Co-overexpression FIT with AtbHLH38 or AtbHLH39 in Arabidopsis-enhanced Cd tolerance via increased Cd sequestration in roots and improved iron homeostasis of shoots. *Plant Physiol.* **2012**, *158*, 790–800. [CrossRef]
23. Xu, Z.; Liu, X.; He, X.; Xu, L.; Huang, Y.; Shao, H.; Zhang, D.; Tang, B.; Ma, H. The Soybean Basic Helix-Loop-Helix Transcription Factor ORG3-Like Enhances Cd Tolerance via Increased Iron and Reduced Cd Uptake and Transport from Roots to Shoots. *Front. Plant Sci.* **2017**, *8*, 1098–1106. [CrossRef] [PubMed]
24. Siedlecka, A.; Kru, P.Z. Cd/Fe interaction in higher plants-its consequences for the photosynthetic apparatus. *Photosynthetica* **1999**, *36*, 321–331. [CrossRef]
25. Gao, C.; Wang, Y.; Xiao, D.S.; Qiu, C.P.; Han, D.G.; Zhang, X.Z.; Wu, T.; Han, Z.H. Comparison of Cd-induced iron-deficiency responses and genuine iron-deficiency responses in Malus xiaojinensis. *Plant Sci.* **2011**, *181*, 269–274. [CrossRef] [PubMed]
26. Mu, N.S.; Hakeem, K.R.; Moha, M.R.; Lee, J.H. Cd toxicity induced alterations in the root proteome of green gram in contrasting response towards iron supplement. *Int. J. Mol. Sci.* **2014**, *15*, 6343–6355. [CrossRef]
27. Mazumder, M.K.; Moul, I.D.; Choudhury, S. Iron (Fe^{3+})-mediated redox responses and amelioration of oxidative stress in Cd (Cd^{2+}) stressed mung bean seedlings: A biochemical and computational analysis. *J. Plant Biochem. Biotechnol.* **2022**, *31*, 49–60. [CrossRef]
28. Mu, N.S.; Kim, T.H.; Qureshi, M.I. Fe modulates Cd-induced oxidative stress and the expression of stress responsive proteins in the nodules of Vigna radiata. *Plant Growth Regul.* **2012**, *68*, 421–433. [CrossRef]
29. Elazab, D.S.; Abdel, W.D.A.; El, M.M.T. Iron and zinc supplies mitigate Cd toxicity in micropropagated banana (*Musa* spp.). *Plant Cell Tissue Organ. Cult.* **2021**, *145*, 367–377. [CrossRef]
30. Kabir, A.H.; Hossain, M.M.; Khatun, M.A.; Man, D.A.; Haider, S.A. Role of Silicon Counteracting Cd Toxicity in Alfalfa (*Medicago sativa* L.). *Front. Plant Sci.* **2016**, *7*, 11–17. [CrossRef]
31. Yao, X.; Cai, Y.; Yu, D.; Liang, G. bHLH104 confers tolerance to Cd stress in Arabidopsis thaliana. *J. Integr. Plant Biol.* **2018**, *60*, 691–702. [CrossRef]
32. Zhu, Y.X.; Du, W.X.; Fang, X.Z.; Zhang, L.L.; Jin, C.W. Knockdown of BTS may provide a new strategy to improve Cd-phytoremediation efficiency by improving iron status in plants. *J. Hazard. Mater.* **2020**, *384*, 121–163. [CrossRef]
33. Inoue, H.; Kob, A.T.; No, Z.T.; Taka, H.M.; Ka, K.Y.; Suzu, K.K.; Naka, Z.M.; Nakani, S.H.; Mori, S.; Nishizawa, N.K. Rice OsYSL15 is an iron-regulated iron(III)-deoxymugineic acid transporter expressed in the roots and is essential for iron uptake in early growth of the seedlings. *J. Biol. Chem.* **2009**, *284*, 3470–3479. [CrossRef]

34. Zheng, L.; Fu, J.M.; Ya, M.N.; Sa, S.A.; Ya, M.M.; Saku, R.I.; Sato, K.; Ma, J.F. Isolation and characterization of a barley yellow stripe-like gene, HvYSL5. *Plant Cell Physiol.* **2011**, *52*, 765–774. [CrossRef] [PubMed]
35. Ka, K.Y.; Ishi, M.Y.; Koba, Y.T.; Yama, K.T.; Naka, N.H.; Nishizawa, N.K. OsYSL16 plays a role in the allocation of iron. *Plant Mol. Biol.* **2012**, *79*, 583–594. [CrossRef]
36. Ao, Y.T.; Koba, Y.T.; Taka, H.M.; Naga, S.S.; Usuda, K.; Ka, K.Y.; Ishi, M.Y.; Naka, N.H.; Mori, S.; Nishizawa, N.K. OsYSL18 is a rice iron (III)-deoxymugineic acid transporter specifically expressed in reproductive organs and phloem of lamina joints. *Plant Mol. Biol.* **2009**, *70*, 681–692. [CrossRef]
37. Zang, J.; Huo, Y.; Liu, J.; Zhang, H.; Liu, J.; Chen, H. Maize YSL2 is required for iron distribution and development in kernels. *J. Exp. Bot.* **2020**, *20*, 31–52. [CrossRef] [PubMed]
38. Di, V.F.; Couch, D.; Con, E.G.; Rosch, Z.H.; Ma, R.S.; Cu, R.C. The Arabidopsis YELLOW STRIPE LIKE4 and 6 transporters control iron release from the chloroplast. *Plant Cell* **2013**, *25*, 1040–1055. [CrossRef]
39. DiDonato, R.J.J.r.; Roberts, L.A.; Sander, S.T.; Eisley, R.B.; Walker, E.L. Arabidopsis Yellow Stripe-Like2 (YSL2): A metal-regulated gene encoding a plasma membrane transporter of nicotianamine-metal complexes. *Plant J.* **2004**, *39*, 403–414. [CrossRef]
40. Waters, B.M.; Chu, H.H.; Didona, T.R.J.; Roberts, L.A.; Eisley, R.B.; Lahner, B.; Salt, D.E.; Walker, E.L. Mutations in Arabidopsis yellow stripe-like1 and yellow stripe-like3 reveal their roles in metal ion homeostasis and loading of metal ions in seeds. *Plant Physiol.* **2006**, *141*, 1446–1458. [CrossRef]
41. Gendre, D.; Czernic, P.; Con, E.G.; Pianelli, K.; Briat, J.F.; Lebrun, M.; Mari, S. TcYSL3, a member of the YSL gene family from the hyper-accumulator *Thlaspi caerulescens*, encodes a nicotianamine-Ni/Fe transporter. *Plant J.* **2007**, *49*, 1–15. [CrossRef]
42. Sasa, K.A.; Ya, M.N.; Xia, J.; Ma, J.F. OsYSL6 is involved in the detoxification of excess manganese in rice. *Plant Physiol.* **2011**, *157*, 1832–1840. [CrossRef]
43. Sheng, H.; Jiang, Y.; Rahmati, M.; Chia, J.C.; Doku, C.T.; Ka, V.Y.; Za, V.T.O.; Mendoza, P.N.; Huang, R.; Smieshka, L.M.; et al. YSL3-mediated copper distribution is required for fertility, seed size and protein accumulation in *Brachypodium*. *Plant Physiol.* **2021**, *186*, 655–676. [CrossRef]
44. Feng, S.; Tan, J.; Zhang, Y.; Liang, S.; Xiang, S.; Wang, H.; Chai, T. Isolation and characterization of a novel Cd-regulated Yellow Stripe-Like transporter (SnYSL3) in *Solanum nigrum*. *Plant Cell Rep.* **2017**, *36*, 281–296. [CrossRef] [PubMed]
45. Wang, J.W.; Li, Y.; Zhang, Y.X.; Chai, T.Y. Molecular cloning and characterization of a *Brassica juncea* yellow stripe-like gene, BjYSL7, whose overexpression increases heavy metal to OPT3lerance of tobacco. *Plant Cell Rep.* **2013**, *32*, 651–662. [CrossRef] [PubMed]
46. Chen, C.; Cao, Q.; Jiang, Q.; Li, J.; Yu, R.; Shi, G. Comparative transcriptome analysis reveals gene network regulating Cd uptake and translocation in peanut roots under iron deficiency. *BMC Plant Biol.* **2019**, *19*, 35. [CrossRef]
47. Meda, A.R.; Scheuermann, E.B.; Prechsl, U.E.; Ere, N.B.; Schaaf, G.; Ha, Y.H.; Weber, G.; von Wi, R.N. Iron acquisition by phytosiderophores contributes to cadmium tolerance. *Plant Physiol.* **2007**, *143*, 1761–1773. [CrossRef]
48. Fang, L.; Ju, W.; Yang, C.; Duan, C.; Cui, Y.; Han, F.; Shen, G.; Zhang, C. Application of signaling molecules in reducing metal accumulation in alfalfa and alleviating metal-induced phytotoxicity in Pb/Cd-contaminated soil. *Ecotoxicol. Environ. Saf.* **2019**, *182*, 109–129. [CrossRef]
49. Gut, S.A.; Hend, R.S.; Guer, R.G.; Ren, A.J.; Lutts, S.; Alse, E.S.; Fernie, A.R.; Hausman, J.F.; Vang, R.J.; Cuy, P.A.; et al. Long-Term Cd Exposure Alters the Metabolite Profile in Stem Tissue of *Medicago sativa*. *Cells* **2020**, *9*, 2707. [CrossRef]
50. Gut, S.A.; Ser, G.K.; Keu, N.E.; Prin, S.E.; Guer, R.G.; Rena, U.J.; Hausman, J.F.; Cuy, P.A. Does long-term cadmium exposure influence the composition of pectic polysaccharides in the cell wall of *Medicago sativa* stems? *BMC Plant Biol.* **2019**, *19*, 271. [CrossRef]
51. Chang, J.D.; Huang, S.; Koni, S.N.; Wang, P.; Chen, J.; Huang, X.Y.; Ma, J.F.; Zhao, F.J. Overexpression of the manganese/Cd transporter OsNRAMP5 reduces Cd accumulation in rice grain. *J. Exp. Bot.* **2020**, *71*, 5705–5715. [CrossRef]
52. Tao, J.; Lu, L. Advances in Genes-Encoding Transporters for Cd Uptake, Translocation, and Accumulation in Plants. *Toxics* **2022**, *10*, 411. [CrossRef] [PubMed]
53. Wani, S.H.; Gai, K.K.; Raz, Z.A.; Saman, T.K.; Ku, M.M.; Govin, D.V. Improving Zinc and Iron Biofortification in Wheat through Genomics Approaches. *Mol. Biol. Rep.* **2022**, *49*, 8007–8023. [CrossRef]
54. Chowdhury, R.; Nallusamy, S.; Shanmugam, V.; Loganathan, A.; Muthurajan, R.; Sivathapandian, S.K.; Paramasivam, J.; Duraialagaraja, S. Genome-wide understanding of evolutionary and functional relationships of rice Yellow Stripe-Like (YSL) transporter family in comparison with other plant species. *Biologia* **2022**, *77*, 39–53. [CrossRef]
55. Lee, G.; Ah, M.H.; Quinta, N.J.; Syll, W.L.; Jani, N.N.; Pre, I.V.; Anderson, J.E.; Piet, Z.B.; Krämer, U. Constitutively enhanced genome integrity maintenance and direct stress mitigation characterize transcriptome of extreme stress-adapted *Arabidopsis halleri*. *Plant J.* **2021**, *108*, 896–911. [CrossRef]
56. Fu, Y.; Zha, T.H.; Li, Y.; Liu, Q.; Trot, S.V.; Li, C. Physiological and Transcriptomic Comparison of Two Sunflower (*Helianthus annuus* L.) Cultivars With High/Low Cadmium Accumulation. *Front. Plant Sci.* **2022**, *13*, 854386. [CrossRef] [PubMed]
57. Curie, C.; Cassin, G.; Couch, D.; Di, V.F.; Higuchi, K.; Le Jean, M.; Misson, J.; Schikora, A.; Czernic, P.; Mari, S. Metal movement within the plant: Contribution of nicotianamine and yellow stripe 1-like transporters. *Ann. Bot.* **2009**, *103*, 1–11. [CrossRef]
58. Koi, K.S.; Inoue, H.; Mizuno, D.; Takahashi, M.; Nakanishi, H.; Mori, S.; Nishizawa, N.K. OsYSL2 is a rice metal-nicotianamine transporter that is regulated by iron and expressed in the phloem. *Plant J.* **2004**, *39*, 415–424. [CrossRef]

59. Liu, Y.; Ji, X.; Nie, X.; Qu, M.; Zheng, L.; Tan, Z.; Zhao, H.; Huo, L.; Liu, S.; Zhang, B.; et al. Arabidopsis AtbHLH112 regulates the expression of genes involved in abiotic stress tolerance by binding to their E-box and GCG-box motifs. *New Phytol.* **2015**, *207*, 692–709. [CrossRef]
60. Balakhnina, T.I.; Kosobryukhov, A.A.; Ivanov, A.A.; Kreslavskii, V.D. The effect of Cd on CO_2 exchange, variable fluorescence of chlorophyll, and the level of antioxidant enzymes in pea leaves. *Russ. J. Plant Physiol.* **2005**, *52*, 15–20. [CrossRef]
61. Zhang, H.; Xu, Z.; Huo, Y.; Guo, K.; Wang, Y.; He, G.; Sun, H.; Li, M.; Li, X.; Xu, N.; et al. Overexpression of Trx CDSP32 gene promotes chlorophyll synthesis and photosynthetic electron transfer and alleviates Cd-induced photoinhibition of PSII and PSI in tobacco leaves. *J. Hazard. Mater.* **2020**, *398*, 122–139. [CrossRef] [PubMed]
62. Han, Y.; Ling, Q.; Dong, F.; de Dios, V.R.; Zhan, X. Iron and copper mi-cronutrients influences cadmium accumulation in rice grains by altering its transport and allocation. *Sci. Total Environ.* **2021**, *777*, 146118. [CrossRef]
63. Zhang, Q.; Huang, D.; Xu, C.; Zhu, H.; Feng, R.W.; Zhu, Q. Fe fortification limits rice Cd accumulation by promoting root cell wall chelation and reducing the mobility of Cd in xylem. *Ecotoxicol. Environ. Saf.* **2022**, *240*, 113700. [CrossRef]
64. Su, Y.; Liu, J.; Lu, Z.; Wang, X.; Shi, G. Effects of iron deficiency on subcellular dis-tribution and chemical forms of cadmium in peanut roots in relation to its translocation. *Environ. Exp. Bot.* **2014**, *97*, 40–48. [CrossRef]
65. Sie, D.A. Interaction between Cd and iron and its effects on photosynthetic capacity of primary leaves of *Phaseolus vulgaris*. *Plant Physiol. Biochem.* **1996**, *34*, 833–841. [CrossRef]
66. Mihucz, V.G.; Csog, Á.; Fodor, F.; Tatár, E.; Szoboszlai, N.; Silaghi-Dumitrescu, L.; Záray, G. Impact of two iron (III) chelators on the iron, Cd, lead and nickel accumulation in poplar grown under heavy metal stress in hydroponics. *J. Plant Physiol.* **2012**, *169*, 561–566. [CrossRef]
67. Wang, C.; Wang, X.; Li, J.; Guan, J.; Tan, Z.; Zhang, Z.; Shi, G. Genome-Wide Identification and Transcript Analysis Reveal Potential Roles of Oligopeptide Transporter Genes in Iron Deficiency Induced Cadmium Accumulation in Peanut. *Front. Plant Sci.* **2022**, *13*, 894848. [CrossRef] [PubMed]
68. Sol, T.A.; Gáspár, L.; Mészáros, I.; Szigeti, Z.; Lévai, L.; Sárvári, E. Impact of iron supply on the kinetics of recovery of photosynthesis in Cd-stressed poplar (*Populus glauca*). *Ann. Bot.* **2008**, *102*, 771–782. [CrossRef]
69. Mu, N.S.; Jeong, B.R.; Kim, T.H.; Lee, J.H.; Soun, D.P. Transcriptional and physiological changes in relation to Fe uptake under conditions of Fe-deficiency and Cd-toxicity in roots of *Vigna radiata* L. *J. Plant Res.* **2014**, *127*, 731–742. [CrossRef]
70. Suzu, K.M.; Ura, B.A.; Sasa, K.S.; Tsugawa, R.; Nishio, S.; Mukai, Y.H.; Murata, Y.; Masuda, H.; Aung, M.S.; Mera, A.; et al. Development of a mugineic acid family phytosiderophore analog as an iron fertilizer. *Nat. Commun.* **2021**, *12*, 1558. [CrossRef]
71. Conte, S.S.; Chu, H.H.; Rodriguez, D.C.; Punshon, T.; Vasques, K.A.; Salt, D.E.; Walker, E.L. Arabidopsis thaliana Yellow Stripe1-Like4 and Yellow Stripe1-Like6 localize to internal cellular membranes and are involved in metal ion homeostasis. *Front. Plant Sci.* **2013**, *4*, 283. [CrossRef]
72. Lichtenthaler, H.K. Chlorophyll fluorescence signatures of leaves during the autumnal chlorophyll breakdown. *J. Plant Physiol.* **1987**, *131*, 101–110. [CrossRef]
73. Reshmi, G.R.; Rajala, K.R. Drought and UV stress response in *Spilanthes acmella* Murr., (tooth-ache plant). *J. Physiol. Biochem.* **2012**, *8*, 110–129.
74. Zhang, X.; Wang, L.; Meng, H.; Wen, H.; Fan, Y.; Zhao, J. Maize ABP9 enhances tolerance to multiple stresses in transgenic Arabidopsis by modulating ABA signaling and cellular levels of reactive oxygen species. *Plant Mol. Biol.* **2011**, *75*, 365–378. [CrossRef] [PubMed]

plants

Article

Amino Acid Residues of the Metal Transporter OsNRAMP5 Responsible for Cadmium Absorption in Rice

Zhengtong Qu and Hiromi Nakanishi *

Department of Global Agricultural Sciences, Graduate School of Agricultural and Life Sciences, The University of Tokyo, Tokyo 113-8657, Japan; kyoku@g.ecc.u-tokyo.ac.jp
* Correspondence: hnak@g.ecc.u-tokyo.ac.jp

Abstract: The transport of metals such as iron (Fe), manganese (Mn), and cadmium (Cd) in rice is highly related. Although Fe and Mn are essential elements for plant growth, Cd is a toxic element for both plants and humans. OsNRAMP5—a member of the same family as the Fe, Mn, and Cd transporter OsNRAMP1—is responsible for the transport of Mn and Cd from soil in rice. Knockout of *OsNRAMP5* markedly reduces both Cd and Mn absorption, and this *OsNRAMP5* knockout is indispensable for the development of low-Cd rice. However, in low-Mn environments, such plants would exhibit Mn deficiency and suppressed growth. We generated random mutations in OsNRAMP5 via error-prone PCR, and used yeast to screen for the retention of Mn absorption and the inhibition of Cd absorption. The results showed that alanine 512th is the most important amino acid residue for Cd absorption and that its substitution resulted in the absorption of Mn but not Cd.

Keywords: OsNRAMP5; cadmium; manganese; rice; transporter; random mutation

Citation: Qu, Z.; Nakanishi, H. Amino Acid Residues of the Metal Transporter OsNRAMP5 Responsible for Cadmium Absorption in Rice. *Plants* **2023**, *12*, 4182. https://doi.org/10.3390/plants12244182

Academic Editor: Ferenc Fodor

Received: 30 October 2023
Revised: 2 December 2023
Accepted: 13 December 2023
Published: 16 December 2023

1. Introduction

The global population surpassed 8 billion in 2023 and is increasing; it is projected to exceed 11.2 billion in 2100 [1]. Worldwide, 750 million people suffer from hunger and undernourishment, a number projected to exceed 840 million in 2030 and 2 billion in 2050 [2]. The total global cultivable area has decreased since 1961 as a result of urbanization [3]. The improvements in crop varieties and techniques resulting from the 'Green Revolution' have increased yields per unit area; however, further increasing yields is problematic. To provide sufficient food for the increasing global population, there is a need to develop plants that are tolerant of poor environments.

Soil contamination by toxic heavy metals precludes its use for agricultural purposes. When crops absorb nutrients such as trace elements from soil, they also take up harmful heavy metals. Among these harmful heavy metals, cadmium (Cd) is an atypical transition heavy metal readily absorbed in conjunction with other minerals required for plant growth (e.g., iron (Fe), zinc (Zn), and manganese (Mn)) [4]. It has a long biological half-life; high mobility, solubility, fluidity, and bioaccumulation; and long-lasting toxicity, irrespective of concentration [5]. Cd is not essential for plant growth or the biological functions of humans and animals. In plants, excess Cd causes growth disorders. Cd contamination is a severe and ubiquitous environmental problem, and Cd enters food chains by being absorbed by plants and then subsequently accumulating in animals and humans. Lifelong intake of Cd, which has a biological half-life of around 10 years, can damage the lungs, kidneys, bones, and reproductive system. In Japan, Itai-Itai disease was first reported in the 1910s, and Cd-exposed miners in Europe suffered lung damage in the 1930s; in both cases, the damage was induced by chronic Cd intoxication [6].

Cd is produced by natural activities (volcanic activity, weathering, and erosion), anthropological activities (smoking, smelting, and fossil fuel combustion), and remobilization of historical sources, including watercourse contamination. Those industrial activities, including mining and smelting, could influence paddy fields to a large extent [7]. As a result, dietary intake accounts for approximately 90% of all Cd intake in the nonsmoking population; other sources include drinking water and exposure to inexpensive jewelry, toys, and plastics [8]. According to the national food survey and estimation of total diet, the Cd intake worldwide is within the range from 0.1 to 0.51 μg/kg of body weight per day, but in comparing the intake of different countries, Asian nations, such as China (0.21–0.51 μg/kg) and Japan (0.31–0.36 μg/kg), showed a higher level of intake than those of the United States (0.13–0.15 μg/kg) and European nations (0.16 μg/kg in Finland, 0.18 μg/kg in Germany, etc.) [9], which could be attributed to the larger consumption of rice in Asian nations [10]. Specifically in China, which is the largest rice producer [11], although the National Standard of the People's Republic of China limits the Cd content in rice to 0.2 mg/kg [12], 10.3% of rice on the Chinese market exceeds this limit [13]. The independent market surveys carried out in six administrative regions in those three major cropping regions showed tested samples from all administrative regions are Cd-contaminated to different extents: the average Cd content ranged from 0.12 to 0.46 mg/kg and 14–100% exceeded the standard limit [14–19]

In rice, Cd is transported within the plant via the apoplastic and symplastic pathways, and both pathways involve transporters of other metallic elements essential for plant growth. Because Cd shares similar chemical properties with Fe, they are closely associated in plants [20]. The mechanisms of the uptake and accumulation of Fe and Cd are somewhat common as a result of similar entry routes within rice. During the vegetative stage, Fe and Cd are absorbed by specific root transporters and transported to the aerial parts via the xylem-to-phloem transfer system, and at grain-filling, grain Fe and Cd are both derived from the phloem [21]. With the presence of Cd, Fe deficiency symptoms could be induced because Cd inhibits not only the absorption of Fe [22], but the transportation of Fe from root to shoot [23]. On the other hand, the addition of Fe could also reduce Cd content in rice [20] and enhance rice growth and yield [24], which suggests that Cd translocation into rice might occur via Fe metabolic pathways [25]. The interaction between Mn and Cd has also been identified, because the accumulation of Cd is reduced in both roots and shoots in the Mn sufficiency environment compared with the Mn deficiency environment [26]. Fe and Mn alleviated Cd toxicity by preventing Cd from being absorbed by forming an Fe plaque on the surface of rice roots [27]. Meanwhile, Fe and Mn could also protect plants from damage induced by Cd on root growth and photosynthesis [28].

Several genes in rice have been reported to take part in xylem loading and phloem redistribution of Fe, Mn, and Cd at different locations in the plants [29]. For example, members of the heavy metal-associated protein (HMA) metal-transporter family transport Cd to the root vascular bundle. Similar to AtHMA4 and AhHMA4, OsHMA2 has also been identified as a transporter of both Zn and Cd, and in OsHMA2-suppressed rice, the concentrations of both Cd and Zn decreased in the leaves and seeds, which suggests that OsHMA2 plays a role in Cd loading to the xylem and participates in root-to-shoot translocation of Cd apart from Zn [7]. Different from OsHMA2, OsHMA3 reportedly does not transport other metals such as Zn [30]. To be specific, OsHMA3, a regulator for Cd transport in the xylem in rice, has the function of mediating vacuolar sequestration of Cd in root cells [31]. The expression of *OsHMA3* was directly proportional to Cd concentration in the environment [32], but with excessive Fe treatment, the expression of *OsHMA3* significantly increased [33]. RNAi-mediated knockdown of *OsHMA3* increased root-to-shoot Cd translocation, and on the other hand, the overexpression of *OsHMA3* reduced shoot Cd accumulation, which indicates that OsHMA3 has the function in vacuolar compartmentation of Cd in roots, which decreases the xylem loading of Cd and subsequent shoot Cd accumulation [34]. Cd is also transported to seeds via the phloem in a manner involving the product of *OsLCT*; the phloem and seeds of *OsLCT1* mutants generated through RNA interference had low levels of Cd [35]. Because Cd is toxic, it is detoxified by inclusion in complexes with thiol compounds such as phytochelatin (PC) and glutathione (GS, a synthetic substrate for PC). In rice, such thiol compounds are synthesized by OsGS and OsPCS, resulting in the extracellular transport of some Cd [36]. Therefore, it is necessary to modify steps in the plant Cd transport pathway—for instance, Cd absorption from soil, transportation from root to leaf, and sequestration into the vacuole—to enhance its detoxification. Doing so would enable the development of low-Cd foods in which Cd is not stored in seeds.

The natural resistance-associated macrophage protein (NRAMP) family is involved in the absorption of metal elements in diverse taxa. NRAMP1 transports divalent metals (e.g., Mn, Fe, and cobalt) across the phagosomal membranes of macrophages, as does divalent metal transporter 1 (DMT1; alternatively, NRAMP2, DCT), which is a transporter of Cd and Fe [37]. The NRAMP family serves as the secondary active transporters with the general features of proton transportation and proton-metal coupling, and the alternating access in the NRAMP family depends largely on the motion and the structure of transmembrane proteins [38]. Rice has seven NRAMP transporters, among which OsNRAMP1 is responsible for the uptake and transport of Cd in plants [39]. Transformation with OsNRAMP1 reduced the Cd tolerance of yeast [40]. However, OsNARMP1 also transports Mn and Fe. Similar interactions between Cd and Fe were also found in both the ferrous Fe transporter iron-regulated transporter 1 (IRT1) and IRT2 in rice. Both OsIRT1 and OsIRT2 are related to Fe uptake in roots and also showed an influx activity of Cd as well as Fe in yeast, showing that OsIRT1 and OsIRT2 are important transporters in roots with the function of the uptake of Cd [41,42]. OsIRTs may contribute to the uptake of Cd in aerobic conditions when water is released. Meanwhile, Cd is absorbed in roots through the OsNRAMP5 transporter, and OsNRAMP5, which has been identified as a transporter of Mn and Cd, is responsible for the absorption of Mn and Cd from soil [43]. The reason that rice accumulates more Cd than other cereal crops may also be related to the *OsNRAMP5* gene having a higher expression in rice [29]. Interestingly, Fe absorption by OsNRAMP5 in root and shoot tissues did not differ significantly between the wild type and an *OsNRAMP5* mutant [44]. Furthermore, knockout of *OsNRAMP5* markedly reduced the amount of Cd in rice by abolishing its uptake from the soil. Therefore, knockout of OsNRAMP5 is a promising trait for producing low-Cd rice. Because OsNRAMP5 transports both Mn and Cd, *OsNRAMP5* knockout also reduced Mn absorption by about 90% [45]; therefore, in low-Mn environments, such plants would exhibit Mn deficiency and suppressed growth.

Mutations in OsIRT1 and AtNRAMP4 alter their metal selectivity [46,47]. Furthermore, the changes in structure of ScaNRAMP also vary the metal transportation [48], which might result from a single amino acid substitution together with the protein stability [49]. Similarly, among the 538 amino acid residues comprising OsNRAMP5, 1 or more may mediate its transport of Mn, Cd, or both. Therefore, substitution of a specific amino acid residue may affect Mn and/or Cd transport in a manner that does not alter the Mn uptake while suppressing Cd uptake by changing the amino acid, the protein structure, or both. Rice with such a mutation could maintain Mn uptake while avoiding Cd accumulation when grown in Cd-contaminated soil with low Mn, with no negative influence on the growth. The development of rice varieties that can absorb Mn but not Cd would enable the cultivation of soil with a greater range of Cd contamination levels than would rice varieties with *OsNRAMP5* knockout, and enlarge the production of sufficient crops with low Cd concentration. To this end, in the present study, we introduced mutations into OsNRAMP5 and evaluated their effects on Mn and Cd transport.

2. Results

2.1. Optimization of Mn and Cd Concentration for Mutant Screening in Yeast

The appropriate conditions of screening were determined by analyzing the transportation of Cd and Mn by OsNRAMP5 in yeast because although when expressed in yeast, OsNRAMP5 functions as a transporter of both Mn and Cd, the growth of the yeast might also depend on the environmental concentrations of the metals [42,44]. In the absence of Mn with different concentrations of EGTA, the growth of the negative control (VC) was inhibited from 10 mM EGTA, and the effect was the most at 20 mM EGTA (Figure 1b,c), but no significant difference was found with 2 mM EGTA (Figure 1a); however, OsNRAMP5-

expressing N5 showed good growth even with 20 mM EGTA as a result of transport of Mn by the OsNRAMP5 (Figure 1a–c). In the presence of different concentrations of Cd, the growth of N5 was impaired from 50 μM Cd, and the most significant difference was found at 100 μM Cd (Figure 1e,f), but the Cd concentration of 10 μM did not influence the growth of the yeast significantly (Figure 1d), and that of VC was unaffected regardless of Cd concentration (Figure 1d–e). Furthermore, both VC and N5 showed inhibited growth in a −Mn/+Cd environment compared with a +Cd environment and −Mn environment, respectively (Figure 1c,e,f,g), and no significant difference was found between the growth of VC and N5 in a −Mn/+Cd environment (Figure 1g). These results indicate that the difference in growth of VC and N5 could be clearly identified in the environment with 20 mM EGTA chelating Mn and 100 μM $CdCl_2$, which demonstrates that the concentrations of EGTA and Cd were appropriate for the later screening.

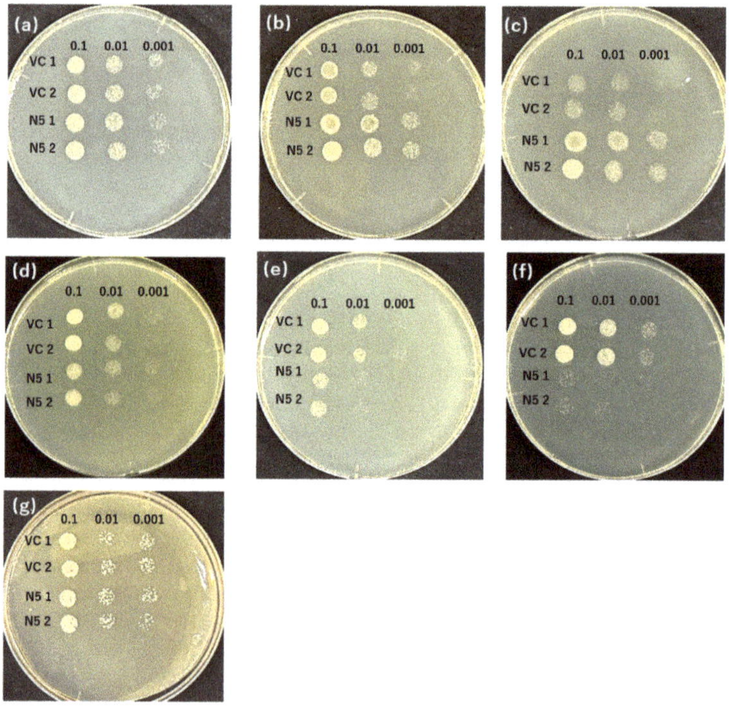

Figure 1. OsNRAMP5 transport activity in yeast. Growth of two individual manganese-absorbing transporter-deficient mutant Δ*smf1*-harboring empty pDR195 vectors (VC 1 and 2) and *OsNRAMP5* (N5 1 and 2) in synthetic defined medium containing (**a**) 2 mM EGTA, (**b**) 10 mM EGTA, (**c**) 20 mM EGTA, (**d**) 10 μM $CdCl_2$, (**e**) 50 μM $CdCl_2$, (**f**) 100 μM $CdCl_2$, (**g**) 20 mM EGTA, and 100 μM $CdCl_2$.

2.2. Patterns of OsNRAMP5 Mutations

We selected 100 colonies from 10 −Mn/+Cd plates in the first screening (200,000 colonies in total) and 20 colonies in the second screening (Table 1). The nucleotide sequences of 20 mutants were classified into 4 patterns: pattern 1, 2 nucleotide mutations corresponding to 1 amino acid mutation (A512T and a silent mutation at the 21st position); pattern 2, 4 nucleotide mutations corresponding to 3 amino acid mutations (S8R, C111Y, A512T, and a silent mutation at the 291st position); pattern 3, 2 nucleotide mutations corresponding to 1 amino acid mutation (A512T and a silent mutation at the 507th position); and pattern 4, 3 nucleotide mutations corresponding to 3 amino acid mutations (S8R, C111Y, and A512T). Only mutants with different amino acids were focused on in this study, so these four patterns were divided into type 1 and type 2 according to their amino acid mutations: patterns 1 and 3 with A512T and patterns 2 and 4 with S8R, C111Y, and A512T. The G to A substitution at position 1534 (alanine to threonine at residue 512) was present in all four patterns.

Table 1. Sequences of mutant *OsNRAMP5* and the locations of the mutations from the start codon of *OsNRAMP5*.

Pattern	Plasmid No.	Mutation Location	Change in Base	Change in a.a.
1	6, 69, 75, 76, 77, 78, 79, 80, 81, 87, 95, 97, 98	#21 #1534	C→T G→A	– A→T
2	9, 73	#22 #291 #332 #1534	A→C T→C G→A G→A	S→R – C→Y A→T
3	35, 36, 70, 82	#507 #1534	T→C G→A	– A→T
4	89	#22 #332 #1534	A→C G→A G→A	S→R C→Y A→T

2.3. *Mutants Absorb Mn but Not Cd*

The absorption of Cd and Mn by the 20 mutants was compared with those of the VC and N5. Under Mn-deficient conditions, all mutants showed growth similar to N5, indicative of similar levels of Mn absorption (Figure 2a). In the presence of Cd, N5 showed little growth, but the mutants—particularly those with pattern 3 (M35, M36, M70, and M82)—showed improved growth compared with the other mutants (Figure 2b). In the presence of Cd but not Mn, the growth of VC and N5 was inhibited by Mn deficiency and Cd toxicity, respectively. However, all mutants showed improved growth, indicating that they could absorb Mn but not Cd in the environment of −Mn/+Cd (Figure 2c).

The patterns with the biggest number of plasmids were selected from each type (pattern 1 from type 1 and pattern 2 from type 2) to test the sensitivity to EGTA and Cd. M6 (mutant from pattern 1) and M9 (mutant from pattern 2) were investigated on media with various concentrations of EGTA and Cd. Apart from the result of VC and N5 similarly to Figure 1, because VC was not sensitive to low concentrations of Mn (2 mM EGTA), the mutants showed only slightly better growth with 10 mM EGTA and a significant difference with 20 mM EGTA (Figure 3a–c). Meanwhile, the growth of the mutant was slightly inhibited with 10 mM of EGTA compared with N5, but when N5 also showed decreased growth with a further increase in EGTA, the mutants had growth similar to it (Figure 3b,c). In the presence of Cd, the mutants were not inhibited in growth as N5 at 10μM Cd (Figure 3d), and even showed slightly better growth at medium concentration (50 μM) compared with VC (Figure 3e). With a high concentration of Cd (100 μM), the growth of the mutants was slightly worse than that of VC (Figure 3f). Meanwhile, the significant difference between the growth of M6 and M9 was not found in either −Mn or +Cd conditions (Figure 3).

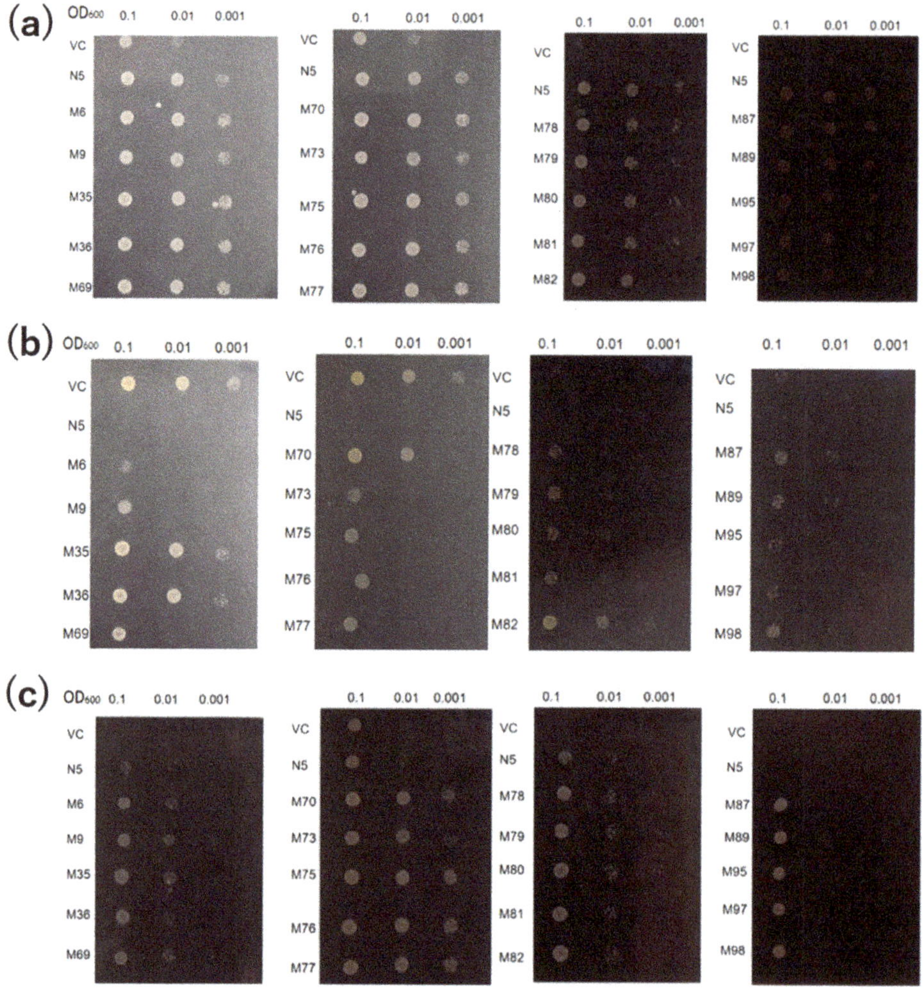

Figure 2. Transport activities of *OsNRAMP5* mutants in Δ*smf1* yeast expressing the *OsNRAMP5* mutants (M), empty pDR195 vector (VC), and wild-type *OsNRAMP5* (N5) in solid synthetic defined medium containing (**a**) 20 mM EGTA, (**b**) 100 μM $CdCl_2$, and (**c**) 20 mM EGTA and 100 μM $CdCl_2$.

In the growth test in liquid medium, the growth rates of M6 and M9 were compared with those of VC and N5. Under Mn-deficient conditions, M6 and M9 grew slightly slower than N5, but much faster than VC (Figure 4a). In the presence of Cd, the growth of N5 was suppressed, and the growth of the two mutants was similar to that of VC (Figure 4b). In the presence of Cd but not Mn, the growth rates of the two mutants were higher than those of VC and N5 (Figure 4c).

The growth rate of VC, N5, M6, and M9 in different concentrations of Cd for 24 h indicated that both mutants were more tolerant in all Cd concentrations compared with the N5 yeast, but similar in growth compared with the VC yeast. Both mutants showed similar growth rates regardless of the concentration of Cd, but N5 showed a decreased growth rate at high Cd concentration (100 μM $CdCl_2$) compared with low Cd concentration (10 μM $CdCl_2$) (Figure 4d).

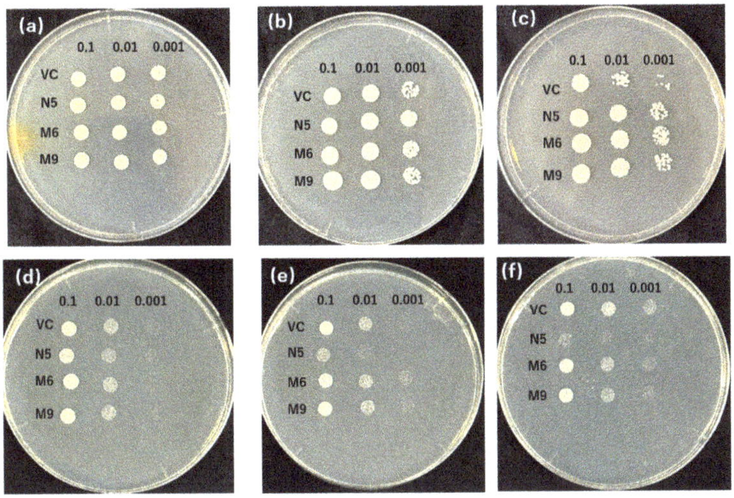

Figure 3. EGTA and Cd sensitivity of *OsNRAMP5* mutants in Δ*smf1* yeast expressing the *OsNRAMP5* mutants (M), empty pDR195 vector (VC), and wild-type *OsNRAMP5* (N5) in solid synthetic defined medium containing (**a**) 2 Mm EGTA, (**b**) 10 Mm EGTA, (**c**) 20 Mm EGTA, (**d**) 10 μM $CdCl_2$, (**e**) 50 μM $CdCl_2$, and (**f**) 100 μM $CdCl_2$.

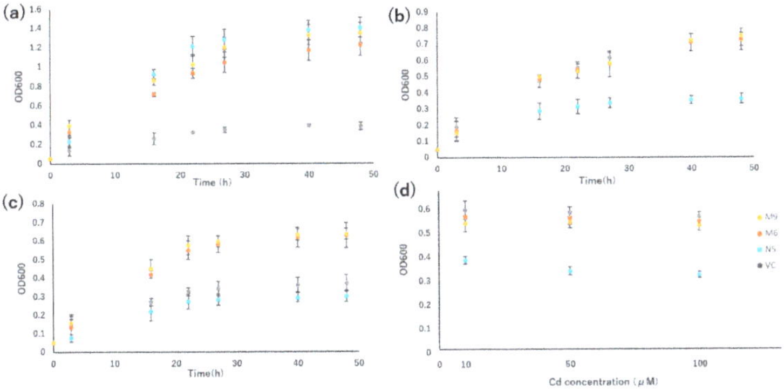

Figure 4. Growth rate of *OsNRAMP5* mutants in Δ*smf1* yeast expressing the *OsNRAMP5* mutants (M6 and M9), empty pDR195 vector (VC), and wild-type *OsNRAMP5* (N5) in liquid synthetic defined medium containing (**a**) 20 mM EGTA, (**b**) 100 μM $CdCl_2$, (**c**) 20 mM EGTA and 100 μM $CdCl_2$, and (**d**) 10 μM, 50 μM, and 100 μM $CdCl_2$.

2.4. Mutants Show Reduced Absorption of Cd but Similar Absorption of Mn Compared with N5

M6 and M9 showed Cd absorption similar to that of VC but significantly different from that of N5 in the presence of 20 μM $CdCl_2$ for 72 h (Figure 5a). In the presence of Cd, M6, M9, VC, and N5 showed similar Mn absorption. In the absence of Cd, M6 showed greater Mn absorption than N5, and M9 showed Mn absorption lower than N5 but similar to VC (Figure 5b).

M6 showed similar Fe absorption to N5 in the absence of Cd, which was significantly higher than in the presence of Cd. However, M9 showed similar absorption of Fe to VC irrespective of the presence of Cd. Meanwhile, the absorption of Fe in both M9 and VC showed lower levels in the absence of Cd, but M6, M9, VC, and N5 showed similar levels of Fe absorption in the presence of Cd (Figure 5c). There was no significant difference in the absorption of Cu compared with M6, M9, VC, and N5, irrespective of the presence of

Cd. Therefore, neither pattern 1 mutants nor the pattern 2 mutants showed an influence in the absorption of Cu. Moreover, M6, M9, VC, and N5 showed similar Zn absorption in the presence of Cd, although in the absence of Cd, only M6 showed elevated Zn absorption (Figure 5e).

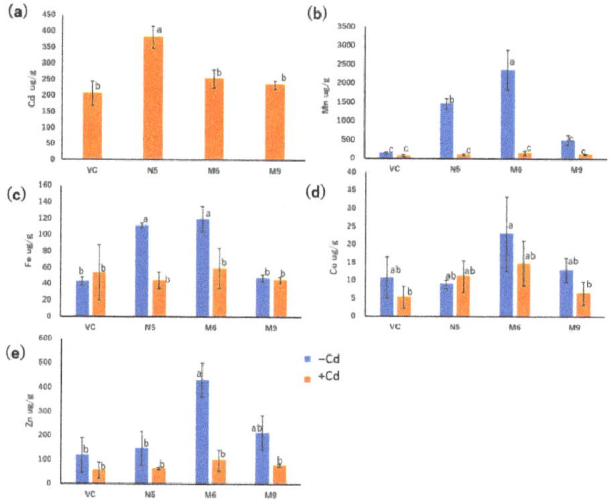

Figure 5. Metal concentrations in empty pDR195 vector (VC), wild-type OsNRAMP5 (N5), and mutants (M6 and M9) in the presence or absence of 20 μM $CdCl_2$ for 72 h followed by drying for 24 h: (**a**) cadmium, (**b**) manganese, (**c**) iron, (**d**) copper, and (**e**) zinc. Data are the means ± standard deviation (n = 3). Different letters indicate significant differences (p < 0.05).

2.5. *Alanine 512 Is Essential for Cd Absorption*

The 20 plasmids harbored a substitution of alanine for threonine at residue 512. To evaluate its importance, we mutated alanine 512 to methionine (A512M), isoleucine (A512I), and aspartic acid (A512D). The metal-transport activities of the mutants were compared with those of VC, N5, and A512T. All mutants had growth rates similar to A512T (Figure 6).

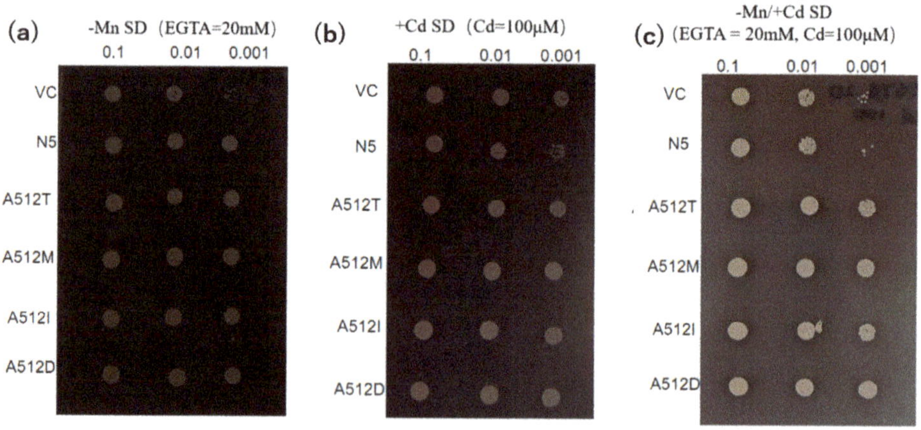

Figure 6. Transport activities of *OsNRAMP5* mutants. Growth of empty pDR195 vector (VC), wild-type *OsNRAMP5* (N5), and amino-acid-512 mutants in synthetic defined medium containing (**a**) 20 mM EGTA, (**b**) 100 μM $CdCl_2$, and (**c**) 20 mM EGTA and 100 μM $CdCl_2$.

3. Discussion

The growth of the mutants in the presence of Cd and the absence of Mn is likely attributable to the mutations (Table 1 and Figure 2a,b, Figures 3 and 4a,b). In the medium containing Cd but not Mn, VC and N5 did not grow, as a result of Mn deficiency and Cd toxicity, respectively (Figures 1 and 2), and in the environment with the presence of Cd and the absence of Mn, the improved growth of N5 compared with that in the environment with the presence of Cd (Figure 1f,g) might be attributed to the chelation by EGTA of Cd, which reduced the free Cd in the medium [50]. However, the obtained *OsNRAMP5* mutants showed good growth (Figure 2). The selected M6 and M9 were investigated and the transport of Mn remained, but they had a low sensitivity to Cd (Figure 3). In liquid medium, the growth rates of the M6 and M9 mutants were similar to that of N5 in the absence of Mn (Figure 4a); both mutants showed similar growth rates under all conditions (Figure 4a–c) and a lower Cd sensitivity in different Cd concentrations compared with N5 (Figure 4d). These findings indicate that M6 and M9 had reduced absorption of Cd in various concentrations of Cd (Figure 2b,c, Figure 3d–f, Figures 4c and 5a) but similar or increased absorption of Mn compared with N5 (Figure 2a,c, Figures 3a–c and 5b), suggesting that the *OsNRAMP5* mutants could mediate the absorption of Mn while suppressing that of Cd. Also, alanine 512, which was common to all mutations, is likely involved in the absorption of Cd (Figure 6). However, even though there was a significant increase in Mn absorption level of M6 and N5 in the absence of Cd, it still showed similar performance with the presence of Cd in M6, M9, VC, and N5 (Figure 5b) due to competitivity of Mn with Cd [51]. Furthermore, because the previous research on OsIRTs indicated that although Cd accumulation in the roots and shoots of *OsIRT1*-overexpression plants was increased under MS medium with excessive Cd, such a phenotype was not shown in the paddy field, which demonstrates that the contribution of the transporters is also affected by the external environmental conditions [52]. The mutations affecting the Mn and Cd transport ability of yeast will be introduced into rice to verify whether rice with mutations can show a higher Mn concentration in the environment with low Mn and lower Cd in the environment with Cd, which is similar to the performance of yeast in this study, in the future.

In the presence of Cd, M6 showed lower Cd absorption, and in the absence of Cd, M6 showed higher Mn (Figure 5a,b), suggesting that the altered Cd and Mn absorption of M6 resulted from the mutation of alanine 512 (Figure 6), and the similar absorption level of Mn in M9 compared to VC might be attributed to the extra mutation of serine 8 and/or cysteine 111 (Table 1 and Figure 5b). Alanine 512 is also important for the absorption of Zn. In the absence of Cd, Zn absorption by N5 was markedly lower than that of M6, and slightly but non-significantly lower than that of M9 (Figure 5e). However, because Cd is more competitive and more easily absorbed than Zn [51], the Zn absorption by M6 and M9 decreased to a level similar to that of VC and N5 in the presence of Cd. Cd is less competitive with Cu [51], which might lead to no significant change in the absorption of Cu with both the presence and absence of Cd in all M6, M9, VC, and N5 (Figure 5d). Moreover, mutations of serine 8, cysteine 111, or both may impede the absorption of Zn and Fe, possibly explaining the similar Zn absorption of N5 and M9 and the lower Fe absorption in M9 than in N5 and M6 in the absence of Cd (Figure 5c,e). For all metals investigated, M9 had similar absorption rates to VC (Figure 7), suggesting that serine 8, cysteine 111, or both are important for metal transport by OsNRAMP5. In M9, the A512T mutation non-significantly enhanced Mn transport compared with VC (Figure 5b). This may explain why the mutations of patterns 2 and 4 were obtained by screening in the absence of Mn.

Particular attention should be paid to the change at nucleotide 507, because yeast with pattern 3 grew better than that with pattern 1 on a Cd-containing medium with the same construction of amino acid (Table 1 and Figure 2b). Meanwhile, the change at nucleotide 21 in M6 might also be important to enhance the absorption of Mn and Zn in the absence of Cd, because M6 showed an increased absorption of Mn and Zn compared with N5, but M9 sharing the mutation at alanine 512 showed a reduced and similar absorption of Mn and Zn, respectively, compared with N5 (Figure 5b,e). The changes from C to T at nucleotide 21 (pattern 1) and T to C at nucleotide 507 (pattern 3) (Table 1) could alter transcriptional efficiency, RNA stability, transfer RNA levels, and protein expression levels even though the amino acid was kept the same [53,54]. It will be very interesting to see if these silent mutations have the same effect in plants as well as in yeast. Whether these are silent mutations that affect absorption could be evaluated by creating a plasmid with the only mutation at nucleotide 21 and a plasmid with the only mutation at nucleotide 507 and evaluating the effect on metal absorption compared with M6, A512T, and yeast with the pattern 3 mutant introduced into both yeasts and plants in a future study. Because all three mutations were not found in other NRAMP proteins, we need further study on whether the influence of these residues is also conserved in other NRAMPs.

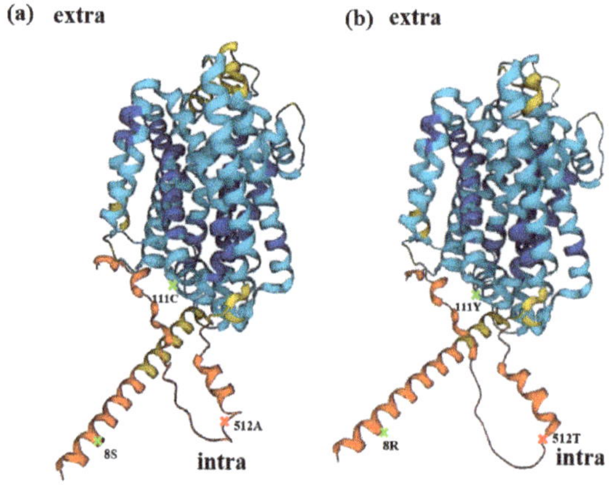

Figure 7. Structural prediction. (**a**) Schematic diagram of wild-type OsNRAMP5 produced with AlphaFold DB version, with alanine 512 (red), which is important for cadmium and manganese transport and affecting the absorption of zinc, and serine 8 and cysteine 111 (green), which influence metal absorption. (**b**) Schematic diagram of OsNRAMP5 with three mutations predicted with AlphaFold DB version, with potentially important locations indicated by the same color as (**a**).

Regarding structural prediction, OsNRAMP5-Q337K, in which a glutamine residue in the eighth transmembrane domain was substituted with a lysine residue, reduced the grain Cd concentration without causing severe Mn deficiency in rice [55]. The three mutations in this study were predicted to be cytoplasmic, and all three mutations slightly changed the structure of the protein (Figure 7). It is necessary to investigate how these residues, which are not extracellular and so cannot interact directly with extracellular metal ions, affect the absorption of metals.

4. Materials and Methods

4.1. Mn and Cd Absorption Assays

The plasmids pDR195 and pDR195 containing *OsNRAMP5* were transformed into the Mn-absorbing transporter-deficient mutant yeast strain Δ*smf1* (MATalpha, his3Δ1; leu2Δ0; meta15Δ0; ura3Δ0; YOL122c:; kanMX4), generating VC and N5, respectively. Metal transport assays were carried out in synthetic defined (SD) medium (2% glucose, 0.5% yeast nitrate base without amino acids, and 2% agar) containing Cd (10 μM, 50 μM, 100 μM $CdCl_2$) but not Mn (2 mM, 10 mM, 20 mM EGTA, pH 5.9) in different concentrations [43,44]. The media were spotted with 8 μL of yeast suspension (OD_{600} = 0.1, 0.01, 0.001), incubated statically at 30 °C for 2 days, and the growth of the yeast strains was monitored.

4.2. Error-Prone PCR

To ligate *OsNRAMP5* into pDR195, *Hind*III and *EcoR*I sites were introduced (Ligation Mighty Mix (TaKaRa)), and the vector was digested with *Bam*HI and *XhoI*I. Mutations were introduced into *OsNRAMP5* via error-prone PCR, which we employed previously [56], in ten separated tubes (using 50 × Titanium Taq DNA Polymerase (TaKaRa)). The PCR conditions were denaturation at 95 °C for 15s, annealing at 55 °C for 15 s, and extension at 68 °C for 2 min for 30 cycles, during which the Mn concentration was changed and random mutations were introduced. OsNRAMP5 has 1614 nucleotides; we used 300 μM Mn to introduce two to five mutations.

4.3. Screening

OsNRAMP5 fragments with random mutations obtained from 10 separated tubes in the error-prone PCR were ligated into pDR195. The vectors with OsNRAMP5 fragments from each error-prone PCR tube were transformed into Δsmf1 and cultured on −Mn/+Cd SD medium (20 mM EGTA, 100 μM $CdCl_2$, pH = 5.9) for 48 h at 30 °C in separated plates for the first screening (10 plates). Next, colonies were diluted and cultured in fresh −Mn/+Cd SD medium for the second screening (100 plates) and the colonies were sequenced.

4.4. Growth Assay

Plasmids harboring *OsNRAMP5* mutants were transformed into Δ*smf1* and cultured on solid −Mn, +Cd, and −Mn/+Cd SD media. The EGTA and Cd sensitivity of the mutants was tested on the solid +Cd media with different conditions (2 mM, 10 mM, 20 mM EGTA and 10 μM, 50 μM, 100 μM $CdCl_2$) using the same method as 4.1. The growth rates of the mutants were analyzed in liquid −Mn, +Cd, and −Mn/+Cd SD media starting with OD_{600} = 0.05 and compared with those of VC and N5 within 48 h. The Cd sensitivity was also analyzed by measuring the growth rate of the mutants in +Cd SD media with different concentrations of Cd and comparing them with those of VC and N5 in 24 h.

4.5. Uptake of Metals

VC, N5, and two *OsNRAMP5* mutant yeast strains (M6 and M9) were cultured in 50 mL of liquid SD medium (20 μM $CdCl_2$, 2.5 μM $MnSO_4$, 0.75 μM $FeCl_3$, 1.5 μM $ZnSO_4$, 0.15 μM $CuSO_4$) for 72 h at 30 °C. The yeasts were centrifuged at 8000× *g* for 5 min, washed with 50 mL of distilled H_2O at 8000× *g* for 5 min 2 times, and washed with 30 mL of distilled H_2O at 8000× *g* for 5 min. The pellets were dried in a warmer at 60 °C for 48 h. The dried pellets were weighed and wet-ashed diluted with 2% HNO_3, and the Fe, Mn, Cu, Zn, and Cd concentrations were analyzed with inductively coupled plasma optical emission spectrometry (ISPS-3500, Seiko Instruments Inc., Chiba, Japan).

4.6. Amino Acid Substitution

The candidate amino acid residue in pDR195 (digested by *Sal*I) was substituted for other amino acid residues to evaluate its function in Mn and Cd transport. The resulting plasmids were transformed into Δ*smf1* and cultured on solid −Mn, +Cd, and −Mn/+Cd SD media.

4.7. Statistical Analysis

The significance of the differences was evaluated using one-way analysis of variance followed by Tukey's multiple comparison test; $p < 0.05$ was considered to indicate significance.

4.8. Protein Structure Analysis

The schematic diagram of the OsNRAMP5 protein structure was produced with AlphaFold DB version.

5. Conclusions

The results in this study indicate that alanine 512 mediates, at least in part, Cd transport by OsNRAMP5; its substitution significantly decreased Cd transport but increased Mn transport. Furthermore, it is also necessary to consider the two amino acid residue changes, S8R and C111Y, other than the 512th alanine for the metal selectivity of the OsNRAMP5 protein. Rice harboring metal-selective OsNRAMP5 could grow in low-Mn soil and/or Cd-contaminated soil, thereby increasing the cultivable land area in the future.

Author Contributions: Conceptualization, H.N. and Z.Q.; methodology, H.N. and Z.Q.; data curation, Z.Q. and H.N.; writing—original draft preparation, Z.Q.; writing—review and editing, H.N.; project administration, H.N; funding acquisition, H.N. and Z.Q. All authors have read and agreed to the published version of the manuscript.

Funding: This work was partially supported by a grant from The Tojuro Iijima Foundation for Food Science and Technology.

Data Availability Statement: All data are contained in the article.

Conflicts of Interest: The authors declare no conflict of interest.

References

1. United Nations Department of Economic and Social Affairs Population Division. *World Population Prospects*; Key Findings & Advance Tables 2015 Revision.; United Nations: New York, NY, USA, 2015.
2. Food and Agriculture Organization of the United Nations. *The State of Food Security and Nutrition in the World;* The State of FOOD Security and Nutrition in the World, Transforming Food Systems for Affordable Healthy Diets; Food and Agriculture Organization of the United Nations: Rome, Italy, 2020.
3. Ritchie, H.; Roser, M. *Land Use*; Our World in Data: Oxford, England, 2013.
4. Haider, F.U.; Cai, L.; Coulter, J.A.; Cheema, S.A.; Wu, J.; Zhang, R.; Ma, W.; Muhammad, F. Cadmium toxicity in plants: Impacts and remediation strategies. *Ecotoxicol. Environ. Saf.* **2021**, *211*, 111887. [CrossRef] [PubMed]
5. Khan, M.A.; Khan, S.; Khan, A.; Alam, M. Soil contamination with cadmium, consequences and remediation using organic amendments. *Sci. Total Environ.* **2017**, *601*, 1591–1605. [CrossRef] [PubMed]
6. Rafati Rahimzadeh, M.; Rafati Rahimzadeh, M.; Kazemi, S.; Moghadamnia, A.A. Cadmium toxicity and treatment: An update. *Casp. J. Intern. Med.* **2017**, *8*, 135–145.
7. Uraguchi, S.; Fujiwara, T. Cadmium transport and tolerance in rice: Perspectives for reducing grain cadmium accumulation. *Rice* **2012**, *5*, 5. [CrossRef] [PubMed]
8. Department of Public Health, Environmental and Social Determinants of Health World Health Organization. *Exposure to Cadmium: A Major Public Health Concern*; WHO: Geneva, Switzerland, 2019.
9. Bellinger, D.; Bolger, M.; Goyer, R.; Barraj, L.; Baines, J. *WHO Food Additives Series: 46, Cadmium*; World Health Organization: Geneva, Switzerland, 2001.
10. China Food and Drug Administration. *National Food Safety Standard Contamination Limit in Food;* China Food and Drug Administration: Beijing, China, 2017; p. 9.
11. European Food Safety Authority. Scientific Opinion of the Panel on Contaminants in the Food Chain. Cadmium in food. *EFSA J.* **2009**, *980*, 1–139.
12. Skorbiansky, S.R.; Childs, N.; Hansen, J. *Rice in Asia's Feed Markets*; RCS-18L-01; USDA-ER: Washington DC, USA, 2018.
13. Liu, B.; Li, T.; Cai, Y. Brief introduction to status quo, hazards and repair methods of cadmium rice. *Mod. Food* **2018**, *21*, 86–89.
14. Li, Q.; Shi, L.; Chen, J.; Tan, H.; Zhou, H.; Yao, L.; Xu, M.; Shang-guan, J.; Lu, K. Investigation on cadmium pollution in foods in country of Jiangxi Province. *Chin. J. Food Hyg.* **2008**, *20*, 330–331.
15. Lei, M.; Zeng, M.; Wang, L.; Williams Paul, N.; Sun, G. Arsenic, lead and cadmium pollution in rice from Hunan markets and contaminated areas and their health risk assessment. *Acta Sci. Circumstantiae* **2010**, *30*, 2314–2320.
16. Wu, D.; Yang, X.; Li, C.; Zhou, C.; Qin, F. Contrentations and health risk assessment of heavy metals in soil and rice and zinc-lead mining area in Guizhou province, China. *J. Agro-Environ. Sci.* **2013**, *32*, 1992–1998.
17. Cai, W.; Su, Z.; Hu, S.; Huang, W.; Xu, X.; Huang, X. Assessment of the content and exposure of lead and cadmium in the major food of Guangdong residents. *Chin. J. Health Lab. Technol.* **2015**, *25*, 2388–2392.
18. Ren, R.; Gong, L.; Wang, S.; Jin, Q. Survey of heavy metal contamination and risk assessment of exposure in Hangzhou indigenous rice. *Chin. J. Health Lab. Technol.* **2020**, *30*, 1516–1520.
19. Jiang, Y.; Ma, N.; Meng, H.; Shi, M.; Zhao, P. Risk assessment on the dietary exposure of cadmium in Guangxi residents. *Chin. J. Food Hyg.* **2021**, *30*, 191–195.
20. Liu, H.; Zhang, C.; Wang, J.; Zhou, C.; Feng, H.; Mahajan, M.D.; Han, X. Influence and interaction of iron and cadmium on photosynthesis and antioxidative enzymes in two rice cultivars. *Chemosphere* **2017**, *171*, 240–247. [CrossRef] [PubMed]
21. Gao, L.; Chang, J.; Chen, R.; Li, H.; Lu, H.; Tao, L.; Xiong, J. Comparison on cellular mechanisms of iron and cadmium accumulation in rice: Prospects for cultivating Fe-rich but Cd-free rice. *Rice* **2016**, *9*, 39. [CrossRef] [PubMed]
22. Sharma, S.S.; Kaul, S.; Metwally, A.; Goyal, K.C.; Finkemeier, I.; Dietz, K.J. Cadmium toxicity to barley (*Hordeum vulgare*) as affected by varying Fe nutritional status. *Plant Sci.* **2004**, *166*, 1287–1295. [CrossRef]
23. Xu, S.S.; Lin, S.; Lai, Z. Cadmium impairs iron homeostasis in *Arabidopsis thaliana* by increasing the polysaccharide contents and the iron-binding capacity of root cell walls. *Plant Soil* **2015**, *392*, 71–85. [CrossRef]

24. Liu, H.J.; Zhang, J.L.; Zhang, F.S. Role of iron plaque in Cd uptake by and translocation within rice (*Oryza sativa* L.) seedlings grown in solution culture. *Environ. Exp. Bot.* **2007**, *59*, 314–320. [CrossRef]
25. Sasaki, A.; Yamaji, N.; Yokosho, K.; Ma, J.F. Nramp5 is a major transporter responsible for manganese and cadmium uptake in rice. *Plant Cell* **2012**, *24*, 2155–2167. [CrossRef]
26. Wang, M.; Ma, W.; Chaney, R.L.; Green, C.E.; Chen, W. Effects of Mn^{2+} on Cd accumulation and ionome in rice and spinach. *J. Environ. Qual.* **2022**, *51*, 890–898. [CrossRef]
27. Liu, W.J.; Zhu, Y.G.; Smith, F.A. Effects of iron and manganese plaques on arsenic uptake by rice seedlings (*Oryza sativa* L.) grown in solution culture supplied with arsenate and arsenite. *Plant Soil* **2005**, *277*, 127–138. [CrossRef]
28. Sebastian, A.; Prasad, M. Iron-and manganese-assisted cadmium tolerance in *Oryza sativa* L.: Lowering of rhizotoxicity next to functional photosynthesis. *Planta* **2015**, *241*, 1519–1528. [CrossRef] [PubMed]
29. Hussain, B.; Li, J.; Ma, Y.; Tahir, N.; Ullah, A. Effects of Fe and Mn cations on Cd uptake by rice plant in hydroponic culture experiment. *PLoS ONE* **2020**, *15*, e0243174. [CrossRef] [PubMed]
30. Takahashi, R.; Ishimaru, Y.; Shimo, H.; Ogo, Y.; Senoura, T.; Nishizawa, N.K.; Nakanishi, H. The OsHMA2 transporter is involved in root-to-shoot translocation of Zn and Cd in rice. *Plant Cell Environ.* **2012**, *35*, 1948–1957. [CrossRef] [PubMed]
31. Miyadate, H.; Adachi, S.; Hiraizumi, A.; Tezuka, K.; Nakazawa, N.; Kawamoto, T.; Katou, K.; Kodama, I.; Sakurai, K.; Takahashi, H.; et al. OsHMA3, a P_{1B}-type of ATPase affects root-to-shoot cadmium translocation in rice by mediating efflux into vacuoles. *New Phytol.* **2011**, *5*, 190–199. [CrossRef] [PubMed]
32. Ueno, D.; Yamaji, N.; Kono, I.; Huang, C.F.; Ando, T.; Yano, M.; Ma, J.F. Gene limiting cadmium accumulation in rice. *Proc. Natl. Acad. Sci. USA* **2010**, *5*, 16500–16505. [CrossRef]
33. Cui, J.; Liu, T.; Li, Y.; Li, F. Selenium reduces cadmium uptake into rice suspension cells by regulating the expression of lignin synthesis and cadmium-related genes. *Sci. Total Environ.* **2018**, *644*, 602–610. [CrossRef]
34. Chen, Z.; Tang, Y.T.; Yao, A.J.; Cao, J.; Wu, Z.H.; Peng, Z.R.; Wang, S.Z.; Xiao, S.; Baker, A.J.M.; Qiu, R.L. Mitigation of Cd accumulation in paddy rice (*Oryza sativa* L.) by Fe fertilization. *Environ. Pollut.* **2017**, *231*, 549–559. [CrossRef]
35. Uraguchi, S.; Kamiya, T.; Sakamoto, T.; Kasai, K.; Sato, Y.; Nagamura, Y.; Yoshida, A.; Kyozuka, J.; Ishikawa, S.; Fujiwara, T. Low-affinity cation transporter (OsLCT1) regulates cadmium transport into rice grains. *Proc. Natl. Acad. Sci. USA* **2011**, *108*, 20959–20964. [CrossRef]
36. Zhong, S.; Li, X.; Li, F.; Huang, Y.; Liu, T.; Yin, H.; Qiao, J.; Chen, G.; Huang, F. Cadmium uptake and transport processes in rice revealed by stable isotope fractionation and Cd-related gene expression. *Sci. Total Environ.* **2021**, *806*, 150633. [CrossRef]
37. Nevo, Y.; Nelson, N. The NRAMP family of metal-ion transporters. *Biochim. Biophys. Acta* **2006**, *1763*, 609–620. [CrossRef]
38. Bozzi, A.T.; Gaudet, R. Molecular Mechanism of Nramp-Family Transition Metal Transport. *J. Mol. Biol.* **2021**, *433*, 166991. [CrossRef] [PubMed]
39. Takahashi, R.; Ishimaru, Y.; Nakanishi, H.; Nishizawa, N.K. Role of the iron transporter OsNRAMP1 in cadmium uptake and accumulation in rice. *Plant Signal. Behav.* **2011**, *6*, 1813–1816. [CrossRef] [PubMed]
40. Takahashi, R.; Ishimaru, Y.; Senoura, T.; Shimo, H.; Ishikawa, S.; Arao, T.; Nakanishi, H.; Nishizawa, N.K. The OsNRAMP1 iron transporter is involved in Cd accumulation in rice. *J. Exp. Bot.* **2011**, *62*, 4843–4850. [CrossRef] [PubMed]
41. Ishimaru, Y.; Suzuki, M.; Tsukamoto, T.; Suzuki, K.; Nakazono, M.; Kobayashi, T.; Wada, Y.; Watanabe, S.; Matsuhashi, S.; Takahashi, M.; et al. Rice plants take up iron as an Fe^{3+}-phytosiderophore and as Fe^{2+}. *Plant J.* **2006**, *5*, 335–346. [CrossRef] [PubMed]
42. Nakanishi, H.; Ogawa, I.; Ishimaru, Y.; Mori, S.; Nishizawa, N.K. Iron deficiency enhances cadmium uptake and translocation mediated by the Fe^{2+} transporters OsIRT1 and OsIRT2 in rice. *Soil Sci. Plant Nutr.* **2006**, *5*, 464–469. [CrossRef]
43. Ishimaru, Y.; Takahashi, R.; Bashir, K.; Shimo, H.; Senoura, T.; Sugimoto, K.; Ono, K.; Yano, M.; Ishikawa, S.; Arao, T.; et al. Characterizing the role of rice NRAMP5 in Manganese, Iron and Cadmium Transport. *Sci. Rep.* **2012**, *2*, 286. [CrossRef]
44. Ishimaru, Y.; Bashir, K.; Nakanishi, H.; Nishizawa, N.K. OsNRAMP5, a major player for constitutive iron and manganese uptake in rice. *Plant Signal. Behav.* **2012**, *7*, 763–766. [CrossRef]
45. Ishikawa, S.; Ishimaru, Y.; Igura, M.; Kuramata, M.; Abe, T.; Senoura, T.; Hase, Y.; Arao, T.; Nishizawa, N.K.; Nakanishi, H. Ion-beam irradiation, gene identification, and marker-assisted breeding in the development of low-cadmium rice. *Proc. Natl. Acad. Sci. USA* **2012**, *109*, 19166–19171. [CrossRef]
46. Rogers, E.E.; Eide, D.J.; Guerinot, M.L. Altered selectivity in an *Arabidopsis* metal transporter. *Proc. Natl. Acad. Sci. USA* **2000**, *97*, 12356–12360. [CrossRef]
47. Pottier, M.; Oomen, R.; Picco, C.; Giraudat, J.; Scholz-Starke, J.; Richaud, P.; Carpaneto, A.; Thomine, S. Identification of mutations allowing Natural Resistance Associated Macrophage Proteins (NRAMP) to discriminate against cadmium. *Plant J. Cell Mol. Biol.* **2015**, *83*, 625–637. [CrossRef]
48. Bozzi, A.T.; Bane, L.B.; Weihofen, W.A.; McCabe, A.L.; Singharoy, A.; Chipot, C.J.; Schulten, K.; Gaudet, R. Conserved methionine dictates substrate preference in *Nramp*-family divalent metal transporters. *Proc. Natl. Acad. Sci. USA* **2016**, *113*, 10310–10315. [CrossRef] [PubMed]
49. Kellogg, E.H.; Leaver-Fay, A.; Baker, D. Role of conformational sampling in computing mutation-induced changes in protein structure and stability. *Proteins* **2011**, *79*, 830–838. [CrossRef] [PubMed]
50. Dai, H.; Wei, S.; Noori, A. The mechanism of chelator improved the tolerance and accumulation of poplar to Cd explored through differential expression protein based on iTRAQ. *J. Hazard. Mater.* **2020**, *393*, 122370. [CrossRef] [PubMed]

51. Jalali, M.; Moradi, F. Competitive sorption of Cd, Cu, Mn, Ni, Pb and Zn in polluted and unpolluted calcareous soils. *Environ. Monit. Assess.* **2013**, *185*, 8831–8846. [CrossRef] [PubMed]
52. Lee, S.; An, G. Over-expression of *OsIRT1* leads to increased iron and zinc accumulations in rice. *Plant Cell Environ.* **2009**, *5*, 408–416. [CrossRef] [PubMed]
53. Chamary, J.V.; Hurst, L.D. The Price of Silent Mutations. *Sci. Am.* **2009**, *300*, 46–53. [CrossRef]
54. Baghban, R.; Farajnia, S.; Rajabibazl, M.; Ghasemi, Y.; Mafi, A.; Hoseinpoor, R.; Rahbarnia, L.; Aria, M. Yeast Expression Systems: Overview and Recent Advances. *Mol. Biotechnol.* **2019**, *61*, 365–384. [CrossRef]
55. Kuramata, M.; Abe, T.; Tanikawa, H.; Sugimoto, K.; Ishikawa, S. A weak allele of *OsNRAMP5* confers moderate cadmium uptake while avoiding manganese deficiency in rice. *J. Exp. Bot.* **2022**, *73*, 6475–6489. [CrossRef]
56. Oki, H.; Yamaguchi, H.; Nakanishi, H.; Mori, S. Introduction of the reconstructed yeast ferric reductase gene, refre1, into tobacco. *Plant Soil* **1999**, *215*, 211–220. [CrossRef]

plants

Article

Evaluation of Siderophores Generated by *Pseudomonas* Bacteria and Their Possible Application as Fe Biofertilizers

José María Lozano-González [1], Silvia Valverde [1,†], Mónica Montoya [2,3], Marta Martín [2], Rafael Rivilla [2], Juan J. Lucena [1] and Sandra López-Rayo [1,*]

1 Department of Agricultural Chemistry and Food Science, Universidad Autónoma de Madrid, Av. Francisco Tomás y Valiente 7, 28049 Madrid, Spain; josem.lozano@uam.es (J.M.L.-G.); silvia.valverde@uva.es (S.V.); juanjose.lucena@uam.es (J.J.L.)
2 Department of Biology, Universidad Autónoma de Madrid, c/Darwin, 2, 28049 Madrid, Spain; monica.montoya@inv.uam.es (M.M.); m.martin@uam.es (M.M.); rafael.rivilla@uam.es (R.R.)
3 Departamento de Química y Tecnología de Alimentos, Escuela Técnica Superior de Ingeniería Agronómica, Alimentaria y de Biosistemas, Universidad Politécnica de Madrid, Ciudad Universitaria, 28040 Madrid, Spain
* Correspondence: sandra.lopez@uam.es
† Current affiliation: I.U. CINQUIMA, Analytical Chemistry Group (TESEA), University of Valladolid, Paseo Belén 7, 47011 Valladolid, Spain.

Abstract: The application of synthetic iron chelates to overcome iron deficiency in crops is leading to a high impact on the environment, making it necessary to find more friendly fertilizers. A promising alternative is the application of biodegradable iron chelates, such as those based on siderophores. In the present work, seven bacterial strains of the genus *Pseudomonas* were selected for their ability to secrete pyoverdine, a siderophore with a high affinity for iron, which could be used as a biofertilizer. The concentration of siderophores secreted by each bacterium expressed as desferrioxamine B equivalents, and the pyoverdine concentration was determined. Their potential as Fe biofertilizers was determined based on their capacity to complex Fe, determining the maximum iron complexation capacity at alkaline pH and selecting the RMC4 strain. The biostimulant capacity of the RMC4 strain was evaluated through the secretion of organic acids such as the hormone Indol-3-acetic acid or glutamic acid, among others, in a kinetic assay. Finally, the genome of RMC4 was determined, and the strain was identified as *Pseudomonas monsensis*. The annotated genome was screened for genes and gene clusters implicated in biofertilization and plant growth promotion. Besides iron mobilization, genes related to phosphorus solubilization, production of phytohormones and biological control, among others, were observed, indicating the suitability of RMC4 as an inoculant. In conclusion, RMC4 and its siderophores are promising sources for Fe biofertilization in agriculture.

Keywords: iron; siderophore; pyoverdine; *Pseudomonas*; biofertilizer

Citation: Lozano-González, J.M.; Valverde, S.; Montoya, M.; Martín, M.; Rivilla, R.; Lucena, J.J.; López-Rayo, S. Evaluation of Siderophores Generated by *Pseudomonas* Bacteria and Their Possible Application as Fe Biofertilizers. *Plants* **2023**, *12*, 4054. https://doi.org/10.3390/plants12234054

Academic Editor: Beatriz Ramos Solano

Received: 30 October 2023
Revised: 24 November 2023
Accepted: 28 November 2023
Published: 2 December 2023

1. Introduction

Iron (Fe) deficiency is a major nutritional disorder in crops causing lower yields and important economic losses [1]. Despite the fact that Fe is the fourth most abundant element on Earth, Fe deficiency has been considered the most common micronutrient deficiency in crops worldwide [1]. This problem is especially relevant in alkaline and calcareous soil conditions characterized by a pH between 7.4–8.5 [2], and a high bicarbonate concentration, buffering the pH and causing the Fe to react with insoluble chemical species, thus limiting its availability for crops [3,4]. To overcome Fe deficiency in crops, the application of synthetic compounds derived from polyaminocarboxylic acids or polyaminophenylcarboxylic acids (commonly known as Fe chelates) such as ethylendiaminetetraacetic acid (EDTA), ethylendiamine-N-N'bis(o-hydroxyphenylacetic) acid (o,o-EDDHA), or N-N'bis(o-hydroxyphenyl)ethylendiamine-N-N'-diacetic acid (HBED) [5] is the most widespread solution [4]. Despite their effectiveness, their high price and environmental impact [6–8] have encouraged the development of new research lines to focus on finding sustainable alternative formulations to these Fe chelates.

One promising line of research in agriculture is the use of siderophore-producing bacteria [9–11]. Under Fe deficiency conditions, these bacteria secrete siderophores. Siderophores are molecules with low molecular weight and high affinity and selectivity for binding and complexing with Fe^{3+} [12]. This high affinity is due to the functional donor groups present in the siderophores (amino, catecholate, hydroxamate, and/or carboxylate), which are able to bind to Fe [13]. Depending on the main functional group present in the siderophore, it can be classified as a catecholate, hydroxamate, carboxylate, or mixed (if they have more than one functional group) type [14]. Currently, many different siderophores of each type are known: catecholate types such as aminochelin, azotochelin, bacillobactin, enterobactin or protochelin; hydroxamate types such as desferrioxamine B (DFOB), putrebactin or vicibactin; carboxylate types such as corynebactin, rhizoferrin or vibrioferrin; and mixed types such as aerobactin, ferribactin, pseudobactin or pyoverdine [9,13,14]. All known siderophores are compiled in a freely usable database [15].

Siderophores exhibit a well-established high-affinity binding with Fe, yet their utilization in biofertilizers remains constrained. Ferreira et al. [16] investigated the efficacy of freeze-dried products derived from siderophore bacterial cultures (*Azotobacter vinelandii* and *Bacillus subtilis*) complexed with Fe in ameliorating Fe deficiency in soybean crops. Results indicated the superior stability of *Azotobacter vinelandii* siderophores in calcareous soils, leading to significant enhancements in dry weight and leaf chlorophyll content. This underscores the potential of *Azotobacter vinelandii* siderophore–Fe complexes as environmentally friendly Fe sources for addressing Fe deficiency in calcareous soils. While the exploration of other bacterial families such as *Bacillus megaterium*, *Pantoea allii*, and *Rhizobium radiobacter* in siderophore production and Fe complexation in calcareous conditions has been studied and dismissed [17], promising attributes have been identified in bacteria from the Pseudomonas family. Therefore, the selection of bacteria with favorable traits for Fe biofertilizers in calcareous environments may hinge on identifying those producing siderophores with strong Fe affinity and high production capacity.

Pseudomonas are Gram-negative bacteria, with well-known plant growth-promoting (PGP) characteristics including the production of siderophores [18], the most significant of which is pyoverdine [19]; production of 1-aminocyclopropane-1-carboxylate deaminase (ACC deaminase) [20]; production of indole acetic acid (IAA) [21]; phosphate solubilization [22]; and nitrogen fixation [7], among others. Furthermore, there are several experiments in the literature testing the effect of the PGP characteristics of *Pseudomonas* on different crops. In the experiment performed by Gusain et al. [23], several bacteria from rainfed agricultural fields of the Garhwal Himalayas were tested for inorganic phosphate solubilization, production of IAA, and production of siderophore. One of the bacteria selected was identified as *Pseudomonas koreensis*, and it promoted plant growth in rice, increasing biomass and phosphorus uptake.

On the other hand, another strategy for the application of these plant growth-promoting bacteria has been found in the literature, which consists of separating the bacterial secretions from the bacteria and isolating the compounds of interest (such as siderophores) for later application. López-Rayo et al. [24] tested the efficacy of ethylenediaminedisuccinic acid ([S,S′]-EDDS) as an Fe fertilizer. [S,S′]-EDDS is a siderophore generated by the actinomycete bacterium *Amycolatpsis japonicum* [25]. López-Rayo et al. [24] observed that the Fe concentration in soybean plants grown in calcareous soil was similar for [S,S′]-EDDS/Fe and EDTA/Fe applications. Nagata et al. [26] applied pyoverdine in Fe deficient tomato plants and observed an improvement in the bioavailability of Fe in tomato plants. Nagata et al. [26] demonstrated the increase in Fe bioavailability in tomato plants due to pyoverdine; however, they used an optimum pH for Fe nutrition (5.75) and a high Fe concentration (100 μM).

This investigation aimed to identify a bacterial strain with a high capacity to produce siderophores from horticultural soils and determine its characterization as an eco-friendly alternative to synthetic ligands for Fe chelation and potential use to alleviate Fe chlorosis in crops. To achieve this objective, wild bacteria of *Pseudomonas* were isolated from soils, and the strain producing a higher concentration of siderophores was selected. The Fe chelating capacity in alkaline soil, the production of organic acids responsible for biostimulant activity in plants, and the plant growth-promoting rhizobacteria (PGPR) characteristics were evaluated in the selected strain.

2. Results

2.1. Bacterial Isolation

Different bacterial strains of the genus *Pseudomonas* were isolated from the rhizosphere of crops. Selected colonies were identified as pseudomonads by their 16S RNA gene sequence. Isolates were also tested for the lack of growth at 37 °C, and siblings were discarded by their BOX pattern. After bacterial isolation, chrome azurol sulphonate (CAS) assay was performed in Petri dishes to determine the amount of siderophore produced by each isolated bacterium. Those strains with more halo formation than the control were selected as possible biofertilizers. The results are shown in Table 1.

Table 1. Measure of halo formation (mm) in the CAS agar assay. Strain F113 was used as a control bacterium.

Strains	Halo Formation (mm)	Strains	Halo Formation (mm)
F113	10		
HFL1	15	RMT4	
HFL3	30	RMT6	13
HFL4	10	RMT7	5
RMC2	10	RMT9	15
RMC4	25	RMT12	20
RMC5	12	RMP5	10
RMC6	7	RMP9	12
RMC8	8	RKP1	7
RMC9	23	RKP2	17
RMT2	10	RKP3	20

Those bacteria that showed more halo formation (mm) than the control bacterium were selected: RMT9 (20 mm), RKP1 (17 mm), RKP2 (20 mm), RKP3 (20 mm), RMC4 (25 mm), RMC9 (23 mm), and HFL3 (30 mm).

2.2. Siderophore and Pyoverdine Production

With the selected bacteria described in 3.1, the CAS liquid assay was performed to quantify siderophore production; the results were expressed as DFOB equivalents (μM) and compared to the strain F113, used as a control. As can be seen in Figure 1, the strain producing the highest concentration of siderophores was RMC4. Also, this strain produced the highest concentration of pyoverdine (Figure 2). Only this strain produced a significantly higher amount of siderophores or pyoverdine than the control strain F113.

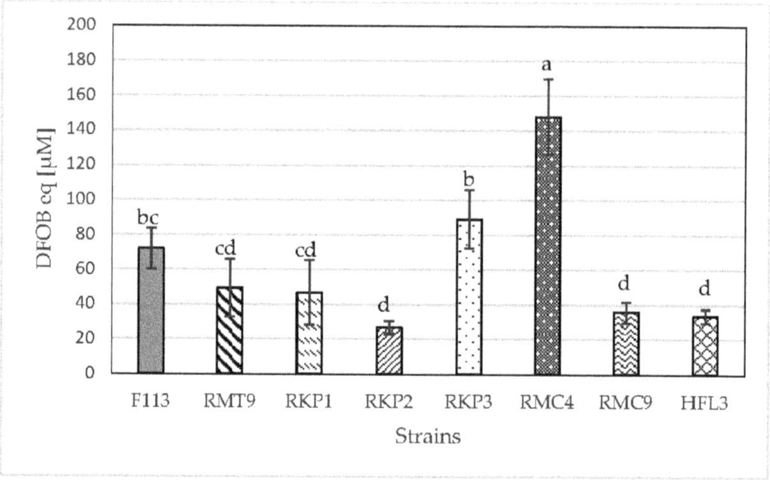

Figure 1. Concentration of siderophore expressed as DFOB equivalents (μM). The data are the mean ± SE (n = 9). Different letters indicate significant differences according to Duncan's test ($p < 0.05$).

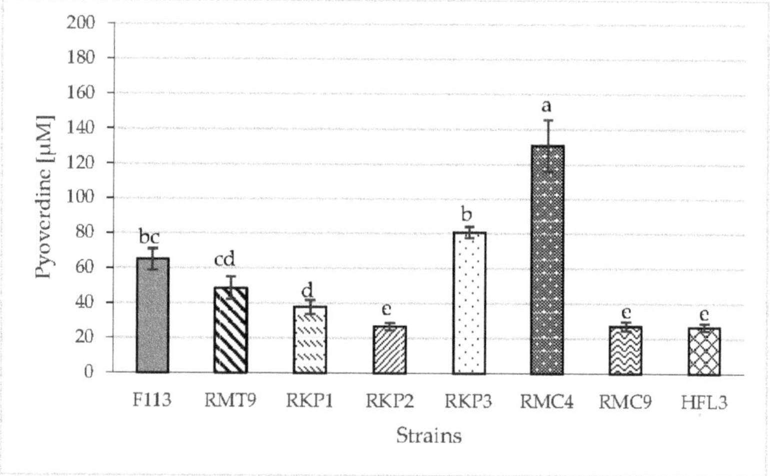

Figure 2. Concentration of pyoverdine (μM) produced by the selected bacterial strains. The data are the mean ± SE (n = 9). Different letters indicate significant differences according to Duncan's test ($p < 0.05$).

2.3. Iron Complexation Capacity Assays

To assess the potential use of the siderophores obtained from the selected strains as Fe biofertilizers, the maximum complexation capacity was determined with the spent media obtained after the growth of the selected strains. To a constant amount of bacterial extract of each bacterial strain, increasing concentrations of Fe^{3+} were added at pH 9, obtaining a curve, where the maximum Fe complexation can be determined (Figure 3). The higher the soluble Fe value, the higher the complexation capacity of the siderophore.

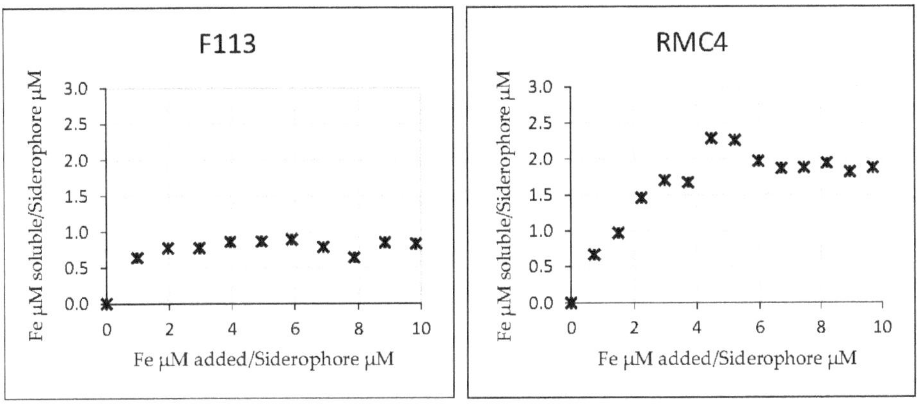

Figure 3. Molar ratio values of Fe: siderophore determined by the representation of the concentration of soluble Fe (µM)/siderophore concentration (µM) vs. concentration of added Fe (µM)/siderophore concentration (µM) for each bacterial strain.

Once the soluble Fe was measured and the siderophore concentration of each spent medium was known, the approximated stoichiometry with which each supernatant would complex Fe was calculated, and the results are shown in Table 2. The RMC4 strain presented the highest value of stoichiometry, which indicated that, at alkaline pHs, the spent medium of the bacterial strain RMC4 could complex a maximum of 2 Fe atoms for each siderophore unit.

Table 2. Maximum number of Fe atoms that can be complexed by each bacterial cell-free supernatant.

Strain	Soluble Fe:Fe Added
F113	1:1
RMT9	1:2
RKP1	1:1
RKP2	1:2
RKP3	3:4
RMC4	2:1
RMC9	2:3
HFL3	1:3

2.4. Titration of Bacterial Secretion

The RMC4 strain was selected because of its high Fe complexation capacity at pH 9. Then, a titration was performed at pH 8.0 and with a fixed wavelength at 480 nm to verify the maximum Fe complexation capacity obtained. The results are shown in Figure 4. The gradual increase in the absorbance of the solution at 480 nm indicated the formation of the pyoverdine/Fe complex attributed to the absorbance of the Fe bond to hydroxamate groups. Once the maximum complexation capacity was reached, the absorbance no longer increased; instead, a slight, gradual decrease was observed. To ascertain the value corresponding to the maximum complexation capacity, a mathematical analysis was performed by calculating the second derivative of the absorbance. The point where the minimum was achieved in this analysis represented the point of maximum complexation. According to this, the complex formed between pyoverdine and Fe had a molar stoichiometry of 1:1 (169 $\pm$ 8 µM of pyoverdine and 168 $\pm$ 5 µM of Fe).

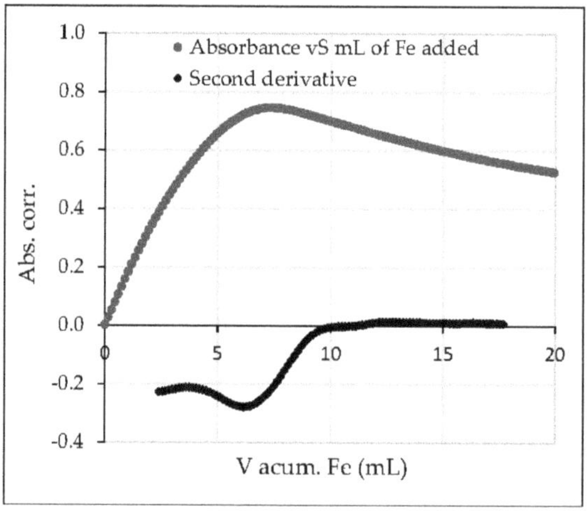

Figure 4. The grey circular dots represent the corrected absorbance for mL of Fe added. The rhomboid dots represent the second derivative of the absorbance versus mL of Fe added.

2.5. Temporal Variation in Organic Acids

Whether the RMC4 bacterial strain produces biostimulant compounds was studied thought the quantification of the organic acids produced in a kinetic assay. Results are shown in Figure 5. The analysis included a large series of acids, but only glutamic acid, acetic acid, aminobutyric acid, IAA, and succinic acid were detected in the secretions. Complementing this study, the concentration of pyoverdine was quantified by spectrophotometric methods (Section 4.2) (Figure 5F). As can be seen in Figure 5, with respect to acetic acid (5A), the concentration remained in a constant range until 72 h, where the concentration increased; aminobutyric acid (5B) showed a peak at 50 h, then its concentration decreased to 0; glutamic acid (5C) did not show marked variation in its concentration during all of the experiment; with IAA (5D), a peak concentration was obtained at 48 h, then it decreased to 0; succinic acid (5E) decreased gradually to 0 concentration at 72 h; and finally, the concentration of pyoverdine (5F) increased steadily, with different slopes up to 150 h. The presence of the IAA among others, indicated that RMC4 could have biostimulant properties.

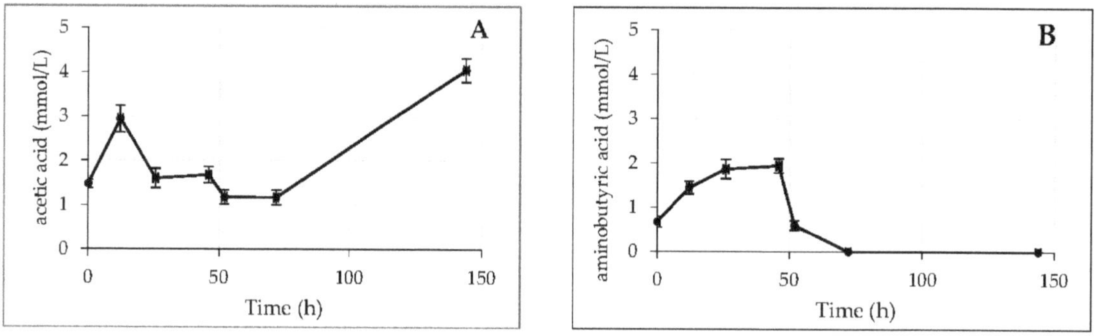

Figure 5. *Cont.*

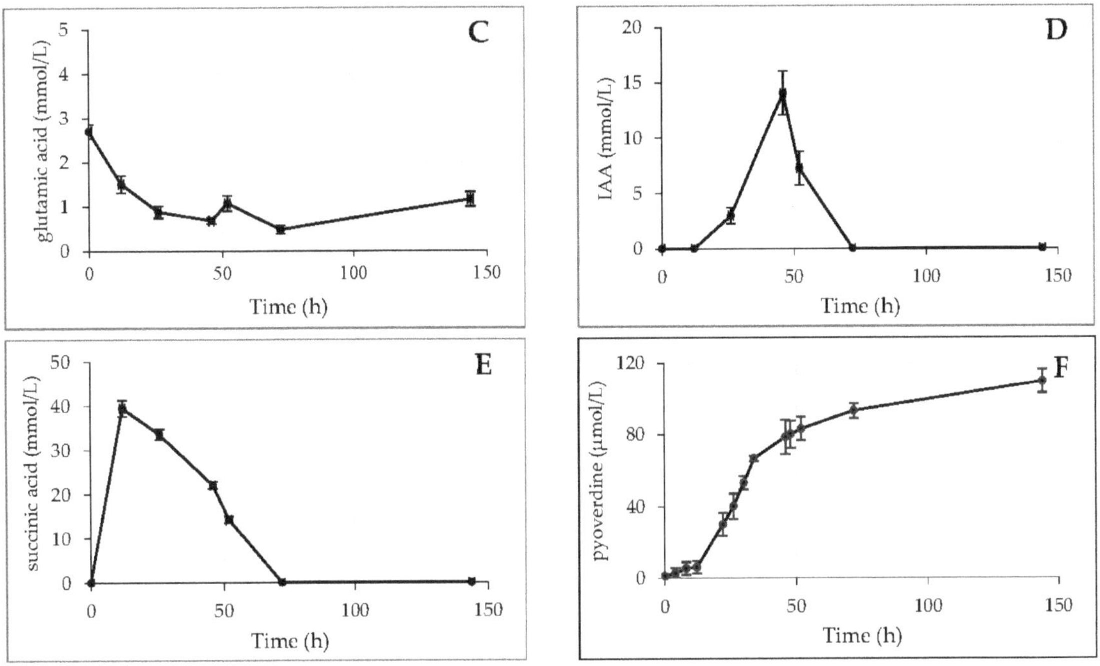

Figure 5. Concentration (mmol/L) of (**A**) acetic acid; (**B**) aminobutyric acid; (**C**) glutamic acid; (**D**) IAA; (**E**) succinic acid, and (**F**) pyoverdine (μmol/L) over time (n = 3) in the bacterial extract of the Pseudomonas RMC4.

2.6. Genomic Analysis of *Pseudomonas monsensis* RMC4

Sequencing of the RMC4 genome resulted in 16 contigs spanning 6,443,679 bp, and the sequencing coverage was estimated at 37×. A completeness of 98.4% was estimated for this genome by the SqueezeMeta pipeline. The genome was shown to belong to a strain of *Pseudomonas monsensis*, a species belonging to the *P. koreensis* subgroup of the *P. fluorescens* cluster of species.

Functional assignment of the genome showed the presence of multiple genes and clusters involved in plant growth promoting activity (Table 3). A cluster was identified in the genome for the biosynthesis of potential antifungals related to lokisin. In addition, the cluster for the hydrocyanic acid production (*hcn* genes) was also found. The potential of the strain RMC4 to produce siderophores was confirmed by the identification of the *fecAR* and *hasDEF* genes that are involved in iron siderophore biosynthesis. The genome also showed the bacteria's ability to produce and secrete the siderophore pyoverdine, the presence of which was observed during growth of the strain under low iron availability conditions. Furthermore, genes and clusters implicated in phosphate mobilization were also found. Regarding the genes involved in phytohormone production by *P. monsensis* RMC4, the cluster composed of the *IaaM* and *IaaH* genes, responsible for the biosynthesis of the auxin IAA, was identified. Likewise, a cluster for the degradation of auxin phenylacetic acid (PAA), which may be involved role in plant–bacteria interaction, was also found. Additionally, the *fitD* gene, which encodes an insect toxin, was also detected in the genome.

Table 3. Identification of the PGPR characteristics of RMC4 strain.

Possible Specie	Genes/Clusters	Function	PGP Category
Pseudomonas monsensis	*Cluster 1*	Type NRPS/lokisin (78%)	Antifungal
	fecAR	Transport of iron dicitrate (III)	Iron siderophore receptor protein
	fitD	Insect toxin	Toxin
	hasDEF	Hemophore biosynthesis	Siderophores
	hcnABC	Hydrocyanic acid biosynthesis	Biocontrol
	hcp (T6SS)	Type VI secretion system	Biocontrol
	iaaHM	Auxin biosynthesis	Phytohormone modulation
	paaFIKY	Phenylacetic acid degradation	Interaction with the environment
	phnBCDENWXZ	Phosphate transport	Nutrient mobilization (P)
	phoBDHH2LPQRU	Phosphate transport	Nutrient mobilization (P)
	pqqABCDE	Pyrroloquinoline quinone biosynthesis	Nutrient solubilization (P)
	pstABCS	Phosphate transport	Nutrient mobilization (P)
	pvdE	Pyoverdine	Nutrient mobilization (Fe)
	ubiA	Production of 4-hydroxybenzoate	Antibiotic

3. Discussion

The main objective of this work was to find and characterize a bacterial strain from horticultural soils with a high capacity to produce siderophores, as an eco-friendly alternative to synthetic ligands to be used as Fe biofertilizer. Besides its Fe complexing capacity, the plant growth-promoting characteristics were also analyzed. A wide variety of naturally produced bacteria were isolated, but only those that fluoresced under UV light were selected, as this is indicative of pyoverdine-producing *Pseudomonas*. The CAS test and the comparison in halo formation were performed on the selected bacteria. The control for comparison was the bacterium *Pseudomonas fluorescens* F113, isolated for the first time from the rhizosphere of sugar beet by Shanahan et al. [27] and well-described. A variant of this bacterium, "F-variant", was generated, which overproduced pyoverdine in an iron-limited medium (SA) and even produced pyoverdine in the iron-rich medium LB [28]. The bacteria that qualitatively generated a larger halo than the control were selected, resulting in a total of seven bacteria: RMT9, RKP1, RKP2, RKP3, RMC4, RMC9, and HFL3 (Table 1). The larger halo formation could be due to a higher production of siderophores. To verify this, a method to quantify the production of siderophores was performed (Figure 1). A colorful dye–iron complex loses its color when a compound with a higher affinity for Fe is added [29]. In an attempt to quantify this process, a known commercial siderophore, DFOB, was used. Thus, the results could be expressed in concentration as DFOB equivalents [30]. The RMC4 bacterium showed a significantly higher concentration of siderophore (DFOB equivalents) than the control bacterium F113. The rest of the bacteria did not show significant differences as compared to the control, except for RKP2, RMC9, and HFL3, which had significantly lower values than the control. Comparing the results in Table 1 and Figure 1, bacteria with a higher halo formation than F113 did not show significantly higher siderophore concentrations than the control bacterium. As one test was qualitative and the other quantitative, it would be expected that concentrations without significant differences could be obtained; however, significantly lower siderophore concentrations than that of the control were obtained. This may have been due to the difference in the culture media used: in the halo formation test, SA medium was used, while in the determination of siderophore concentration, MMS was used. MMS has been shown to cause a significant increase in pyoverdine production by *Pseudomonas fluorescens* [31]. However, this increase has not been observed in other bacteria of the *Pseudomonas* genus, such as *P. aeruginosa*, *chlororaphis*, *pertucinogena*, *putida*, *stutzeri*, and *syringae*. Sasirekha and Srividya [32] observed that the bacterium *Pseudomonas aeruginosa* FP6 produced more siderophores in a culture medium with mannitol or sucrose as a carbon source (instead of succinic acid) or with yeast extract or urea as a nitrogen source (instead of ammonium sulfate). Murugappan et al. [33] optimized the culture medium for maximum siderophore production from the bacterium *Pseudomonas putida* (CMMB2) and observed that, for this bacterium, the culture medium for maximum production of siderophores was the MM9 medium, the best carbon source was succinate (as used in this paper), and the nitrogen source was NH_4Cl, with a pH adjusted to 8 (7 was the pH used in this work). The use of MMS may have resulted in the optimization of pyoverdine production by those bacteria belonging to the *fluorescens* species, while the maximum yield was not obtained from those not belonging to this group.

The quantification of siderophores by CAS assay may have presented interference due to compounds present in the culture medium that could also have complexed Fe (such as phosphates) and cause discoloration before the addition of DFOB. However, as the same culture medium was used for all bacteria, in combination with the application of a blank using only the culture medium, this was not likely to interfere with the results. The CAS assay is the most widely used method for the quantification of siderophores; however, due to the large number of siderophores that could be secreted by the same bacterium and the complexity of the culture supernatant, it is not possible to accurately determine the real concentration of siderophores. For the identification of each siderophore, advanced analytical techniques, such as ultraperformance liquid chromatography coupled to tandem mass spectrometry [34], would have to be used. Nevertheless, the CAS assay is a very fast, reliable, and inexpensive method of approximating the concentration of siderophores.

The RMC4 bacterium produced a significantly higher concentration of pyoverdine than the control bacterium, while RKP1, RKP2, RMC9, and HFL3 bacteria produced significantly lower concentrations of pyoverdine than F113 (Figure 2). These results are possibly related to the group of *Pseudomonas* to which each bacterium belongs, as explained above. Related to RMC4, a comparison between Figures 1 and 2 showed that the percentage of siderophores corresponding to pyoverdine was $88.1 \pm 8.8\%$; this means that the major component of the segregated siderophores in RMC4 bacteria was pyoverdine. This result is consistent with that described in the literature. Under Fe deficiency conditions, bacteria could synthesize siderophores to bind Fe^{3+} and load into the cytoplasm via highly specific transport systems [35]. If the Fe limitation conditions are moderated, bacteria could produce siderophores with low affinity for Fe but metabolically inexpensive to produce, although under extremely Fe deficient conditions, bacteria could produce highly efficient but metabolically expensive siderophores [36,37]. In the case of *Pseudomonas aeruginosa*, in situations of moderate Fe deficiency, pyochelin is secreted, an inefficient but metabolically profitable siderophore; under severe Fe deficiency, pyoverdine, a siderophore very efficient ($K_{LFe(III)} = 10^{30.8}$; $K_{LHFe(III)} = 10^{43.0}$) [38] but metabolically very costly, is secreted [39]. In the present experiment, severe Fe deficiency conditions were induced, causing the bacteria to tend to secrete pyoverdine rather than other siderophores with less affinity for Fe; this was consistent with the fact that a higher percentage of the siderophores excreted corresponded to pyoverdine.

A complexation capacity assay was performed, and the stoichiometry of the Fe complex was determined (Table 2). The resulting stoichiometry for RMC4 (Figure 3) was around 2, showing that the maximum Fe complexation capacity of the siderophores secreted by this bacterium is double that of F113, RKP3, and RMC9, which resulted in a 1:1 stoichiometry.

In the experiment conducted by Ferreira et al. [17] to determine the Fe complexation capacity of other bacterial strains, the method described by Villén was also used to determine the maximum Fe complexation capacity of different siderophores under alkaline conditions. Ferreira et al. [17] studied the siderophores secreted by the bacteria *Azotobacter vinelandii*, *Bacillus megaterium*, *Bacillus subtilis*, *Pantoaea allii*, and *Rhizobium radiobacter* and the maximum Fe complexation capacity by the abovementioned method, but ambiguous values were obtained. This occurrence could be ascribed to the diversity of bacteria studied, which may secret various siderophores with differing iron affinities, and also to the highly alkaline conditions in which the maximum complexation capacity test was carried out. Ferreira et al. [17] argued that pH 9 could be used to test the maximum complexation capacity in basicity ranges typical of alkaline soils; however, it would be advisable to carry out a study of the stability of the siderophore–Fe complexation at different pH values. The pH will define the type of siderophore–Fe bonding. Hydroxamate functional groups bind to Fe^{3+} through the loss of a proton, which is conditioned by the pH of the medium [40]. Catecholate functional groups bind Fe^{3+} after the loss of two protons in neutral-alkaline pH via phenolic oxygens [41]. The pH could also affect the stability of the siderophore, and even competition for binding different metals, as siderophores such as DFOB are more likely to bind to Co^{2+} than Fe^{3+} at alkaline pH [42]. Therefore, pH is a limiting parameter when analyzing the maximum complexation capacity; the state of the functional groups and the stability of the compound at the pH of the study must be considered. The study of maximum Fe complexation capacity in alkaline soils should be studied at pHs where Fe chlorosis problems may occur. Iron precipitates as (hydr)oxides in soils at pH 7.5 in the presence of $CaCO_3$ [24]. It would, therefore, be interesting to carry out the study starting from this pH, and not directly from pH 9.

Due to the results obtained in the maximum complexation capacity assay (Figure 3/Table 2), the bacterial strain RMC4 was chosen to perform an assessment and calculate more accurately the maximum Fe complexation capacity. The titration yielded a maximum complexation capacity of molar pyoverdine:Fe ratio of 1:1. A priori, these results may seem to contradict the results obtained previously; however, there are several factors to be considered. The titration was performed at 480 nm to observe the absorbance of the hydroxamate–Fe bond. This indicated that this group has a 1:1 iron complexing capacity; however, the rest of the possible complexing functional groups that may be present in the peptide chain of pyoverdine were not being observed. Furthermore, in Figure 4, which shows the formation of the complex, the increase in absorbance does not have a linear trend; two trend lines can be observed in the gradual increase in absorbance with different slopes, which could indicate the formation of different pyoverdine–Fe complexes where, probably, different functional groups would be involved. Also, once the maximum complexation point was reached, a decrease in absorbance was observed, which could be indicative of the degradation of the pyoverdine–Fe complex or the formation of another, more-stable pyoverdine–Fe complex not visible at the selected wavelength. Finally, it was shown that the selected strain is able to biosynthesize hemophore groups (Table 3), which indicates that pyoverdine was probably not the only complexing agent present in the medium. If the hypotheses presented were confirmed, we could potentially develop a promising iron biofertilizer with multiple complexing groups. Some of these groups might form more-stable iron complexes, while others could form less-stable ones. Consequently, when this iron biofertilizer is applied to plants, it could release the iron complexed in the less-stable groups more rapidly compared to the iron complexed in the more-stable groups. This might result in a slow release of iron, ensuring a continuous supply of iron to the plants even after the faster-release iron has been utilized.

The sequencing of the RMC4 genome revealed that the bacterium sequenced belonged to the species *P. monsensis*. The genome of *P. monsensis* RMC4 confirmed that this strain had potential as a plant growth-promoting rhizobacterium (PGPR) because it contains several beneficial genes for plants (Table 3). In fact, it is known that *P. fluorescens* strains are involved in plant growth-promoting activity by several mechanisms, for example, the production of siderophores and nutrient solubilization and mobilization, or in biological control through the production of antibiotics and fungicides [43–45]. Regarding the biocontrol potential of these rhizobacteria, Sehrawat et al. [46] reported on the beneficial effect of using antagonistic microorganisms (e.g., HCN producers) against pathogens, indicating they can be employed as a sustainable strategy, thus avoiding the use of pesticides. Type six secretion systems of pseudomonads have been proposed as a biocontrol trait against phytopathogenic bacteria [47]. The gene cluster for biosynthesis of the antifungal lokisin and the production of hydrocyanic acid identified in the genome of the RMC4 strain could give the bacteria the capacity for biological control. These results have also been previously reported in the genus *Pseudomonas* [48–50]. In addition, this strain has been shown to have insecticidal activity, which has also been observed in *P. fluorescens* towards agricultural pests [51]. To promote plant growth, microorganisms have also developed several mechanisms to mobilize and mineralize nutrients, such as iron or phosphate, that are not available for plant uptake [52,53]. In this sense, soil bacteria are known to produce small organic molecules, siderophores, under iron-limiting conditions through high-affinity interactions [54,55]. Our results indicate that *P. monsensis* RMC4 is involved in the biosynthesis of several iron siderophores, as previously found by Gu et al. [56] in the *P. koreensis* group. In addition, Fernandez et al. [57] informed that pseudomonads also could mobilize phosphate in the soil, and this was indeed observed in the RMC4 strain. Furthermore, phytohormones such as IAA and PAA, through catabolism, play an important role in plant growth and development and are involved in the interactions between microorganisms and plant roots [58–60]. The presence in the *P. monsensis* RMC4 strain of both IAA biosynthesis and PAA degradation pathways could be important for the PGPR activity. Therefore, the present study provides a more comprehensive view of the capacity of the *P. monsensis* RMC4 strain as a plant growth-promoting biofertilizer and its potential as a biostimulant.

The results described above showing the highest production of siderophores and pyoverdine and higher iron complexation capacity show that *Pseudomonas monsensis* RMC4 is an excellent candidate to be used as an Fe biofertilizer and able to mobilize iron. Furthermore, the finding that the genome encodes many other plant growth-promoting traits highlights its use as a polyvalent agricultural inoculant. A kinetic study was performed (144 h) where some carboxylic acids with biostimulant properties [22] were measured (glutamic acid, acetic acid, aminobutyric acid, IAA, and succinic acid); data are shown in Figure 5. The limiting factor in the production of the different acids was the Fe deficiency of the bacteria, which affected many metabolic processes of the bacteria, such as protein and nucleic acid synthesis. Glutamic acid is an amino acid with biostimulant properties, possibly related to the fact that it is the central product in the nitrogen metabolism pathway [61], is involved in chlorophyll biosynthesis [62], and has been shown to exert a positive effect on Fe uptake in tomato plants with lime-induced Fe deficiency [63]. As shown in Figure 5C, the initial concentration of glutamic acid decreased until 48 h and then increased in a linear progression until the end of the experiment. Glutamic acid is known to be involved in the biosynthesis of pyoverdine, being a component of the side chain, and could be modified to succinimide, catalyzed by pyoverdine I decarboxylase (PvdN), or to α-ketoglutarate, catalyzed by pyoverdine aminotransferase (PtaA) [64]. The concentration of pyoverdine increased until 48 h, then remained stable until the end of the experiment (Figure 5F). As mentioned above, the biosynthesis of pyoverdine requires a very high energy expenditure for the organism, probably after 48 h, and not having obtained Fe, it is likely that the bacteria are no longer producing pyoverdine, keeping its concentration stable and possible causing metabolites used for its biosynthesis (such as glutamic acid) to increase in concentration. Acetic acid can be produced in the metabolic pathway of the bacteria. High concentrations of acetic acid are an important physiological stress factor in cells [65]. Acetic acid had a maximum at 12 h (Figure 5A), then remained stable with values around 2 $mmol \cdot L^{-1}$, and after 72 h, its values increased to a maximum of 4 $mmol \cdot L^{-1}$. The initial values were probably due to natural generation of the bacteria´s metabolism, and the maximum value in the final time could be indicative of physiological damage caused by severe Fe deficiency and failure of Fe acquisition strategies to work. Gamma-aminobutyric acid (GABA) is a phytohormone secreted by plants with abiotic stress regulation functions [66]. Several studies have reported its positive effects in horticultural crops under abiotic stresses; in melon plants with saline-alkaline stress, it induced increased growth, reduced oxidative stress levels, and increased antioxidant enzymes [67]. In bacteria, GABA is synthesized from the α-decarboxylation of L-glutamic acid. The GABA concentration value was around 2 $mmol \cdot L^{-1}$ until 26 h, then the concentration decreased until it was not detected after 52 h. The variation in values was similar to that observed for glutamic acid, under Fe deficiency conditions; the normal functioning of metabolism was disrupted, leading to the inability to metabolize GABA. Indol-3-acetic acid is a well-known phytohormone involved in the regulation of growth, stem elongation, and seed germination, among other functions. In bacteria, IAA synthesis plays a key role in plant–microorganism interaction. These interactions could promote plant phytostimulation but also could be pathogenic [56]. In the present experiment, high concentrations of IAA were detected at 26 h, then ceased to be produced after 34 h. It was demonstrated that the RMC4 bacterium was able to produce IAA, a compound with PGP characteristics that could have a biostimulant effect on the plant. This result agrees with that obtained for the PGPR characterization (Table 3): the bacterium possessed the *iaaHM* gene, whose function is the biosynthesis of auxins (such as IAA). Succinic acid was the carbon source used by the bacteria in this experiment. The concentration of this compound decreased until it was not detected after 30 h (Figure 5E). Perhaps, as it was the only carbon source available to the bacteria, this compound was consumed by the bacteria and once completely consumed, the bacterial metabolism was deregulated, causing the chain reaction that was observed with the other compounds (it stopped producing GABA, IAA, and pyoverdine). Regarding pyoverdine, initially, its production was slow. After 12 h, the pyoverdine concentration was less than 10 $\mu mol \cdot L^{-1}$. Probably due to the activation of the Fe acquisition mechanism, the production of pyoverdine increased substantially until 48 h, when it reached a concentration of 85 $\mu mol \cdot L^{-1}$, which means that in the first 12 h, the production of 0.83 $\mu mol \cdot L^{-1} \cdot h^{-1}$ was observed, and in the following 36 h, the pyoverdine production rate was 2.08 $\mu mol \cdot L^{-1} \cdot h^{-1}$, an increase of 2.5 fold in the production rate. Possibly due to the deregulation of the bacteria´s metabolism (as explained before), from 48 h until the end of the experiment (144 h), pyoverdine production remained stable, obtaining a concentration of 108 $\mu mol \cdot L^{-1}$ at the end of the experiment. A further 23 $\mu mol \cdot L^{-1}$ was obtained from 48 h to 144 h, resulting in a production rate of 0.24 $\mu mol \cdot L^{-1} \cdot h^{-1}$ of pyoverdine in the last part of the experiment. With these results, three phases of pyoverdine production by the bacterium were observed. The first phase may correspond to metabolic activation; the bacterium was aware of the need for Fe and activated the strategies for acquiring this element. The second phase corresponds to the metabolic zenith; the production of pyoverdine is the fastest, as a result of the activation of Fe acquisition strategies. Finally, the last phase corresponds to metabolic collapse, since the Fe acquisition strategies have not worked and the bacterial metabolism has collapsed, which will probably lead to the death of the bacterium. For the biotechnology industry interested in optimizing pyoverdine production, the challenge to overcome is to optimize the parameters that affect pyoverdine production (temperature conditions, carbon source, nitrogen source, iron concentration, pH, etc.) in order to extend the metabolic zenith phase as much as possible and try to increase the production per hour in this phase.

4. Materials and Methods

4.1. Bacterial Isolation

Bacteria were isolated from the rhizosphere of different horticultural plant species: pepper (*Capsicum annuum*), tomato (*Solanum lycopersicum*), and pumpkin (*Curcubita* sp.). Roots were weighed and immersed in saline solution (8.5% NaCl). After vigorously shaking and serial decimal dilutions, the supernatants were used to inoculate plates of Sucrose-Asparagine (SA) culture medium, a medium that is selective for pseudomonads and stimulates pyoverdine production [68]. Ampilicin (100 $\mu g \cdot mL^{-1}$) and Cycloheximide (100 $\mu g \cdot mL^{-1}$) were added to the medium to increase selectivity and to avoid the growth of eukaryotes. After two days of incubation, colonies showing a yellow/green fluorescence were selected as potential *Pseudomonas* isolates. Colonies that did not show fluorescent pigment were discarded. Those that survived temperatures of 37 °C were discarded to avoid potential pathogens. In addition, using the Box-PCR technique as described by Gutiérrez-Barranquero et al. [69], sibling strains were discarded. Those bacteria showing higher siderophore production than the control bacterium *Pseudomonas fluorescens* F113 [28] according to the halo formation in the CAS agar assay [29] were selected as potential biofertilizer candidates. Strains were identified as *Pseudomonas* spp. by amplification and sequencing of the 16S RNA gene.

4.2. Culture Conditions and Siderophore and Pyoverdine Production

The selected bacterial strains were grown in 10% phosphorus minimal medium succinate (MMS), described as optimized for siderophore production by bacteria, with the following composition (reagents obtained from Panreac (Barcelona, Spain) except where indicated): 0.6 $g \cdot L^{-1}$ dipotassium hydrogenphosphate trihydrate; 0.3 $g \cdot L^{-1}$ monopotassium phosphate; 0.2 $g \cdot L^{-1}$ magnesium sulfate heptahydrate; 1.0 $g \cdot L^{-1}$ ammonium sulfate; and 4.0 $g \cdot L^{-1}$ succinic acid (Sigma-Aldrich (St. Louis, MO, USA)), pH fixed at 7.00 ± 0.01 [19,31]. Culture conditions were as described by Vindeirinho et al. [31]. In brief, starter cultures were obtained by inoculating 2 loops of culture in MMS with ~3.7 $\mu mol \cdot L^{-1}$ $FeCl_3 \cdot 6\ H_2O$ (Panreac), at 30 °C, 150 rpm, for ~8 h. Pre-cultures were prepared also in MMS with Fe 0.37 $\mu mol \cdot L^{-1}$, by inoculating an appropriate volume of the starter culture, and incubated under the same conditions described above, for ~16 h until the bacteria reached exponential phase (optical density $OD_{600} \sim 1.0$). Finally, cultures designated for siderophore production were prepared in 200 mL MMS without Fe, by inoculating an appropriate volume of the pre-culture until an initial OD_{600} of ~ 0.1. Bacteria were incubated for ~48 h under the same conditions described above. Subsequently, samples were centrifuged (3000× *g*, 15 min, at 25 °C), filtered using a 0.45 μm cellulose nitrate membrane filter (Labbox Labware S.L. Barcelona, Spain) and stored protected from light at −20 °C until siderophore quantification. The quantification of siderophores was conducted using the CAS liquid assay method originally outlined by Schwyn and Neilands [29], as amended by Mehnert et al. [30]. In brief, 1.5 mL of 1 $mol \cdot L^{-1}$ $FeCl_3 \cdot 6\ H_2O$ dissolved in 10 $mol \cdot L^{-1}$ HCl (Merck, suprapur) was mixed with 7.5 mL of 2 $mol \cdot L^{-1}$ CAS solution; then, the mixture was added slowly to 6 mL of 10 $mol \cdot L^{-1}$ cetyltrimethylammonium bromide (purchased from Sigma-Aldrich). Simultaneously, 9.76 g 2-(N-morpholino)ethanesulfonic acid (MES) (Sigma-Aldrich) was dissolved in 80 mL water, and pH was adjusted to 5.6 with 50% KOH. Water was added to attain a final volume of 85 mL, and this MES buffer solution was then combined with the dye solution. After 4 h, 150 μL of the culture supernatant and 150 μL of metal CAS solution were mixed, and OD was measured at 630 nm. To quantify siderophores, a commercial siderophore known as DFOB (Sigma-Aldrich) was utilized as standard, and the concentrations within culture supernatants were expressed in "DFOB-equivalents". A standard curve was constructed by plotting the discolorization (*d*) of the metal CAS solution at 630 nm as a function of siderophore concentration (Equation (1)). Sterile culture medium was used as reference solution (A_{ref}), and zero absorbance (A_0) was performed using a mixture of the metal CAS solution and 2 $mol \cdot L^{-1}$ DFOB. Presuming an association/dissociation equilibrium, data were fitted using (Equation (2)) with y_{max} set at 100. Both the methodology and the equations were obtained from the protocol described by Mehnert et al. [30].

$$d = \frac{A_{ref} - A_{supernatant}}{A_{ref} - A_0} \tag{1}$$

$$y = y_{max} \cdot \left(1 - e^{-kx}\right) \tag{2}$$

The concentration of pyoverdine was measured by UV-VIS spectroscopy. The chromophore group, common in the pyoverdines produced by *Pseudomonas* genera, has a maximum absorption peak between 380–400 nm; thus, the concentration was determined using the Lambert–Beer law, using the molar extinction coefficient (ε) of 16,000 $L \cdot mol^{-1} \cdot cm^{-1}$ [19].

To avoid Fe contamination, all glassware was soaked with HCl (VWR, Normapur) 6M overnight and, afterwards, washed with ultrapure water (Milli-Q system, Bedford, MA, USA).

4.3. Iron Complexation Capacity Assays

The complexation capacity assay was performed as described by Villén et al. [70]. To a fixed volume of the supernatant, increasing concentrations (from 0 to 16 $mg \cdot L^{-1}$) of $FeCl_3 \cdot 6\ H_2O$ were added, then the pH was adjusted to 9.00 ± 0.01. The solution was allowed to stand for 3 h, then the pH was adjusted again to 9. Subsequently, the solution was left in the dark for 24 h. Afterwards, pH was adjusted once more. Then, the mixture was centrifuged (10,000× *g*, 15 min) and filtered by 0.45 μm pore-size nylon membrane. Finally, the final volume was adjusted to 50 mL, and the concentration of Fe was determined by atomic absorption spectroscopy with flame atomization (Perkin-Elmer Analyst 800; Shelton, CT, USA).

4.4. Titration of Bacterial Secretion

The assessment protocol described by Yunta et al. [71] was followed with modifications. Pyoverdine was obtained as described in Section 4.2. The pH was adjusted to 8, and the concentration of pyoverdine was measured as described above. The experimental solution (10 mL) was placed in a 50 mL thermostatic (25.0 ± 0.5 °C) jacketed reaction vessel provided with an airtight cap fitted with a gas inlet and outlet tubes, combined pH glass electrode, a spectrode, two piston burets (tips placed below the surface of the solution), and a magnetic stirrer. The photometric titration consisted of the gradual addition of Fe^{3+} standard solution to the pyoverdine until the absorbance at 480 nm remained constant. Potentiometric measurements were performed with Metrohm 719 and 721 potentiometers (precision of 0.1 mV) combined with a pH glass electrode that kept the pH constant at 8.0, while a NaOH 0.2 mol·L^{-1} solution was automatically added if necessary. Photometric titrations were carried out using a Metrohm 662 photometer (resolution of 10 ± 0.1 nm) with a light spectrode of path length 2 × 10 nm. Both potentiometers were controlled by the software for PC Tiamo 2.5 (Metrohm AG, Switzerland). The titration was performed in triplicate.

4.5. Evaluation of Temporal Variation in Organic Acid Concentrations Secreted by Bacteria

A kinetic assay was carried out to evaluate temporal changes in organic acid concentrations secreted by the selected bacteria. The analysis was focused on the identification of 11 carboxylic acids, including monocarboxylic acids (MCAs) and polycarboxylic acids (PCAs): acetic, lactic, oxalic, citric, aminobutyric, succinic, malic, gluconic, fumaric, pyruvic acids, and IAA. For this, the experiment was carried out for 144 h (6 days), and samples were taken every 4 h. Zero time corresponded to the cultures to produce siderophores, after performing the initial pre-culture and culture procedures described in Section 4.2. For its evaluation, a previous sample treatment was necessary. Briefly, 1.0 mL of sample was loaded onto a Strata-X-AW 33 μm polymeric weak anion cartridge (SPE) (Phenomenex, Torrance, USA), previously conditioned with 5 mL of methanol (MeOH) (Sigma-Aldrich) and 5 mL of 25 mM tris(hydroxymethyl)aminomethane-acetate (Tris-OAc) (Sigma-Aldrich) (pH = 7.5) for MCAs and 25 mM MES (Sigma-Aldrich) (pH = 4.5) for PCAs at about 1 mL·min^{-1} by means of a suction system. The SPE cartridge was then washed with 5 mL of a mixture of 25 mM Tris-OAc:MeOH (90:10, v/v) for MCAs and 25 mM Tris-OAc:MeOH (90:10, v/v) in the case of PCAs. The rinse was discarded, and after 10 min of drying time, the analytes were eluted with 2 mL of 25 mM Tris-OAc: 0.01M HCl (10:90, v/v) mixture for MCAs and 0.1 M HCl for PCAs. The resulting solution was passed through a nylon 0.45 μm syringe filter, and 10 μL was injected into the chromatographic system. Analysis was achieved by high-performance liquid chromatography coupled to a refractive index detector (HPLC-RID) system (1260 Infinity model Agilent Technologies, Waldbronn, Germany). A Bio-Rad Aminex HPX-87 H column (300 × 7.8 mm, 9 μm) was used, protected by a guard column from Phenomenex. Analysis conditions were set as follows: the mobile phase was sulfuric acid 5 mM, the flow rate was 0.5 mL·min^{-1}, the column temperature was set at 60 °C, and the temperature of the refractive index detector was at 50 °C in positive polarity mode.

During sampling times, the concentration of pyoverdine was also measured by UV-Vis spectrophotometry, as explained in Section 4.2. The experiment was conducted in triplicate.

4.6. Sequencing and Analysis of the Genome of Strain RMC4

The genomic DNA of strain RMC4 was extracted using the NucleoSpig® Microbial DNA kit (Macherey-Nagel, Düren. Germany), and the quality was determined on an agarose gel and quantified using a Qubit fluorimeter (Invitrogen, Carlsbad, CA, USA). The genome was analyzed on a MinION sequencer (Oxford Nanopore Technologies, UK). The library was prepared with 1 µg of DNA and the Nanopore Ligation Sequencing kit (SQK-LSK-110). The library was loaded onto a MinION flow cell (R9.4.1 pores) following the manufacturer's recommendations. Subsequently, sequencing was performed via MinKNOW software (v22.03.6). Guppy (v6.0.7) was used to carry out the base calling step, which consists of translating raw signal data into nucleotide sequences in FASTQ format, and reads shorter than 5000 nts were excluded before further analysis [72]. For bioinformatic analysis, NanoPack [73] was applied to evaluate the read length and ASE calling quality. Additionally, the ONT reads were filtered, and the adapter sequences were removed with Porechop (v0.2.4). The novo assembly was performed using Flye (v2.9, [74]), and SquezzeMeta was used to annotate the genome and determine the taxonomic classification [75]. In addition, the Type-Strain Genome Server (TYGS) platform [76] was used to determine whether the genome sequenced corresponded or not to a known bacterial species, through the percentage of DNA–DNA hybridization (%DDH). The *P. monsensis* RMC4 genome has been submitted to the NCBI database and is available under BioProiect accession number PRJNA1028413. Secondary metabolite biosynthesis clusters were identified using the AntiSMASH (v7.0) web application [77]. The genome was also searched using subsystems technology (RAST server, [78]) in order to identify genes and gene clusters implicated in plant growth promotion.

4.7. Statistical Analysis

IBM SPSS Statistics 24.0 software (SPSS Inc., Chicago, IL, USA) was used for one-way analysis of variance (ANOVA). Siderophore and pyoverdine production were compared using Duncan's test for $p < 0.05$.

5. Conclusions

In the present work, RMC4 was selected from a large group of bacteria isolated from horticultural soils according to their siderophore production, corresponding mainly to pyoverdine. The Fe chelating capacity was evaluated at alkaline pH, elucidating that the bacterial secretion had the capacity to form an Fe complex in a 1:2 molar ratio (secretion:Fe), which was explained not only by the pyoverdine but also by the presence of other siderophores of compounds able to complex Fe in the studied conditions. In addition, RMC4 showed plant biostimulant characteristics according to its high production of IAA or glutamic acid and the gene clusters related to phosphorus mobilization/solubilization and the production of antibiotics and antifungals detected by genomic analysis. These results contribute to the existing knowledge of *Pseudonomas* as a siderophore-producing bacterium and bring novel insights to the identification of eco-friendly alternatives to synthetic ligands for Fe chelation and their potential use to alleviate Fe chlorosis in crops.

Future experiments involving plants can contribute to advancing our comprehension of the effectiveness of Fe complexed by RMC4 secretions (mainly pyoverdine–Fe) as a biofertilizer in calcareous environments and assessing the biostimulant impact of RMC4 on plant growth.

Author Contributions: Conceptualization, S.L.-R., R.R., J.J.L., M.M. (Marta Martín) and J.M.L.-G.; methodology, S.L.-R., R.R., J.J.L., M.M. (Marta Martín), S.V. and J.M.L.-G.; formal analysis; S.L.-R., R.R., J.J.L., M.M. (Marta Martín), S.V., M.M. (Mónica Montoya) and J.M.L.-G.; investigation, S.V., M.M. (Mónica Montoya) and J.M.L.-G.; writing—original draft preparation, S.L.-R., R.R., J.J.L., M.M. (Marta Martín) and J.M.L.-G.; writing—review and editing, S.L.-R., R.R., J.J.L., M.M. (Marta Martín), S.V. and J.M.L.-G.; project administration, S.L.-R., R.R., J.J.L. and M.M. (Marta Martín); funding acquisition, S.L.-R., R.R., J.J.L. and M.M. (Marta Martín). All authors have read and agreed to the published version of the manuscript.

Funding: The authors gratefully acknowledge the financial support of the Ministerio de Economía y Competitividad through project RTI2018-096268-B-I00, and the Ministerio de Ciencia e Innovación, the Agencia Estatal de Investigación, and Feder funds through the projects PID2022-141721OB-C21 and PID2021-125070OB-I00. J. M. Lozano is recipient of the FPI grant from the Ministerio de Ciencia e Innovación (PRE-2019-091246). M. Montoya is recipient of the Margarita Salas grant of the Ministerio de Universidades and Universidad Politécnica de Madrid (RD 289/2021) supported by the European Union Next-GenerationEU.

Data Availability Statement: The data presented in this study are openly accessible at e-cienciaDatos, https://edatos.consorciomadrono.es/.

Acknowledgments: The authors acknowledge S. García Méndez, M. Torres de la Casa, P. Jareño, and M. Pérez-Cerrato for their contribution to the experiments, and David Durán for his training on the techniques used.

Conflicts of Interest: The authors declare no conflict of interest.

References

1. Briat, J.; Dubos, C.; Gaymard, F. Iron nutrition, biomass production, and plant product quality. *Trends Plant Sci.* **2015**, *20*, 33–40. [CrossRef]
2. Lindsay, W.L.; Schwab, A.P. The chemistry of iron in soils and its availability to plants. *J. Plant Nutr.* **1982**, *5*, 821–840. [CrossRef]
3. Chaney, R.L. Diagnostic practices to identify iron deficiency in higher plants. *J. Plant Nutr.* **1984**, *7*, 47–67. [CrossRef]
4. Lucena, J.J. Effects of bicarbonate, nitrate and other environmental factors on iron deficiency chlorosis. A review. *J. Plant Nutr.* **2000**, *23*, 1591–1606. [CrossRef]
5. Nadal, P.; Hernández-Apaolaza, L.; Lucena, J.J. Effectiveness of N,N′-Bis(2-hydroxy-5-methylbenzyl) ethylenediamine-N,N′-diacetic acid (HJB) to supply iron to dicot plants. *Plant Soil* **2009**, *325*, 65. [CrossRef]
6. Bloem, E.; Haneklaus, S.; Haensch, R.; Schnug, E. EDTA application on agricultural soils affects microelement uptake of plants. *Sci. Total Environ.* **2017**, *577*, 166–173. [CrossRef] [PubMed]
7. Nowack, B. Environmental Chemistry of Aminopolycarboxylate Chelating Agents. *Environ. Sci. Technol.* **2002**, *36*, 4009–4016. [CrossRef]
8. Schenkeveld, W.D.C.; Hoffland, E.; Reichwein, A.M.; Temminghoff, E.J.M.; van Riemsdijk, W.H. The biodegradability of EDDHA chelates under calcareous soil conditions. *Geoderma* **2012**, *173*, 282–288. [CrossRef]
9. Soares, E.V. Perspective on the biotechnological production of bacterial siderophores and their use. *Appl. Microbiol. Biotechnol.* **2022**, *106*, 3985–4004. [CrossRef]
10. Ghosh, S.K.; Bera, T.; Chakrabarty, A.M. Microbial siderophore–A boon to agricultural sciences. *Biol. Control.* **2020**, *144*, 104214. [CrossRef]
11. Singh, P.; Chauhan, P.K.; Upadhyay, S.K.; Singh, R.K.; Dwivedi, P.; Wang, J.; Jain, D.; Jiang, M. Mechanistic insights and potential use of siderophores producing microbes in rhizosphere for mitigation of stress in plants grown in degraded land. *Front. Microbiol.* **2022**, *13*, 898979. [CrossRef]
12. Ahmed, E.; Holmström, S.J.M. Siderophores in environmental research: Roles and applications. *Microb. Biotechnol.* **2014**, *7*, 196–208. [CrossRef] [PubMed]
13. Hofmann, M.; Retamal-Morales, G.; Tischler, D. Metal binding ability of microbial natural metal chelators and potential applications. *Nat. Prod. Rep.* **2020**, *37*, 1262–1283. [CrossRef] [PubMed]
14. Ferreira, C.M.H.; Soares, H.M.V.M.; Soares, E.V. Promising bacterial genera for agricultural practices: An insight on plant growth-promoting properties and microbial safety aspects. *Sci. Total Environ.* **2019**, *682*, 779–799. [CrossRef] [PubMed]
15. Siderophore Base—The Web Database of Microbial Siderophores. Available online: http://bertrandsamuel.free.fr/siderophore_base/siderophores.php (accessed on 20 October 2022).
16. Ferreira, C.M.H.; López-Rayo, S.; Lucena, J.J.; Soares, E.V.; Soares, H.M.V.M. Evaluation of the Efficacy of Two New Biotechnological-Based Freeze-Dried Fertilizers for Sustainable Fe Deficiency Correction of Soybean Plants Grown in Calcareous Soils. *Front. Plant Sci.* **2019**, *10*, 1335. [CrossRef]
17. Ferreira, C.M.H.; Vilas-Boas, Â.; Sousa, C.A.; Soares, H.M.V.M.; Soares, E.V. Comparison of five bacterial strains producing siderophores with ability to chelate iron under alkaline conditions. *AMB Express* **2019**, *9*, 78. [CrossRef] [PubMed]
18. Cornelis, P.; Matthijs, S. *Pseudomonas Siderophores and Their Biological Significance*; Varma, A., Chincholkar, S.B., Eds.; Springer: Berlin/Heidelberg, Germany, 2007; pp. 193–203. [CrossRef]
19. Meyer, J.M.; Abdallah, M.A. The Fluorescent Pigment of Pseudomonas fluorescens: Biosynthesis, Purification and Physicochemical Properties. *Microbiology* **1968**, *107*, 319–328. [CrossRef]
20. Shaharoona, B.; Arshad, M.; Zahir, Z.A.; Khalid, A. Performance of Pseudomonas spp. containing ACC-deaminase for improving growth and yield of maize (*Zea mays* L.) in the presence of nitrogenous fertilizer. *Soil Biol. Biochem.* **2006**, *38*, 2971–2975. [CrossRef]
21. Ahmad, F.; Ahmad, I.; Khan, M. Indole Acetic Acid Production by the Indigenous Isolates of Azotobacter and Fluorescent Pseudomonas in the Presence and Absence of Tryptophan. *Turk. J. Biol.* **2005**, *29*, 29–34.

22. Rodríguez, H.; Fraga, R. Phosphate solubilizing bacteria and their role in plant growth promotion. *Biotechnol. Adv.* **1999**, *17*, 319–339. [CrossRef]
23. Gusain, Y.S.; Kamal, R.; Mehta, C.M.; Singh, U.S.; Sharma, A.K. Phosphate solubilizing and indole-3-acetic acid producing bacteria from the soil of Garhwal Himalaya aimed to improve the growth of rice. *J. Environ. Biol.* **2015**, *36*, 301.
24. López-Rayo, S.; Sanchis-Pérez, I.; Ferreira, C.M.H.; Lucena, J.J. [S,S]-EDDS/Fe: A new chelate for the environmentally sustainable correction of iron chlorosis in calcareous soil. *Sci. Total Environ.* **2019**, *647*, 1508–1517. [CrossRef]
25. López-Rayo, S.; Valverde, S.; Lucena, J.J. [S, S]-EDDS Ligand as a Soil Solubilizer of Fe, Mn, Zn, and Cu to Improve Plant Nutrition in Deficient Soils. *J. Agric. Food Chem.* **2023**, *71*, 9728–9737. [CrossRef]
26. Nagata, T.; Oobo, T.; Aozasa, O. Efficacy of a bacterial siderophore, pyoverdine, to supply iron to *Solanum lycopersicum* plants. *J. Biosci. Bioeng.* **2013**, *115*, 686–690. [CrossRef]
27. Shanahan, P.; O'Sullivan, D.J.; Simpson, P.; Glennon, J.D.; O'Gara, F. Isolation of 2, 4-diacetylphloroglucinol from a fluorescent pseudomonad and investigation of physiological parameters influencing its production. *Appl. Environ. Microbiol.* **1992**, *58*, 353–358. [CrossRef] [PubMed]
28. Sánchez-Contreras, M.; Martín, M.; Marta, V.; O'Gara, F.; Ildefonso, B.; Rafael, R. Phenotypic Selection and Phase Variation Occur during Alfalfa Root Colonization by *Pseudomonas fluorescens* F113. *J. Bacteriol.* **2002**, *184*, 1587–1596. [CrossRef] [PubMed]
29. Schwyn, B.; Neilands, J.B. Universal chemical assay for the detection and determination of siderophores. *Anal. Biochem.* **1987**, *160*, 47–56. [CrossRef] [PubMed]
30. Mehnert, M.; Retamal-Morales, G.; Schwabe, R.; Vater, S.; Heine, T.; Levicán, G.J.; Schlömann, M.; Tischler, D. Revisiting the Chrome Azurol S Assay for Various Metal Ions. *Solid State Phenom.* **2017**, *262*, 509–512. [CrossRef]
31. Vindeirinho, J.M.; Soares, H.M.V.M.; Soares, E.V. Modulation of Siderophore Production by *Pseudomonas fluorescens* Through the Manipulation of the Culture Medium Composition. *Appl. Biochem. Biotechnol.* **2021**, *193*, 607–618. [CrossRef] [PubMed]
32. Sasirekha, B.; Srividya, S. Siderophore production by *Pseudomonas aeruginosa* FP6, a biocontrol strain for *Rhizoctonia solani* and *Colletotrichum gloeosporioides* causing diseases in chilli. *Agric. Nat. Resources.* **2016**, *50*, 250–256. [CrossRef]
33. Murugappan, R.M.; Aravinth, A.; Rajaroobia, R.; Karthikeyan, M.; Alamelu, M.R. Optimization of MM9 Medium Constituents for Enhancement of Siderophoregenesis in Marine *Pseudomonas putida* Using Response Surface Methodology. *Indian J. Microbiol.* **2012**, *52*, 433–441. [CrossRef] [PubMed]
34. Giannelli, G.; Bisceglie, F.; Pelosi, G.; Bonati, B.; Cardarelli, M.; Antenozio, M.L.; Degola, F.; Visioli, G. Phyto-Beneficial Traits of Rhizosphere Bacteria: In Vitro Exploration of Plant Growth Promoting and Phytopathogen Biocontrol Ability of Selected Strains Isolated from Harsh Environments. *Plants* **2022**, *11*, 230. [CrossRef] [PubMed]
35. Braun, V.; Hantke, K. Recent insights into iron import by bacteria. *Curr. Opin. Chem. Biology.* **2011**, *15*, 328–334. [CrossRef] [PubMed]
36. Cornelis, P. Iron uptake and metabolism in pseudomonads. *Appl. Microbiol. Biotechnol.* **2010**, *86*, 1637–1645. [CrossRef] [PubMed]
37. Mossialos, D.; Meyer, J.; Budzikiewicz, H.; Wolff, U.; Koedam, N.; Baysse, C.; Anjaiah, V.; Cornelis, P. Quinolobactin, a New Siderophore of *Pseudomonas fluorescens* ATCC 17400, the Production of Which Is Repressed by the Cognate Pyoverdine. *Appl. Environ. Microbiol.* **2000**, *66*, 487–492. [CrossRef] [PubMed]
38. Albrecht-Gary, A.; Blanc, S.; Rochel, N.; Ocaktan, A.Z.; Abdallah, M.A. Bacterial Iron Transport: Coordination Properties of Pyoverdin PaA, a Peptidic Siderophore of *Pseudomonas aeruginosa. Inorg. Chem.* **1994**, *33*, 6391–6402. [CrossRef]
39. Dumas, Z.; Ross-Gillespie, A.; Kümmerli, R. Switching between apparently redundant iron-uptake mechanisms benefits bacteria in changeable environments. *Proc. R. Soc. B Biol. Sci.* **2013**, *280*, 20131055. [CrossRef]
40. Crumbliss, A.L. Iron bioavailability and the coordination chemistry of hydroxamic acids. *Coord. Chem. Rev.* **1990**, *105*, 155–179. [CrossRef]
41. Raymond, K.N.; Müller, G.; Matzanke, B.F. Complexation of Iron by Siderophores a Review of Their Solution and Structural Chemistry and Biological Function. In *Structural Chemistry*; Springer: Berlin/Heidelberg, Germany, 1984; pp. 49–102.
42. Neubauer, U.; Nowack, B.; Furrer, G.; Schulin, R. Heavy Metal Sorption on Clay Minerals Affected by the Siderophore Desferrioxamine B. *Environ. Sci. Technol.* **2000**, *34*, 2749–2755. [CrossRef]
43. Gross, H.; Loper, J.E. Genomics of secondary metabolite production by *Pseudomonas* spp. *Nat. Prod. Rep.* **2009**, *26*, 1408–1446. [CrossRef]
44. Raaijmakers, J.M.; De Bruijn, I.; Nybroe, O.; Ongena, M. Natural functions of lipopeptides from Bacillus and Pseudomonas: More than surfactants and antibiotics. *FEMS Microbiol. Rev.* **2010**, *34*, 1037–1062. [CrossRef]
45. Benaissa, A. Plant growth promoting rhizobacteria a review. *Alger. J. Environ. Sci. Technol.* **2019**, *5*.
46. Sehrawat, A.; Sindhu, S.S.; Glick, B.R. Hydrogen cyanide production by soil bacteria: Biological control of pests and promotion of plant growth in sustainable agriculture. *Pedosphere* **2022**, *32*, 15–38. [CrossRef]
47. Bernal, P.; Allsopp, L.P.; Filloux, A.; Llamas, M.A. The Pseudomonas putida T6SS is a plant warden against phytopathogens. *ISME J.* **2017**, *11*, 972–987. [CrossRef] [PubMed]
48. Omoboye, O.O.; Oni, F.E.; Batool, H.; Yimer, H.Z.; De Mot, R.; Höfte, M. Pseudomonas cyclic lipopeptides suppress the rice blast fungus *Magnaporthe oryzae* by induced resistance and direct antagonism. *Front. Plant Sci.* **2019**, *10*, 901. [CrossRef]
49. Gu, S.; Yang, T.; Shao, Z.; Wang, T.; Cao, K.; Jousset, A.; Friman, V.P.; Pommier, T. Siderophore-mediated interactions determine the disease suppressiveness of microbial consortia. *Msystems* **2020**, *5*, 10–1128. [CrossRef]

50. Anand, A.; Chinchilla, D.; Tan, C.; Mène-Saffrané, L.; L'Haridon, F.; Weisskopf, L. Contribution of hydrogen cyanide to the antagonistic activity of Pseudomonas strains against *Phytophthora infestans*. *Microorganisms* **2020**, *8*, 1144. [CrossRef]
51. Kupferschmied, P.; Maurhofer, M.; Keel, C. Promise for plant pest control: Root-associated pseudomonads with insecticidal activities. *Front. Plant Sci.* **2013**, *4*, 287. [CrossRef]
52. Richardson, A.E.; Barea, J. -M.; Mcneill, A.M.; Prigent-Combaret, C. Acquisition of phosphorus and nitrogen in the rhizosphere and plant growth promotion by microorganisms. *Plant Soil* **2009**, *321*, 305–339. [CrossRef]
53. Etesami, H.; Adl, S.M. Plant Growth-Promoting Rhizobacteria (PGPR) and Their Action Mechanisms in Availability of Nutrients to Plants. In *Phyto-Microbiome in Stress Regulation*; Springer: Berlin/Heidelberg, Germany, 2020; pp. 147–203.
54. Hider, R.C.; Kong, X. Chemistry and biology of siderophores. *Nat. Prod. Rep.* **2010**, *27*, 637–657. [CrossRef]
55. Saha, M.; Sarkar, S.; Sarkar, B.; Sharma, B.K.; Bhattacharjee, S.; Tribedi, P. Microbial siderophores and their potential applications: A review. *Environ. Sci. Pollut. Res.* **2016**, *23*, 3984–3999. [CrossRef] [PubMed]
56. Gu, Y.; Ma, Y.N.; Wang, J.; Xia, Z.; Wei, H.L. Genomic insights into a plant growth-promoting *Pseudomonas koreensis* strain with cyclic lipopeptide-mediated antifungal activity. *Microbiol. Open* **2020**, *9*, e1092. [CrossRef] [PubMed]
57. Fernández, L.; Agaras, B.; Zalba, P.; Wall, L.G.; Valverde, C. *Pseudomonas* spp. isolates with high phosphate-mobilizing potential and root colonization properties from agricultural bulk soils under no-till management. *Biol Fertil. Soils* **2012**, *48*, 763–773. [CrossRef]
58. Spaepen, S.; Vanderleyden, J.; Remans, R. Indole-3-acetic acid in microbial and microorganism-plant signaling. *FEMS Microbiol. Rev.* **2007**, *31*, 425–448. [CrossRef] [PubMed]
59. Akram, W.; Anjum, T.; Ali, B. Phenylacetic acid is ISR determinant produced by *Bacillus fortis* IAGS162, which involves extensive re-modulation in metabolomics of tomato to protect against Fusarium wilt. *Front. Plant Sci.* **2016**, *7*, 498. [CrossRef]
60. Kunkel, B.N.; Harper, C.P. The roles of auxin during interactions between bacterial plant pathogens and their hosts. *J. Exp. Bot.* **2018**, *69*, 245–254. [CrossRef]
61. Lea, P.J.; Miflin, B.J. Alternative route for nitrogen assimilation in higher plants. *Nature* **1974**, *251*, 614–616. [CrossRef]
62. Porra, R.J. Recent Progress in Porphyrin and Chlorophyll Biosynthesis. *Photochem. Photobiol.* **1997**, *65*, 492–516. [CrossRef]
63. Cerdán, M.; Sánchez-Sánchez, A.; Jordá, J.D.; Juárez, M.; Sánchez-Andreu, J. Effect of commercial amino acids on iron nutrition of tomato plants grown under lime-induced iron deficiency. *J. Plant Nutr. Soil Sci.* **2013**, *176*, 859–866. [CrossRef]
64. Ringel, M.T.; Brüser, T. The biosynthesis of pyoverdines. *Microbial. Cell.* **2018**, *5*, 424–437. [CrossRef]
65. Adams, M.R. *Fermented Weaning Foods. Microbiology of Fermented Foods*; Wood, B.J.B., Ed.; Springer: Berlin/Heidelberg, Germany, 1998; pp. 790–811. [CrossRef]
66. Zheng, Y.; Wang, X.; Cui, X.; Wang, K.; Wang, Y.; He, Y. Phytohormones regulate the abiotic stress: An overview of physiological, biochemical, and molecular responses in horticultural crops. *Front. Plant Sci.* **2023**, *13*, 1095363. [CrossRef] [PubMed]
67. Xiang, L.; Hu, L.; Xu, W.; Zhen, A.; Zhang, L.; Hu, X. Exogenous γ-Aminobutyric Acid Improves the Structure and Function of Photosystem II in Muskmelon Seedlings Exposed to Salinity-Alkalinity Stress. *PLoS ONE* **2016**, *11*, 0164847. [CrossRef] [PubMed]
68. Scher, F.M.; Baker, R. Effect of *Pseudomonas putida* and a synthetic iron chelator on induction of soil suppressiveness to *Fusarium* wilt pathogens. *Phytopathology* **1982**, *72*, 1567–1573. [CrossRef]
69. Gutiérrez-Barranquero, J.A.; Carrión, V.J.; Murillo, J.; Arrebola, E.; Arnold, D.L.; Cazorla, F.M.; De Vicente, A. A *Pseudomonas syringae* Diversity Survey Reveals a Differentiated Phylotype of the Pathovar syringae Associated with the Mango Host and Mangotoxin Production. *Phytopathology* **2013**, *103*, 1115–1129. [CrossRef] [PubMed]
70. Villén, M.; Lucena, J.J.; Cartagena, M.C.; Bravo, R.; García-Mina, J.; de la Hinojosa, M.I.M. Comparison of Two Analytical Methods for the Evaluation of the Complexed Metal in Fertilizers and the Complexing Capacity of Complexing Agents. *J. Agric. Food Chem.* **2007**, *55*, 5746–5753. [CrossRef] [PubMed]
71. Yunta, F.; García-Marco, S.; Lucena, J.J.; Gómez-Gallego, M.; Alcázar, R.; Sierra, M.A. Chelating agents related to ethylenediamine bis (2-hydroxyphenyl) acetic acid (EDDHA): Synthesis, characterization, and equilibrium studies of the free ligands and their Mg^{2+}, Ca^{2+}, Cu^{2+}, and Fe^{3+} chelates. *Inorg. Chem.* **2003**, *42*, 5412–5421. [CrossRef] [PubMed]
72. Wick, R.R.; Judd, L.M.; Holt, K.E. Performance of neural network basecalling tools for Oxford Nanopore sequencing. *Genome Biol.* **2019**, *20*, 129. [CrossRef]
73. De Coster, W.; D'Hert, S.; Schultz, D.T.; Cruts, M.; Van Broeckhoven, C. NanoPack: Visualizing and processing long-read sequencing data. *Bioinformatics* **2018**, *34*, 2666–2669. [CrossRef]
74. Kolmogorov, M.; Bickhart, D.M.; Behsaz, B.; Gurevich, A.; Rayko, M.; Shin, S.B.; Kuhn, K.; Pevzner, P.A. metaFlye: Scalable long-read metagenome assembly using repeat graphs. *Nat. Methods* **2020**, *17*, 1103–1110. [CrossRef]
75. Tamames, J.; Puente-Sánchez, F. SqueezeMeta, a highly portable, fully automatic metagenomic analysis pipeline. *Front. Microbiol.* **2019**, *9*, 3349. [CrossRef]
76. Meier-Kolthoff, J.P.; Carbasse, J.S.; Peinado-Olarte, R.L.; Göker, M. TYGS and LPSN: A database tandem for fast and reliable genome-based classification and nomenclature of prokaryotes. *Nucleic Acids Res.* **2022**, *50*, D801–D807. [CrossRef] [PubMed]

77. Blin, K.; Shaw, S.; Augustijn, H.E.; Reitz, Z.L.; Biermann, F.; Alanjary, M.; Fetter, A.; Weber, T. antiSMASH 7.0: New and improved predictions for detection, regulation, chemical structures and visualisation. *Nucleic Acids Res.* **2023**, *2023*, gkad344. [CrossRef] [PubMed]
78. Overbeek, R.; Olson, R.; Pusch, G.D.; Olsen, G.J.; Davis, J.J.; Disz, T.; Edwards, D.A.; Stevens, R. The SEED and the Rapid Annotation of microbial genomes using Subsystems Technology (RAST). *Nucleic Acids Res.* **2014**, *42*, D206–D214. [CrossRef] [PubMed]

plants

MDPI

Article

Effect of the Nonpathogenic Strain *Fusarium oxysporum* FO12 on Fe Acquisition in Rice (*Oryza sativa* L.) Plants

Jorge Núñez-Cano [1], Francisco J. Romera [1], Pilar Prieto [2], María J. García [1], Jesús Sevillano-Caño [1], Carlos Agustí-Brisach [1], Rafael Pérez-Vicente [3], José Ramos [4] and Carlos Lucena [1,*]

[1] Departamento de Agronomía (Unit of Excellence 'María de Maeztu' 2020-24), Edificio Celestino Mutis (C-4), Campus de Excelencia Internacional Agroalimentario de Rabanales (ceiA3), Universidad de Córdoba, 14071 Córdoba, Spain; jorgenunezcano@gmail.com (J.N.-C.); ag1roruf@uco.es (F.J.R.); b92gadem@uco.es (M.J.G.); o42secaj@uco.es (J.S.-C.); cagusti@uco.es (C.A.-B.)

[2] Departamento de Mejora Genética, Instituto de Agricultura Sostenible (IAS), Consejo Superior de Investigaciones Científicas (CSIC), 14004 Córdoba, Spain; pilar.prieto@ias.csic.es

[3] Departamento de Botánica, Ecología y Fisiología Vegetal, Edificio Celestino Mutis (C-4), Campus de Excelencia Internacional Agroalimentario de Rabanales (ceiA3), Universidad de Córdoba, 14071 Córdoba, Spain; bv1pevir@uco.es

[4] Departamento de Química Agrícola, Edafología y Microbiología, Edificio Severo Ochoa (C-6), Campus de Excelencia Internacional Agroalimentario de Rabanales (ceiA3), Universidad de Córdoba, 14071 Córdoba, Spain; mi1raruj@uco.es

* Correspondence: b42lulec@uco.es; Tel.: +34-957218488

Abstract: Rice (*Oryza sativa* L.) is a very important cereal worldwide, since it is the staple food for more than half of the world's population. Iron (Fe) deficiency is among the most important agronomical concerns in calcareous soils where rice plants may suffer from this deficiency. Current production systems are based on the use of high-yielding varieties and the application of large quantities of agrochemicals, which can cause major environmental problems. The use of beneficial rhizosphere microorganisms is considered a relevant sustainable alternative to synthetic fertilizers. The main goal of this study was to determine the ability of the nonpathogenic strain *Fusarium oxysporum* FO12 to induce Fe-deficiency responses in rice plants and its effects on plant growth and Fe chlorosis. Experiments were carried out under hydroponic system conditions. Our results show that the root inoculation of rice plants with FO12 promotes the production of phytosiderophores and plant growth while reducing Fe chlorosis symptoms after several days of cultivation. Moreover, Fe-related genes are upregulated by FO12 at certain times in inoculated plants regardless of Fe conditions. This microorganism also colonizes root cortical tissues. In conclusion, FO12 enhances Fe-deficiency responses in rice plants, achieves growth promotion, and reduces Fe chlorosis symptoms.

Keywords: biostimulant; Fe deficiency; phytosiderophores; rhizosphere microorganisms; graminaceous plants

Citation: Núñez-Cano, J.; Romera, F.J.; Prieto, P.; García, M.J.; Sevillano-Caño, J.; Agustí-Brisach, C.; Pérez-Vicente, R.; Ramos, J.; Lucena, C. Effect of the Nonpathogenic Strain *Fusarium oxysporum* FO12 on Fe Acquisition in Rice (*Oryza sativa* L.) Plants. *Plants* **2023**, *12*, 3145. https://doi.org/10.3390/plants12173145

Academic Editor: Ferenc Fodor

Received: 19 July 2023
Revised: 29 August 2023
Accepted: 30 August 2023
Published: 31 August 2023

1. Introduction

It is estimated that the world population will reach approximately 9 billion inhabitants by the year 2050, with an increase in food demand of 70% [1]. Rice cultivation is a very important cereal throughout the world, since it is the staple food for more than half of the world's population. It is cultivated in more than 100 countries and provides more than 20% of the calories consumed worldwide [2]. Currently, production systems are mainly based on the use of high-yield varieties and the application of large amounts of agrochemicals, which leads to unsustainable agriculture [3]. Among the problems caused by these production systems, there is soil and groundwater contamination, imbalance of soil nutrients and reduction of soil biodiversity [4,5]. Given this situation, it is necessary to change to a sustainable production system which is more environmentally friendly and has less dependence on chemical fertilizers [6,7]. For this reason, the development of crop varieties more efficient in nutrient acquisition and a better management of the rhizosphere are necessary [8]. The rhizosphere is the fraction of the soil close to the roots, rich in energy and in which a large number of microbes, like rhizobacteria and fungi, live [9]. Some of these microbes associated with plants could be exploited to achieve promotion of growth and greater plant productivity [10]. Many of these mutualistic microbes release nutrient solubilizing compounds or modify the physiology and architecture of the roots in order to help plants to obtain nutrients like iron (Fe) and others [11–15].

Iron (Fe) is one of the most abundant elements in the earth's lithosphere, but its solubility and availability for plants is low in calcareous soils with pH ranging from 7.4 to 8.5 [16,17]. In this case, plants may suffer from Fe deficiency, showing chlorosis in the youngest leaves [18–20]. Fe is a redox-active metal that is involved in hemoproteins related to electron transfer in photosynthesis and mitochondrial respiration, as well as protection against reactive oxygen species [21]. This nutrient is also involved in other key processes in plant physiology, such as chlorophyll biosynthesis, nitrogen assimilation and the biosynthesis of hormones like gibberellic acid, ethylene and jasmonic acid [21–23].

Plants have evolved two distinct strategies, namely Strategy I and Strategy II, to facilitate the uptake of Fe from the soil. Strategy I is employed by non-graminaceous plants, such as dicots, and involves the reduction of Fe^{3+} to Fe^{2+} before absorption [24,25]. This reduction process is facilitated by a ferric reductase located at the root surface, which is encoded by the *AtFRO2* gene in *Arabidopsis thaliana*. Subsequently, Fe^{2+} is taken up through an Fe^{2+} transporter, which is encoded by the *AtIRT1* gene in *Arabidopsis thaliana* [24,25]. When facing Fe deficiency, these Strategy I plants activate several physiological and morphological responses in their roots. These responses include an increased ferric reductase activity, enhanced capacity for Fe^{2+} uptake, acidification of the rhizosphere (due to H^+-ATPases encoded by *AtAHA* genes in *Arabidopsis*), as well as escalated synthesis and release of organic acids (e.g., malate and citrate) and phenolic compounds (such as coumarins and others) [17,19]. Morphologically, noteworthy adaptations include the formation of subapical root hairs, cluster roots, and transfer cells, all aimed at increasing the root's contact surface with the soil. The enhancement of both morphological and physiological responses is particularly significant in the subapical region of the roots [19,26].

To obtain Fe from the soil, Strategy II plant species release PhytoSiderophores (PS) from their roots, through transporters like the one encoded by the *TOM1* gene in rice, which form stable Fe^{3+} chelates with Fe^{3+} ions in the soil [27]. These Fe^{3+} chelates (Fe^{3+}-PS) are then taken up by specific epidermal root cell plasma membrane transporters, like the one encoded by the *YSL15* gene in rice [25,28,29]. Under Fe-deficient conditions, Strategy II species greatly increase the production and release of PSand the number of Fe^{3+}-PS transporters, and develop other physiological responses [29]. The increased production of PS is related to a higher expression of genes encoding transcription factors, like *IRO2*, which upregulate the expression of genes implicated in PS synthesis, like the *NAAT* gene, encoding the enzyme nicotianamine aminotransferase [30]. Rice, traditionally considered a Strategy II species [29], also presents some characteristics of Strategy I species, such as an enhanced Fe^{2+} uptake through a Fe^{2+} transporter, encoded by the *OsIRT1* gene [27,31,32]. For this reason, some authors consider it a plant species that uses a combined strategy [33–35].

Our group has demonstrated a role for ethylene, whose production increases in Fe-deficient roots, in the regulation of Fe-deficiency responses by Strategy I plant species [25,36]. However, there are few publications relating ethylene to Fe deficiency responses in Strategy II plant species [19,25]. In fact, Romera et al. [37] found that there were no differences in ethylene production between Fe-sufficient and Fe-deficient roots of several Strategy II plant species, like maize, wheat and barley. However, in the roots of rice plants that possesses a combined strategy, ethylene production is also higher under Fe-deficient conditions [38]. These results are reinforced by the discovery that, under Fe deficiency, ethylene synthesis genes, such as *OsACS*, *OsACO*, *OsSAMS* and *OsMTK*, are upregulated in rice roots [29,38,39]. *SAMS* and *MTK* genes are also involved in nicotianamine (NA) and PS synthesis [40,41].

It has been well demonstrated that some beneficial rhizosphere microorganisms, i.e., bacteria and fungi, are able to enhance plant nutrition and growth. These kinds of beneficial microorganisms are called Plant-Growth-Promoting Bacteria or Plant-Growth-Promoting Rhizobacteria (PGPB or PGPR, respectively) and Plant-Growth-Promoting Fungi (PGPF) [9]. Some of them can also boost plant defenses, rendering the entire plant more resistant to pathogens and pests, through a phenomenon called Induced Systemic Resistance (ISR) [9]. Some of the ISR-eliciting microorganisms are also able to enhance plant Fe nutrition by inducing Fe-deficiency responses because both processes (ISR and the responses) are modulated by similar hormones and signaling molecules, like ethylene and NO [19]. In Strategy I plants, the effects of these microorganisms on Fe nutrition are associated with their capacity to upregulate many key Fe-related genes, like *FIT*, *MYB72*, *IRT1*, *FRO2*, and others [9,19]. Nonetheless, their effects on Strategy II plants have been less studied [19].

In some cases, nonpathogenic strains of *Fusarium oxysporum* have been found to trigger Induced Systemic Resistance (ISR) [42,43], providing protection against soilborne pathogens like *Fusarium* spp. wilt and *Verticillium dahliae*-induced wilt, effectively reducing disease symptoms [42–44]. However, it has been proposed that these nonpathogenic strains might also induce a different type of resistance known as Endophytic Mediated Resistance (EMR). EMR occurs when a plant gains resistance against pathogens after being colonized by an endophytic microorganism, such as *Fusarium* spp. [45].

This form of resistance (EMR), unlike ISR, is characterized by endophytic microorganisms that typically do not provide protection against pathogens in the above-ground tissues [46]. Moreover, some studies suggest that ethylene does not play a role in this type of resistance [47]. However, the claim that nonpathogenic strains of *Fusarium oxysporum* induce this resistance (EMR) is a subject of debate due to conflicting evidence. For instance, research has shown that when *Capsicum annuum* plants are inoculated with the nonpathogenic strain of *F. oxysporum* FO47, they gain resistance against *Verticillium dahliae* and experience reduced foliar damage [48]. Constantin et al. [47] have proposed that endophytic microorganisms can induce EMR, a resistance mechanism distinct from ISR, with the intriguing feature of ethylene independence. This assertion implies that the introduction of nonpathogenic strains of F. oxysporum to plant hosts would activate resistance pathways not reliant on ethylene signaling. However, our own research, documented and published by our group, has revealed a contrary finding. Specifically, the inoculation of plants with the FO12 strain resulted in a noticeable augmentation of ethylene-related gene expression at specific intervals [49].

Within this framework, most recently, the nonpathogenic strain *F. oxysporum* FO12 has been characterized not only as a potential biological control agent of Verticillium wilt of olive, a disease caused by the soilborne pathogen *Verticllium dahliae* [44,50–53], but also as a resistance host inducer modulating in parallel the Fe acquisition in Arabidopsis and cucumber plant models [49]. In order to go ahead in generating knowledge on this last evidence, in this study, we have conducted experiments with rice plants (*Oryza sativa* L. cv. Puntal) using the nonpathogenic strain FO12 of *F. oxysporum* to demonstrate whether FO12 could induce Fe-deficiency responses in rice, as other ISR-eliciting fungi do in dicot plants [19].

2. Results

2.1. Effect of the Inoculation with FO12 on Fe Chlorosis

The main symptom of Fe deficiency in plants is an interveinal yellowing of the youngest leaves, known as Fe chlorosis. It is produced because Fe plays an important role in the functioning of some enzymes involved in the synthesis of chlorophyll [23,54]. Rice plants cultivated with Fe presented higher SPAD values than those without Fe (Figure 1). The inoculation caused a clear promotive effect on the SPAD index of plants grown under Fe deficiency but had almost no effect in those grown under Fe sufficiency (Figure 1).

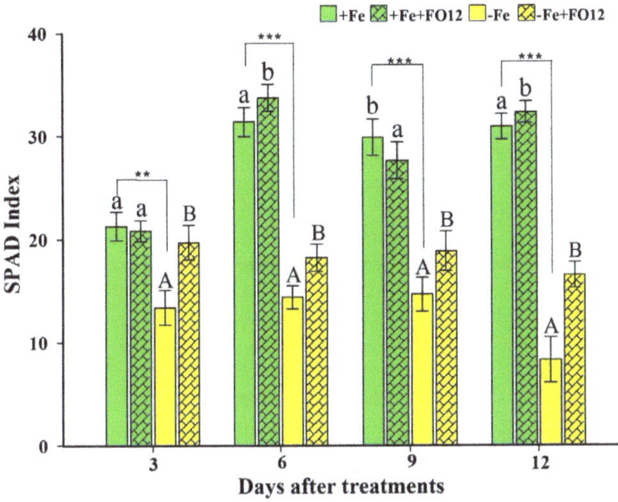

Figure 1. Effect of the inoculation with the nonpathogenic strain *Fusarium oxysporum* FO12 on the SPAD index of rice plants grown under Fe sufficiency or Fe deficiency. SPAD index determinations were carried out at 3, 6, 9 and 12 d after treatments. Treatments: –Fe, –Fe+FO12, +Fe and +Fe+FO12. The values represented are mean ± ES (n = 8). Within each time, ** $p < 0.01$ or *** $p < 0.001$ indicate significant differences between treatments. For each evaluation moment, different lowercase or capital letters indicate significant differences between non-inoculated or inoculated plants with FO12 for +Fe or –Fe treatments, respectively.

2.2. Effect of the Inoculation with FO12 on Growth Promotion

To determine growth promotion, half of the 22 d old rice plants were inoculated with the FO12 strain, and then both inoculated and control plants were cultivated for 12 additional days, either under Fe deficiency (–Fe) or Fe 70 μM (+Fe). After 6 d of the inoculation with FO12, there was a significant growth-promoting effect both in shoots and roots just under Fe sufficiency (Figure 2a,b). However, no changes were observed both in shoots and roots relative to inoculation in Fe-deficient plants (Figure 2a,b).

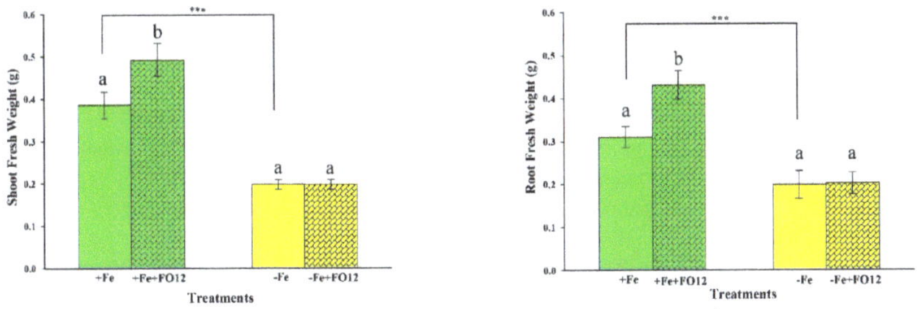

(a) (b)

Figure 2. Effect of Fe deficiency and inoculation with FO12 in the growth of rice plants. (**a**) Shoot fresh weight. (**b**) Root fresh weight. To determine this effect, half of the 22 d old plants were inoculated. Then, both inoculated and control plants were cultivated for 12 additional days, either under Fe sufficiency (+Fe) or Fe deficiency (–Fe). After that time, roots and shoots were excised and weighed separately. The values represented are mean ± ES (n = 8). Different letters indicate significant differences according to Duncan's multiple range test ($p < 0.05$). Similarly, *** $p < 0.001$ indicate significant differences between treatments.

2.3. Effect of the Inoculation with FO12 on Phytosiderophores Production

Regarding PS production, an increasing trend is observed up to 24 h, then it decreases substantially (Figure 3). In the first sampling carried out at 6 h, there are statistical differences between the treatments without Fe. The highest average value is observed in the treatment that was inoculated with FO12. In the case of treatments with Fe, differences between treatments are also observed. However, the plants that were inoculated with FO12 showed a lower average than those that were not inoculated. At 12 h, the plants of the treatments without Fe and inoculated with FO12 presented the highest mean PS production among all the treatments. Likewise, it was the only treatment where statistical differences were observed. The sampling carried out at 24 h was the one where the highest values were obtained in terms of PS production. The plants of the treatments without Fe plus FO12 had the highest means of this sampling. In the same way, statistical differences were observed both in the treatments with and without Fe, the plants of the treatments inoculated with FO12 being the ones that presented the highest values. At 48 h, PS production by the plants declined in all treatments (Figure 3).

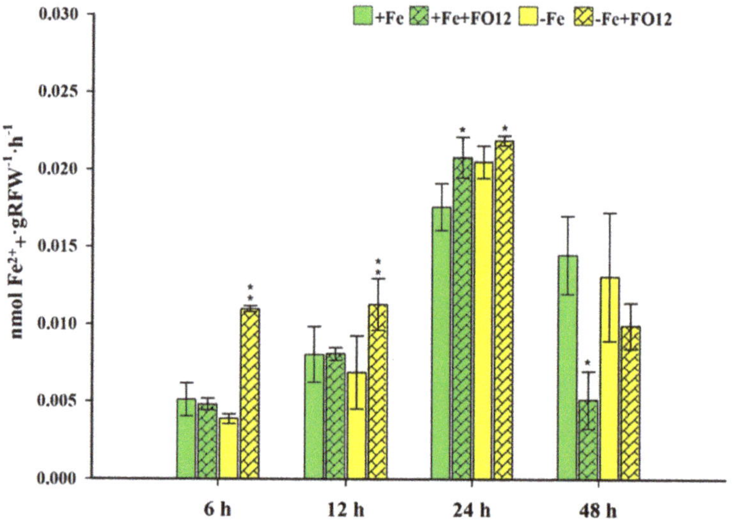

Figure 3. Evolution of phytosiderophore production in rice plants during 48 h of treatments. Four treatments were carried out: plants with Fe (+Fe), plants with Fe and inoculated with FO12 (+Fe+FO12), plants without Fe (−Fe), plants without Fe and inoculated with FO12 (−Fe+FO12). The inoculation was carried out the same day the Fe-deficiency treatment was applied. Within each sampling time, * or ** indicate significant differences ($p < 0.05$ or $p < 0.01$) relative to their respective non-inoculated control plants.

2.4. Effect of the Inoculation with FO12 on Expression of Fe-Related Genes

The relative expressions of PS-related genes (*TOM1*, *IRO2*, *NAAT* and *YSL15*) by the roots of the rice plants, as well as that of the *IRT1* gene, were also analysed (Figure 4). Inoculation with FO12 strain had a clear inductor effect over the expression of all genes analysed in rice plants grown under Fe-deficient conditions. Similarly, most of the genes, except *NAAT* and *IRT1*, presented higher expression in the +Fe+FO12 treatment in relation to the +Fe treatment (control plants) (Figure 4).

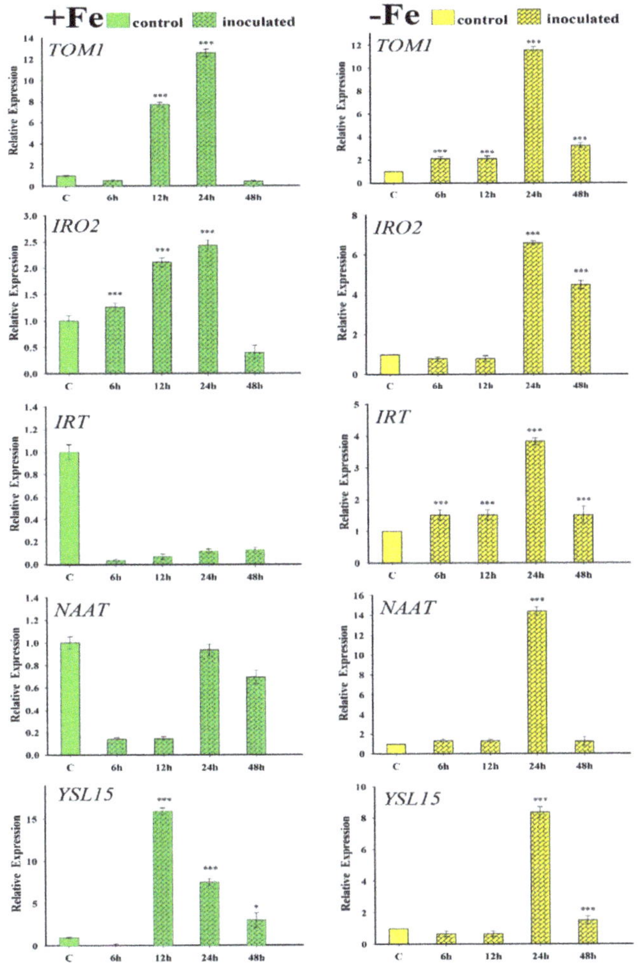

Figure 4. Effect of FO12 on the relative expression of PS-related genes (*TOM1*, *IRO2*, *IRT1*, *NAAT* and *YSL15*) in roots of rice plants. Four treatments were carried out: plants with Fe (+Fe), plants with Fe and inoculated with FO12 (–Fe+FO12), plants without Fe (–Fe), and plants without Fe and inoculated with FO12 (+Fe+FO12). The data represent the mean ± SE of three independent biological replicates and two technical replicates 2 d after treatments. Within each time, * $p < 0.05$ or *** $p < 0.001$ indicate significant differences in relation to the control treatment.

2.5. Iron Concentration in the Plant Substrate Affects FO12 Colonization in Rice Root Tissues

With the aim of assessing the effect of Fe on the colonization process of rice root tissues by FO12, the GFP-tagged *F. oxysporum* was visualized using CLSM in the absence of Fe and compared with the colonization process observed in the presence of Fe at 70 μM concentration. Using the GFP-FO12 in CLSM experiments allowed the in situ visualization of the fungus on/in rice roots without any tissue manipulation. Rice roots were already colonized by conidia of *F. oxysporum* 1 dai (days after inoculation) in both treatments. Conidia were observed already germinating at this time point and did germinate similarly over rice roots in both treatments (–Fe and +Fe). After 4 dai, differences were already observed in GFP-FO12 colonization progress on rice roots, with the colonization of the rice root surface being more profuse in the absence of Fe than in the presence of Fe (Figure 5a,b).

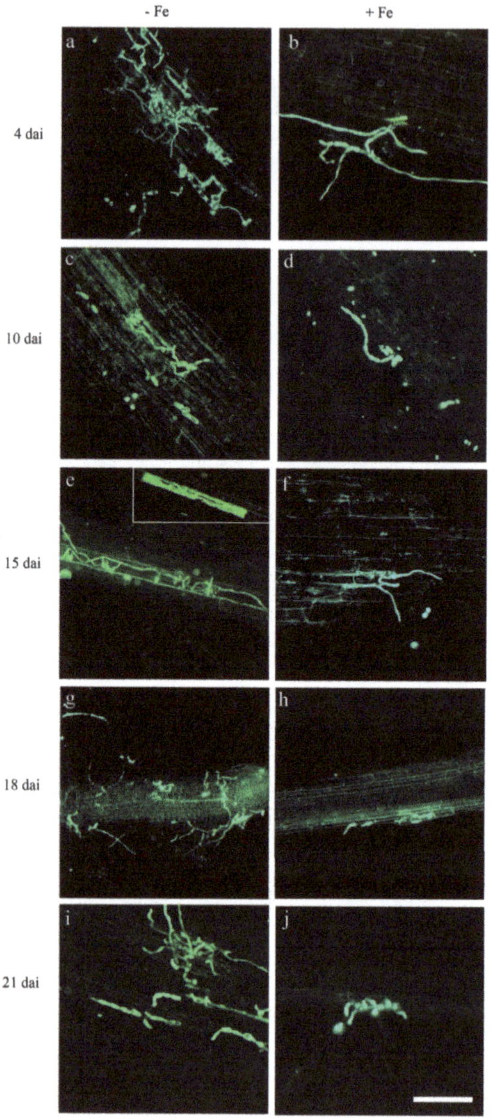

Figure 5. CLSM images of the time-course colonization processes of rice roots by the GFP-FO12 (in green). Confocal analysis was carried out on 4–5 cm long roots to show surface GFP-FO12 colonization. Images are projections of 20 adjacent confocal optical sections. The focal step between confocal optical sections was 0.5 µm. (**a**,**b**) Surface colonization at 4 dai by GFP-FO12 on rice roots of plants (**a**) without Fe, and with a supplement of (**b**) 70 µM Fe. (**c**,**d**) Surface colonization at 10 dai by GFP-FO12 on rice roots of plants growing (**c**) without Fe and (**d**) with an addition of 70 µM Fe. (**e**,**f**) Surface and internal (inset) colonization at 15 dai by GFP-FO12 on rice roots of plants (**e**) without Fe and (**f**) with a supplement of 70 µM Fe. (**g**,**h**) Surface colonization at 18 dai by GFP-FO12 on rice roots of plants (**g**) without Fe and (**h**) with a supplement of 70 µM Fe. (**i**,**j**) Surface colonization at 21 dai by GFP-FO12 on rice roots of plants (**i**) without Fe and (**j**) with a supplement of 70 µM Fe. Rice roots were colonized by GFP-FO12 hyphae in plants growing both in the presence and in the absence of Fe during the whole bioassay.

During the progression of the experiment, GFP-FO12 developed on the root surface at a later stage (10 dai) both in the presence and in the absence of Fe (Figure 5c,d). Fungus conidia showed a slight reduction in the development of hyphae on the root surface of plants growing in the presence of Fe at 15 dai (Figure 5e,f). From this time point, we kept analysing plants by CLSM until 21 dai, although no substantial variations in plants were detected from 10 dai until the end of the experiment. Thus, GFP-FO12 reached a similar degree of hyphae development on the root surface of rice plants from 10 dai until the end of the bioassay. In fact, the visualization of the fungus at 18 dai (Figure 5g,h) showed an equivalent colonization degree to those of the preceding and following days until the end of the experiment (21 dai, Figure 5i,j). As with the observations taken previously, the proportion of hyphae were slightly lower in the root surface of plants growing with Fe at the end of the experiment (Figure 5i,j).

The colonization progress of GFP-FO12 on the rice root surface displayed non-uniformity, both in the presence and absence of Fe. Some regions of the roots exhibited a significant abundance of hyphal colonization, while other areas remained entirely devoid of the fungus. Utilizing Confocal Laser Scanning Microscopy (CLSM), we were able to identify internal colonization of rice roots by GFP-FO12. This internal colonization by GFP-FO12 was evident in the cortical tissue of rice roots starting from the early days of the bioassay and continued until the end, particularly in plants growing without Fe (Figure 5e and inset).

It did not detect GFP-FO12 hyphae proliferating in the vascular tissue at any time. In contrast, GFP-FO12 was only detected on the rice root surface up to the end of the experiment in the roots of rice plants growing with Fe, although we cannot rule out that internal colonization can also occur in plants in the presence of Fe.

Most of the conidia germinated and the proliferation of hyphae could be observed on the root surface in plants growing both in the presence and in the absence of Fe. A slightly higher proliferation of hyphae could be observed on the root surface in plants growing with no concentration of Fe, in the absence of iron from the beginning of the bioassay.

3. Discussion

To obtain Fe from the soil, Strategy II plant species release PhytoSiderophores (PS) from their roots, which form stable Fe^{3+} chelates (Fe^{3+}-PS). PS are released through transporters, like the one encoded by the *TOM1* gene in rice, while the Fe^{3+}-PS are taken up by transporters, like the one encoded by the *YSL15* gene in rice [25,27–29]. Under Fe-deficient conditions, Strategy II species greatly increase the production and release of PS and the number of Fe^{3+}-PS transporters, and develop other physiological and regulatory responses [29]. Rice, traditionally considered a Strategy II species [29], also presents some characteristics of Strategy I species, such as an enhanced Fe^{2+} uptake through a Fe^{2+} transporter, encoded by the *OsIRT1* gene [27,31,32]. For this reason, some authors consider it a plant species that uses a combined strategy [33–35]. Despite the activation of these Fe-deficiency responses, crops can need an added contribution of fertilizers to withstand the enormous pressure that current agriculture exerts on crop production.

Searching for production strategies focused on gradually reducing the dependence on the application of large amounts of chemical products is one of the main challenges of current agriculture. The use of beneficial microorganisms is one of the strategies that is becoming more and more established every day. Microorganisms contribute to enhancing the tolerance of plants to abiotic stresses and to increasing their resistance to pathogens. In the same way, they can promote plant growth and increase the acquisition of nutrients through different mechanisms, such as changes in the soil structure and nutrient solubility, and changes in root morphology and physiology [55–58].

The results of this study show some positive effect of FO12 on plant growth and Fe acquisition by rice plants. The primary indication of Fe deficiency in plants manifests as interveinal yellowing in the most juvenile leaves, a condition referred to as Fe chlorosis. This discoloration arises due to iron's pivotal function within several enzymes engaged in chlorophyll synthesis [23,54]. FO12 can affect the photosynthetic activity of the plants since those growing without Fe that were inoculated presented higher SPAD values (Figure 1). These results are similar to those obtained with inoculation with the fungus *Trichoderma asperellum* SL2 in rice [59]. However, the SPAD values did not show significant differences between control and FO12 inoculated treatments in cucumber plants cultured in a calcareous soil [49].

The effect on the development of the plants was reflected in a greater weight of both the aerial part and the roots (Figure 2). These positive effects have also been identified in various studies with the use of endophytic microbes [60–62]. FO12, in addition to inducing Fe-deficiency responses, exhibits growth-promoting properties, much like other Plant-Growth-Promoting Fungi (PGPF) and Plant-Growth-Promoting Bacteria (PGPB) [9]. Numerous studies have reported similar findings with other beneficial rhizosphere microorganisms. For instance, [63] utilized a combination of four microorganisms to inoculate soybean seeds, resulting in a noticeable growth-promoting effect on the plants. Similarly, Fontenelle et al. [64] observed a highly significant growth promotion when various isolates of *Trichoderma* spp. were applied to tomato plants under greenhouse conditions.

The application of PGPR or PGPF has proven to be an effective method in improving Fe chlorosis in calcareous soils [65–67]. Liu et al. [66] demonstrated the promotion of growth and enhanced mineral uptake in strawberry plants by inoculating them with isolates like *Agrobacterium, Bacillus* and *Alcaligenes.* Likewise, Liu et al. [66] achieved growth promotion in alfalfa plants by inoculating them with *Pseudomonas aeruginosa* and *Enterobacter aerogenes* cultured under saline–alkali conditions in a greenhouse. Moreover, El_Komy et al. [67] ameliorated the symptoms caused by *Fusarium solani, Macrophomina phaseolina* and *Rhizoctonia solani* in sunflower plants cultivated in a calcareous soil under field conditions, showing a clear growth-promoting effect with the inoculation of a mixture of rhizobacteria.

Many microbes produce signals that induce in the plant PS production and hormones that favor Fe acquisition [68]. In this study, the results obtained show that FO12 can induce a greater PS production by rice plants (Figure 3). This positive effect was higher when plants were growing without Fe (Figure 3). In order to summarize our results and the results of other authors [19,49] it seems that FO12 induces Fe-deficiency responses like other ISR-eliciting microorganisms.

According to Constantin et al. [47], endophytic micro-organisms induce EMR. These authors suggested that this type of resistance is independent of ethylene, in contrast to ISR. This statement means that the colonization of the plant by nonpathogenic strains of *F. oxysporum* would induce resistance in an ethylene-independent manner. However, Aparicio et al. [49] showed that the expression of ethylene-related genes was evidently enhanced at certain times under inoculation with FO12 in cucumber plants. Additionally, NO levels were also increased with the inoculation in the subapical region of the roots. Kavroulakis et al. [69] demonstrated that ethylene-deficient mutants of tomato inoculated with *F. solani* were more susceptible to pathogen attack, supporting a role for ethylene in the acquisition of resistance against *Fusarium*. Furthermore, NO and ethylene enhance the expression of several Fe acquisition genes in *Arabidopsis thaliana* [70,71]. Given that FO12 has been shown to upregulate ethylene-related genes and enhance NO production [49], it is plausible to consider that FO12 might induce iron deficiency responses in an ethylene/NO-dependent manner. Consequently, this implies that the FO12 strain could trigger Induced Systemic Resistance (ISR) rather than Endophytic-Mediated Resistance (EMR). Something similar could happen with rice plants, traditionally considered a Strategy II species but that possesses some characteristics of Strategy I species, in which several authors have shown that ethylene can also play a role in the regulation of some of its Fe-deficiency responses [25].

The expression of Fe-related genes in rice plants was evaluated (Figure 4). Results indicated a more enhanced expression in the presence of FO12. In cucumber plants, the FO12 strain induces the upregulation of Fe-related genes, including *CsFRO1, CsIRT1,* and *CsHA1* [49]. Similar effects have been observed in *Arabidopsis thaliana* when root-inoculated with the WCS417 strain of *Pseudomonas simiae*, leading to the increased expression of *MYB72*, *FRO2* and *IRT1* [11,72]. Moreover, the *Paenibacillus polymyxa* BFKC01 strain was found to promote growth and enhance the expression of *FRO2*, *FIT* and *IRT1* in *A. thaliana* [73]. Additionally, exposure of tomato plants to *Trichoderma* volatiles induced the expression of Fe-deficiency genes, such as *LeFRO*, *LeIRT*, and *LeFER* [74].

Fusarium oxysporum, being an endophytic microorganism, can grow inside the plant but typically does not colonize the vascular system. Instead, it forms fungal hyphae along the root cortex and endodermis [45]. Our findings align with this understanding, as the FO12 strain was observed to colonize the intercellular spaces of the cortical cells (Figure 5), supporting its endophytic nature. Notably, the inoculation of rice plants with the FO12 strain is facilitated by depleting Fe from the nutrient solution. On the other hand, the cortical tissue of rice roots was internally colonized by GFP-FO12 from early days until the end of the bioassay when plants were growing without Fe (Figure 5e, inset). These results agree with those obtained by Guerra & Anderson [75], showing that *Phaseolus vulgaris* plants growing in a hydroponic system with restricted Fe and B were more susceptible against *Fusarium* wilt. Similar results were obtained recently by Aparicio et al. [49], analyzing the time-course colonization processes of cucumber roots by the FO12-GFP-tagged strain. Collectively, these findings suggest that the depletion of Fe in the plant promotes endophytic colonization by non-pathogenic strains, such as FO12.

4. Materials and Methods

4.1. Seed Germination and Plant Cultivation

Experiments were carried out with rice (*Oriza sativa* L. var. 'Puntal') plants. Seeds were surface sterilized as described by Aparicio et al. [49]. Then, seedlings were transferred to a hydroponic system. Each of eight seedling groups were inserted in plastic lids and held in the holes of a thin polyurethane raft floating on an aerated nutrient solution containing 2 mM $Ca(NO_3)_2$, 0.75 mM K_2SO_4, 0.65 mM $MgSO_4$, 0.5 mM KH_2PO_4, 50 µM KCl, 10 µM H_3BO_3, 1 µM $MnSO_4$, 0.5 µM $CuSO_4$, 0.5 µM $ZnSO_4$, 0.05 µM $(NH_4)_6Mo_7O_{24}$ and 45 µM Fe-EDTA. When the plants were 22 d old, different treatments were applied: (1) +Fe (complete nutrient solution with 70 µM Fe-EDTA); (2) +Fe +FO12 (+Fe treatment + FO12 inoculum); (3) –Fe (nutrient solution without Fe-EDTA); and (4) –Fe +FO12 (–Fe treatment + FO12 inoculum). Plants were maintained in the treatments from 4 to 21 d, depending on the experiments.

4.2. Cultivation of Fungus and Inoculum Preparation

The FO12 strain and GFP-FO12, which was transformed with GFP (Green Fluorescent Protein) [76], were generously provided by the 'Patología Agroforestal' group from the 'Universidad de Córdoba'. The FO12 strain was cultured in 250 mL of Potato Dextrose Broth (PDB, Scharlau) in 1-L flasks. Following a 4-day incubation period at 28 °C with continuous shaking at 110 rpm, the culture was filtered using a sterile nylon filter with a pore size of 10 µM (NY-LON_10, *Filtra Vibración*) to separate the spores from the mycelia. The spores were then centrifuged at 10,000 rpm for 10 min and subsequently resuspended in 4 mL of sterile distilled water. Prior to inoculation, the spore concentration was determined by counting them in a Neubauer chamber. As for GFP-FO12, the same culture method was employed, with the addition of 25 ppm of Hygromycin to maintain selective pressure.

4.3. Plant Inoculation

Following the method by Navarro-Velasco et al. [77], 22-day-old rice plants underwent root immersion for inoculation. Initially, a solution containing 10^7 spores/mL was prepared using distilled water and poured into a 1.5 L capacity tray. To prevent contact between the solution and shoots, certain lanes were covered with insulating tape. The roots of the plants were then submerged in the solution for a duration of 30 min. Throughout the process, constant gentle shaking of the system was maintained to prevent spore precipitation. After inoculation, the plants were subsequently subjected to the mentioned treatments: +Fe or –Fe. For each treatment, there was a corresponding control group that was not inoculated.

4.4. Physiological and Morphological Assessments

Plants were periodically harvested (at 4, 8, 12 and 16 d after treatments) to observe growth promotion, by determining shoot height and both shoot and root fresh weights. Furthermore, in this study, Fe chlorosis was evaluated in the rice plants at 3, 6, 9 and 12 d after treatments by determining the chlorophyll level of the rice plants by the SPAD index using a portable chlorophyll meter, Minolta SPAD-502. Four readings were made per plant, taken over the center area of fully developed apical leaves, assigning the average value to each plant.

Phytosiderophore (PS) assessments were carried out at 6, 12, 24, 48 and 72 h after the treatment´s application, using the methodology described by Inal et al. [78] and Reichman et al. [79]. Rice plants were removed from the treatment and washed three times with deionized water. They were then placed in containers with 70 mL of deionized water for 3 h with constant aeration. From the root exudates produced in these 3 h, 9 mL aliquots were taken in vials, where 0.5 mL of 0.6 μM $FeCl_3$ was added. They were stirred for 1 h to form the Fe (III)-PS compounds. Immediately afterwards, 1 mL of 1.0 M sodium acetate buffer (pH 7.0) was added to the solutions, and the mixture was stirred for 15 min to precipitate the remaining Fe (III). Next, the solutions were filtered through a 0.2 μm filter to remove any solid particles, and then 0.25 mL of 6 M HCL and 0.5 mL of 80 g L^{-1} hydroxylamine hydrochloride were added to reduce Fe (III) to Fe (II). The solutions were then placed in an oven and maintained at a temperature between 50–60 °C for 30 min. After the incubation, 0.25 mL of 2.5 g L^{-1} and 1 mL of 2.0 M sodium acetate buffer (pH 4.7) were introduced to the mixture. Finally, the contents of the tubes were mixed by shaking them briefly for 5 min.

The absorbance was determined at 562 nm. PS release rates were calculated as Fe equivalents.

After PS determination, roots were collected and kept at −80 °C for gene expression determination using primer pairs shown on Table 1.

Table 1. Primer pairs used for rice gene expression analysis.

Gene	Sequence 5-3
OsNAAT1	Forward: TAAGAG GATAATTGATTTGCTTAC
	Reverse: CTG ATCATTCCAATCCTAGTACAAT
OsYSL15	Forward: AACATAAGGGGGACTG GTAC
	Reverse: TGATTACCGCAATGATGCTTAG
OsIRO2	Forward: CTCCCATCGTTTCGGCTACCT
	Reverse: GCTGGGCACTCCTCGTTGATC
OsTOM1	Forward: GCCCAAGAACGCCAAAATGA
	Reverse: GGCTTGAAGGTCAACGCAAG
OsIRT1	Forward: CGTC TTCTTCTTCTCCACCACGAC
	Reverse: GCAGCTGATGATCGAGTCTG ACC
OsActin	Forward: TGCTATGTACGTCGC CATCCAG
	Reverse: AATGAGTAACCACGCTCCGTCA

4.5. *Fusarium oxysporum* Colonization Studies in Rice Roots

The GFP-tagged *F. oxysporum* FO12 transformant (GFP-FO12) was used to monitor the infection and colonization process of an entire rice plant growing in hydroponic system. Fifty plants were inoculated with the GFP-FO12, and the colonization of root tissue samples was visualized by CLSM (confocal laser scanner microscopy) over the 21 d post-inoculation period.

Two different treatments were set up in the presence of the GFP-FO12 to assess the effect of the concentration of Fe in the GFP-FO12 colonization process of root tissues. Thus, 25 inoculated plants were growth in the absence of Fe and 25 inoculated plants were grown with 70 μM Fe concentration. Two plants per treatment were analysed per day until 7 d after inoculation (7 dai). From this time point, two plants per treatment were analysed every 2 days until the end of the bioassay (21 dai). All the inoculated plants were examined by CLSM.

Rice tissue samples for microscopic studies were prepared according to the protocol previously described by Prieto et al. [79]. Rice roots were thin enough to perform CLSM analysis without vibratome sectioning. Therefore, whole roots were used to visualize the *F. oxysporum* colonization process. At least ten different roots per plant were mounted in a slice with distilled water to perform CLSM analysis.

Whole root tissues from the different treatments were used to collect single confocal optical sections using an Axioskop 2 MOT microscope (Carl Zeiss, Jena, Germany) equipped with a krypton and an argon laser, controlled by Carl Zeiss Laser Scanning System LSM5 PASCAL software (Carl Zeiss). GFP-FO12 was visualized using a 488 nm argon laser light (detection at 500–520 nm). Finally, data were recorded and the images relocated for analysis to Zeiss LSM Image Browser version 4.0 (Carl Zeiss). Confocal stacks were mounted and analysed to assess colonization of GFP-FO12. Images included in Figure 5 were obtained from projections of adjacent confocal optical sections. Final figures were handled with PhotoShop 10.0 software (Adobe Systems, San Jose, CA, USA).

4.6. *qRT-PCR Analysis*

Genes related to PS production by root cells of Fe-deficient rice plants were analysed. The following genes were analysed: *OsTOM1*, a deoxymugineic acid (DMA) efflux transporter; *OsNAAT*, which participates in DMA biosynthesis for Fe (III)-DMA uptake and translocation; *OsYSL15*, a Fe (III)-DMA transporter; *OsIRT1*, a Fe^{2+} transporter; and *OsIRO2*, which is an essential regulator involved in mediation of Fe uptake.

Roots were first ground into a fine powder using a mortar and pestle in liquid nitrogen. Total RNA was then extracted from the powdered roots using Tri Reagent solution (Molecular Research Center, Inc., Cincinnati, OH, USA), following the manufacturer's instructions. To generate cDNA, 3 μg of DNase-treated root RNA was reverse transcribed using M-MLV reverse transcriptase (Promega, Madison, WI, USA) and random hexamers for amplification. For the study of gene expression, quantitative real-time polymerase chain reaction (qRT-PCR) was performed on a qRT-PCR Bio-Rad CFX connect thermal cycler. The amplification profile involved cycles with the following conditions: initial denaturation and polymerase activation at 95 °C for 3 min, followed by amplification and quantification at 90 °C for 10 s, 57 °C for 15 s, and 72 °C for 30 s. A final melting curve stage was performed from 65 to 95 °C with an increment of 0.5 °C for 5 s to ensure the absence of primer dimer or nonspecific amplification products. The PCR reactions were set up in 20 μL of SYBR Green Bio-RAD PCR Master Mix, following the manufacturer's instructions. To detect any contamination in the reaction components, controls containing water instead of cDNA were included. For normalization of gene expression, a reference gene (*OsActin*) was used. The specific primers utilized in the qRT-PCR analysis are listed in Table 1.

4.7. *Statistical Analysis*

The statistical analyses were carried out using IBM SPSS Statistics 25. To assess the normal distribution of the variables studied, the Shapiro–Wilk normality test was applied. If the significance value was greater than 0.05, the data were considered to follow a normal distribution (parametric). Conversely, if the significance value was below 0.05, the data were considered non-parametric. For comparisons between the inoculated and control treatments, either Student's *t*-test (parametric) or the Mann–Whitney test (non-parametric) was used to determine significant differences ($p < 0.05$). To compare gene expression between control and inoculated treatments at different times, analysis of variance (One-way ANOVA) and Dunnett's test were utilized, setting the threshold for statistical significance at $p < 0.05$.

5. Conclusions

In conclusion, the results of this study demonstrate that the FO12 strain exhibits promising characteristics as a biofertilizer for rice plants. Its ability to colonize rice roots under Fe-deficient conditions, induce the expression of Fe-relative genes, increase phytosiderophore production and promote plant growth and development highlights its potential as an effective Fe biofertilizer. However, further research is necessary to fully understand its mechanisms of action and to optimize its application for agricultural purposes. Continued investigation and experimentation will be crucial in harnessing the full potential of the FO12 strain as a biofertilizer to enhance Fe uptake and improve crop productivity in rice and potentially other agricultural systems. Our results also indicate that the FO12 strain can indeed induce Fe-deficiency responses in rice plants, similarly to other ISR-eliciting microbes.

Author Contributions: J.N.-C. performed experiments and wrote the paper. C.L. and F.J.R. helped with the experimental procedures. J.S.-C. and C.A.-B. helped with statistical analysis. P.P. performed colonization analysis of roots. R.P.-V. and M.J.G. conducted analysis of gene expression. C.L., J.R. and F.J.R. helped with the design of experiments and revised the manuscript. All authors have read and agreed to the published version of the manuscript.

Funding: Open Access funding provided thanks to the CRUE-CSIC and Universidad de Cordoba/CBUA agreement with the Review Plants.

Institutional Review Board Statement: Not applicable.

Data Availability Statement: The original contributions presented in the study are included in the article.: All the data included in this article are publicly available. Further inquiries can be directed to the corresponding author.

Acknowledgments: This work was possible thanks to the support of the 'Plan propio de investigación de la Universidad de Córdoba', the 'programa operativo de fondos FEDER Andalucía', and the 'Secretaría Nacional de Ciencia, Tecnología e Innovación' (SENACYT) de Panamá. We also acknowledge the financial support of MICINN, the Spanish State Research Agency, through the Severo Ochoa and María de Maeztu Program for Centres and Units of Excellence in R&D (Ref. CEX2019-000968-M).

Conflicts of Interest: The authors declare no conflict of interest.

References

1. United Nations Department of Economic and Social Affairs Population Division. *World Population Prospects*; Summary of Results; United Nations: New York, NY, USA, 2022.
2. Fukagawa, N.K.; Ziska, L.H. Rice: Importance for Global Nutrition. *J. Nutr. Sci. Vitaminol.* **2019**, *65*, S2–S3. [CrossRef] [PubMed]
3. Horrigan, L.; Lawrence, R.S.; Walker, P. How sustainable agriculture can address the environmental and human health harms of industrial agriculture. *Environ. Health Perspect.* **2002**, *110*, 445–456. [CrossRef] [PubMed]
4. Dubey, V. Impact of continuous use of chemical fertilizer. *Int. J. Eng. Res. Dev.* **2012**, *3*, 13–16.
5. Savci, S. Investigation of effect of chemical fertilizers on environment. *APCBEE Procedia* **2012**, *1*, 287–292. [CrossRef]
6. Bhardwaj, D.; Ansari, M.W.; Sahoo, R.K.; Tuteja, N. Biofertilizers function as key player in sustainable agriculture by improving soil fertility, plant tolerance and crop productivity. *Microb. Cell Factories* **2014**, *13*, 66. [CrossRef]
7. FAO. *The Future of Food and Agriculture—Alternative Pathways to 2050*; FAO: Rome, Italy, 2018; (CC BY-NC-SA 3.0 IGO).
8. Shen, J.; Li, C.; Mi, G.; Li, L.; Yuan, L.; Jiang, R.; Zhang, F. Maximizing root/rhizosphere efficiency to improve crop productivity and nutrient use efficiency in intensive agriculture of China. *J. Exp. Bot.* **2013**, *64*, 1181–1192. [CrossRef]
9. Pieterse, C.M.J.; Zamioudis, C.; Berendsen, R.L.; Weller, D.M.; Van Wees, S.C.; Bakker, P.A. Induced systemic resistance by beneficial microbes. *Annu. Rev. Phytopathol.* **2014**, *52*, 347–375. [CrossRef]

10. Pathania, P. Role of plant growth—Promoting bacteria in sustainable agriculture. *Biocatal. Agric. Biotechnol.* **2020**, *30*, 101842. [CrossRef]
11. Zamioudis, C.; Hanson, J.; Pieterse, C.M.J. β-Glucosidase BGLU42 is a MYB72-dependent key regulator of rhizobacteria-induced systemic resistance and modulates iron deficiency responses in *Arabidopsis* roots. *New Phytol.* **2014**, *204*, 368–379. [CrossRef]
12. Contreras-Cornejo, H.A.; López-Bucio, J.S.; Méndez-Bravo, A.; Macías-Rodríguez, L.; Ramos-Vega, M.; Guevara-García, Á.A.; López-Bucio, J. Mitogen-Activated protein kinase 6 and ethylene and auxin signaling pathways are involved in *Arabidopsis* root-system architecture alterations by *Trichoderma atroviride*. *Mol. Plant-Microbe Interact.* **2015**, *28*, 701–710. [CrossRef]
13. García-López, A.M.; Avilés, M.; Delgado, A. Effect of various microorganisms on phosphorus uptake from insoluble Ca-phosphates by cucumber plants. *J. Plant Nutr. Soil Sci.* **2016**, *179*, 454–465. [CrossRef]
14. Garnica-Vergara, A.; Barrera-Ortiz, S.; Muñoz-Parra, E.; Raya-González, J.; Méndez-Bravo, A.; Macías-Rodríguez, L.; Ruiz-Herrera, L.F.; López-Bucio, J. The volatile 6-pentyl-2H-pyran-2-one from *Trichoderma atroviride* regulates *Arabidopsis thaliana* root morphogenesis via auxin signaling and ETHYLENE INSENSITIVE 2 functioning. *New Phytol.* **2016**, *209*, 1496–1512. [CrossRef]
15. Verbon, E.H.; Trapet, P.L.; Stringlis, I.A.; Kruijs, S.; Bakker, P.A.; Pieterse, C.M.J. Iron and Immunity. *Annu. Rev. Phytopathol.* **2017**, *55*, 355–375. [CrossRef]
16. Briat, J.F.; Dubos, C.; Gaymard, F. Iron nutrition, biomass production, and plant product quality. *Trends Plant Sci.* **2015**, *20*, 33–40. [CrossRef] [PubMed]
17. Vélez-Bermúdez, I.C.; Schmidt, W. Plant strategies to mine iron from alkaline substrates. *Plant Soil* **2022**, *483*, 1–25. [CrossRef]
18. Loeppert, R.H. Reactions of iron and carbonates in calcareous soils. *J. Plant Nutr.* **1986**, *9*, 195–214. [CrossRef]
19. Romera, F.J.; García, M.J.; Lucena, C.; Martínez-Medina, A.; Aparicio, M.A.; Ramos, J.; Alcántara, E.; Angulo, M.; Pérez-Vicente, R. Induced systemic resistance (ISR) and Fe deficiency responses in dicot plants. *Front. Plant Sci.* **2019**, *10*, 287. [CrossRef]
20. Taalab, A.; Ageeb, G.; Siam, H.S.; Mahmoud, S.A. Some Characteristics of Calcareous soils. *Middle East J.* **2019**, *8*, 96–105.
21. Marschner, P. Rhizosphere Biology. In *Marschner's Mineral Nutrition of Higher Plants*; Elsevier: Amsterdam, The Netherlands, 2012; pp. 369–388.
22. Hänsch, R.; Mendel, R.R. Physiological functions of mineral micronutrients (Cu, Zn, Mn, Fe, Ni, Mo, B, Cl). *Curr. Opin. Plant Biol.* **2009**, *12*, 259–266. [CrossRef]
23. Kroh, G.E.; Pilon, M. Regulation of Iron Homeostasis and Use in Chloroplasts. *Int. J. Mol. Sci.* **2020**, *21*, 3395. [CrossRef]
24. Brumbarova, T.; Bauer, P.; Ivanov, R. Molecular mechanisms governing Arabidopsis iron uptake. *Trends Plant Sci.* **2015**, *20*, 124–133. [CrossRef]
25. Lucena, C.; Romera, F.J.; García, M.J.; Alcántara, E.; Pérez-Vicente, R. Ethylene participates in the regulation of Fe deficiency responses in Strategy I plants and in rice. *Front. Plant Sci.* **2015**, *6*, 1056. [CrossRef]
26. Li, W.; Lan, P. The understanding of the plant iron deficiency responses in Strategy I plants and the role of ethylene in this process by omic approaches. *Front. Plant Sci.* **2017**, *8*, 40. [CrossRef]
27. Kobayashi, T.; Nakanishi Itai, R.; Nishizawa, N.K. Iron deficiency responses in rice roots. *Rice* **2014**, *7*, 27. [CrossRef]
28. Mori, S. Iron acquisition by plants. *Curr. Opin. Plant Biol.* **1999**, *2*, 250–253. [CrossRef]
29. Kobayashi, T.; Nishizawa, N.K. Iron uptake, translocation, and regulation in higher plants. *Annu. Rev. Plant Biol.* **2012**, *63*, 131–152. [CrossRef]
30. Inoue, H.; Takahashi, M.; Kobayashi, T.; Suzuki, M.; Nakanishi, H.; Mori, S.; Nishizawa, N.K. Identification and localisation of the rice nicotianamine aminotransferase gene OsNAAT1 expression suggests the site of phytosiderophore synthesis in rice. *Plant Mol. Biol.* **2008**, *66*, 193–203. [CrossRef]
31. Ishimaru, Y.; Suzuki, M.; Tsukamoto, T.; Suzuki, K.; Nakazono, M.; Kobayashi, T.; Wada, Y.; Watanabe, S.; Matsuhashi, S.; Takahashi, M.; et al. Rice plants take up iron as an Fe^{3+}-phytosiderophore and as Fe^{2+}. *Plant J.* **2006**, *45*, 335–346. [CrossRef]
32. Ishimaru, Y.; Kakei, Y.; Shimo, H.; Bashir, K.; Sato, Y.; Sato, Y.; Uozumi, N.; Nakanishi, H.; Nishizawa, N.K. A Rice Phenolic Efflux Transporter Is Essential for Solubilizing Precipitated Apoplasmic Iron in the Plant Stele. *J. Biol. Chem.* **2011**, *286*, 24649–24655. [CrossRef]
33. Ricachenevsky, F.K.; Sperotto, R.A. There and back again, or always there? The evolution of rice combined strategy for Fe uptake. *Front. Media SA* **2014**, *5*, 189. [CrossRef]
34. Wairich, A.; De Oliveira, B.H.N.; Arend, E.B.; Duarte, G.L.; Ponte, L.R.; Sperotto, R.A.; Ricachenevsky, F.K.; Fett, J.P. The combined strategy for iron uptake is not exclusive to domesticated rice (*Oryza sativa*). *Sci. Rep.* **2019**, *9*, 16144. [CrossRef] [PubMed]
35. Rai, S.; Singh, P.K.; Mankotia, S.; Swain, J.; Satbhai, S.B. Iron homeostasis in plants and its crosstalk with copper, zinc, and manganese. *Plant Stress* **2021**, *1*, 100008. [CrossRef]
36. Romera, F.J.; Lucena, C.; García, M.J.; Alcántara, E.; Pérez-Vicente, R. *Stress Signaling in Plants: Genomics and Proteomics Perspectives*; Springer Science and Business Media LLC.: Berlin/Heidelberg, Germany, 2017.
37. Romera, F.J.; Alcantara, E.; De La Guardia, M.D. Ethylene production by Fe-deficient roots and its involvement in the regulation of Fe-deficiency stress responses by Strategy I plants. *Ann. Bot.* **1999**, *83*, 51–55. [CrossRef]
38. Wu, J.; Wang, C.; Zheng, L.; Wang, L.; Chen, Y.; Whelan, J.; Shou, H. Ethylene is involved in the regulation of iron homeostasis by regulating the expression of iron-acquisition-related genes in *Oryza sativa*. *J. Exp. Bot.* **2011**, *62*, 667–674. [CrossRef]
39. Itai, R.N.; Ogo, Y.; Kobayashi, T.; Nakanishi, H.; Nishizawa, N.K. Rice genes involved in phytosiderophore biosynthesis are synchronously regulated during the early stages of iron deficiency in roots. *Rice* **2013**, *6*, 16. [CrossRef]

40. Suzuki, M.; Takahashi, M.; Tsukamoto, T.; Watanabe, S.; Matsuhashi, S.; Yazaki, J.; Kishimoto, N.; Kikuchi, S.; Nakanishi, H.; Mori, S. Biosynthesis and secretion of mugineic acid family phytosiderophores in zinc-deficient barley. *Plant J.* **2006**, *48*, 85–97. [CrossRef]
41. Li, Y.; Wang, N.; Zhao, F.; Song, X.; Yin, Z.; Huang, R.; Zhang, C. Changes in the transcriptomic profiles of maize roots in response to iron-deficiency stress. *Plant Mol. Biol.* **2014**, *85*, 349–363. [CrossRef]
42. Alabouvette, C.; Olivain, C. Modes of action of non-pathogenic strains of *Fusarium oxysporum* in controlling Fusarium wilts. *Plant Prot. Sci.* **2002**, *38*, 195–199. [CrossRef]
43. Patil, S.; Sriram, S.; Savitha, M.J.; Arulmani, N. Induced systemic resistance in tomato by non-pathogenic *Fusarium* species for the management of *Fusarium* wilt. *Arch. Phytopathol. Plant Prot.* **2011**, *44*, 1621–1634. [CrossRef]
44. Varo, A.; Raya-Ortega, M.C.; Trapero, A. Selection and evaluation of micro-organisms for biocontrol of *Verticillium dahliae* in olive. *J. Appl. Microbiol.* **2016**, *121*, 767–777. [CrossRef]
45. Constantin, M.E.; Vlieger, B.V.; Takken, F.L.W.; Rep, M. Diminished pathogen and enhanced endophyte colonization upon coinoculation of endophytic and pathogenic *Fusarium* strains. *Microorganisms* **2020**, *8*, 544. [CrossRef]
46. De Lamo, F.J.; Takken, F.L. Biocontrol by *Fusarium oxysporum* using endophyte-mediated resistance. *Front. Plant Sci.* **2020**, *11*, 37. [CrossRef]
47. Constantin, M.E.; de Lamo, F.J.; Vlieger, B.V.; Rep, M.; Takken, F.L.W. Endophyte-mediated resistance in tomato to *Fusarium oxysporum* is independent of ET, JA, and SA. *Front. Plant Sci.* **2019**, *10*, 979. [CrossRef] [PubMed]
48. Veloso, J.; Díaz, J. *Fusarium oxysporum* Fo47 confers protection to pepper plants against *Verticillium dahliae* and *Phytophthora capsici*, and induces the expression of defence genes. *Plant Pathol.* **2012**, *61*, 281–288. [CrossRef]
49. Aparicio, M.A.; Lucena, C.; García, M.J.; Ruiz-Castilla, F.J.; Jiménez-Adrián, P.; López-Berges, M.S.; Prieto, P.; Alcántara, E.; Pérez-Vicente, R.; Ramos, J.; et al. The nonpathogenic strain of *Fusarium oxysporum* FO12 induces Fe deficiency responses in cucumber (*Cucumis sativus* L.) plants. *Planta* **2023**, *257*, 50. [CrossRef] [PubMed]
50. Mulero-Aparicio, A.; Agustí Brisach, C.; Varo, A.; López-Escudero, F.J.; Trapero, A. A non-pathogenic strain of *Fusarium oxysporum* as a potential biocontrol agent against Verticillium wilt of olive. *Biol. Control* **2019**, *139*, 104045. [CrossRef]
51. Mulero-Aparicio, A.; Cernava, T.; Turrà, D.; Schaefer, A.; Di Pietro, A.; López-Escudero, F.J.; Trapero, A.; Berg, G. The Role of Volatile Organic Compounds and Rhizosphere Competence in Mode of Action of the Non-pathogenic Fusarium oxysporum FO12 Toward Verticillium Wilt. *Front. Microbiol.* **2019**, *10*, 1808. [CrossRef]
52. Mulero-Aparicio, A.; Trapero, F.J.; López-Escudero, F.J. A nonpathogenic strain of *Fusarium oxysporum* and grape marc compost control Verticillium wilt of olive. *Phytopathol. Mediterr.* **2020**, *59*, 159–167. [CrossRef]
53. Mulero-Aparicio, A.; Varo, A.; Agustí-Brisach, C.; López-Escudero, F.J.; Trapero, A. Biological control of Verticillium wilt of olive in the field. *Crop Prot.* **2020**, *128*, 104993. [CrossRef]
54. Tottey, S.; Block, M.A.; Allen, M.; Westergren, T.; Albrieux, C.; Scheller, H.V.; Merchant, S.; Jensen, P.E. *Arabidopsis* CHL27, located in both envelope and thylakoid membranes, is required for the synthesis of protochlorophyllide. *Proc. Natl. Acad. Sci. USA* **2003**, *100*, 16119–16124. [CrossRef]
55. Velazquez, E.; Peix, A.; Zurdo-Pineiro, J.L.; Palomo, J.L.; Mateos, P.F.; Rivas, R.; Munoz-Adelantodo, E.; Toro, N.; Garcia-Benavides, P.; Martinez-Molina, E. The coexistence of symbiosis and pathogenicity-determining genes in *Rhizobium rhizogenes* strains enables them to induce nodules and tumors or hairy roots in plants. *Mol. Plant-Microbe Interact.* **2005**, *18*, 1325–1332. [CrossRef] [PubMed]
56. Berg, G. Plant–microbe interactions promoting plant growth and health: Perspectives for controlled use of microorganisms in agriculture. *Appl. Microbiol. Biotechnol.* **2009**, *84*, 11–18. [CrossRef] [PubMed]
57. Richardson, A.E.; Barea, J.M.; McNeill, A.M.; Prigent-Combaret, C. Acquisition of phosphorus and nitrogen in the rhizosphere and plant growth promotion by microorganisms. *Plant Soil.* **2009**, *321*, 305–339. [CrossRef]
58. Halpern, M.; Bar-Taly, A.; Ofeky, M.; Minzy, D.; Mullerx, T.; Yermiyahu, U. The Use of biostimulants for enhancing nutrient uptake. *Adv. Agron.* **2015**, *130*, 141–174. [CrossRef]
59. Doni, F.; Fathurrahman, F.; Mispan, M.S.; Suhaimi, N.S.M.; Yusoff, W.M.W.; Uphoff, N. Transcriptomic profiling of rice seedlings inoculated with the symbiotic fungus *Trichoderma asperellum* SL2. *J. Plant Growth Regul.* **2019**, *38*, 1507–1515. [CrossRef]
60. Pal, G.; Kumar, K.; Verma, A.; White, J.F.; Verma, S.K. Functional Roles of Seed-Inhabiting Endophytes of Rice. In *Seed Endophytes: Biology and Biotechnology*; Verma, S.K., White, J.J.F., Eds.; Springer International Publishing: Berlin/Heidelberg, Germany, 2019; pp. 213–236. [CrossRef]
61. Hossain, G.M.A.; Ghazali, A.H.; Islam, T.; Mia, M.A.B. Enhanced Nutrient Accumulation in Non-Leguminous Crop Plants by the Application of Endophytic Bacteria *Bacillus* Species. In *Bacilli in Agrobiotechnology: Plant Stress Tolerance, Bioremediation, and Bioprospecting*; Islam, M.T., Rahman, M., Pandey, P., Eds.; Springer International Publishing: Berlin/Heidelberg, Germany, 2022; pp. 349–364. [CrossRef]
62. Verma, N.; Narayan, O.P.; Prasad, D.; Jogawat, A.; Panwar, S.L.; Dua, M.; Johri, A.K. Functional characterization of a high-affinity iron transporter *PiFTR*) from the endophytic fungus *Piriformospora indica* and its role in plant growth and development. *Environ. Microbiol.* **2022**, *24*, 689–706. [CrossRef]
63. Bakhshandeh, E.; Gholamhosseini, M.; Yaghoubian, Y.; Pirdashti, H. Plant growth promoting microorganisms can improve germination, seedling growth and potassium uptake of soybean under drought and salt stress. *Plant Growth Regul.* **2020**, *90*, 123–136. [CrossRef]

64. Fontenelle, A.D.B.; Guzzo, S.D.; Lucon, C.M.M.; Harakava, R. Growth promotion and induction of resistance in tomato plant against *Xanthomonas euvesicatoria* and *Alternaria solani* by *Trichoderma* spp. *Crop Prot.* **2011**, *30*, 1492–1500. [CrossRef]
65. Ipek, M.; Pirlak, L.; Esitken, A.; Figen Dönmez, M.; Turan, M.; Sahin, F. Plant growth-promoting rhizobacteria (Pgpr) increase yield, growth and nutrition of strawberry under high-calcareous soil conditions. *J. Plant Nutr.* **2014**, *37*, 990–1001. [CrossRef]
66. Liu, J.; Tang, L.; Gao, H.; Zhang, M.; Guo, C. Enhancement of alfalfa yield and quality by plant growth-promoting rhizobacteria under saline-alkali conditions. *J. Sci. Food Agric.* **2019**, *99*, 281–289. [CrossRef]
67. El_Komy, M.H.; Hassouna, M.G.; Abou-Taleb, E.M.; Al-Sarar, A.S.; Abobakr, Y. A mixture of *Azotobacter*, *Azospirillum*, and *Klebsiella* strains improves root-rot disease complex management and promotes growth in sunflowers in calcareous soil. *Eur. J. Plant Pathol.* **2020**, *156*, 713–726. [CrossRef]
68. Roesti, D.; Gaur, R.; Johri, B.; Imfeld, G.; Sharma, S.; Kawaljeet, K.; Aragno, M. Plant growth stage, fertiliser management and bio-inoculation of arbuscular mycorrhizal fungi and plant growth promoting rhizobacteria affect the rhizobacterial community structure in rain–Fed wheat fields. *Soil Biol. Biochem.* **2006**, *38*, 1111–1120. [CrossRef]
69. Kavroulakis, N.; Ntougias, S.; Zervakis, G.I.; Ehaliotis, C.; Haralampidis, K.; Papadopoulou, K.K. Role of ethylene in the protection of tomato plants against soil-borne fungal pathogens conferred by an endophytic *Fusarium solani* strain. *J. Exp. Bot.* **2007**, *58*, 3853–3864. [CrossRef] [PubMed]
70. Garcia, M.J.; Lucena, C.; Romera, F.J.; Alcantara, E.; Perez-Vicente, R. Ethylene and nitric oxide involvement in the up-regulation of key genes related to iron acquisition and homeostasis in *Arabidopsis*. *J. Exp. Bot.* **2010**, *61*, 3885–3899. [CrossRef]
71. Garcia, M.J.; Romera, F.J.; Alcantara, E.; Perez-Vicente, R. A new model involving ethylene, nitric oxide and Fe to explain the regulation of Fe-acquisition genes in Strategy I plants. *Plant Physiol. Biochem.* **2011**, *49*, 537–544. [CrossRef] [PubMed]
72. Zamioudis, C.; Korteland, J.; Van Pelt, J.A.; van Hamersveld, M.; Dombrowski, N.; Bai, Y.; Hanson, J.; Van Verk, M.C.; Ling, H.Q.; Schulze-Lefert, P.; et al. Rhizobacterial volatiles and photosynthesis-related signals coordinate MYB72 expression in *Arabidopsis* roots during onset of induced systemic resistance and iron-deficiency responses. *Plant J.* **2015**, *84*, 309–322. [CrossRef]
73. Zhou, C.; Guo, J.; Zhu, L.; Xiao, X.; Xie, Y.; Zhu, J.; Ma, Z.; Wang, J. *Paenibacillus polymyxa* BFKC01 enhances plant iron absorption via improved root systems and activated iron acquisition mechanisms. *Plant Physiol. Biochem.* **2016**, *105*, 162–173. [CrossRef]
74. Martínez-Medina, A.; Van Wees, S.C.M.; Pieterse, C.M.J. Airborne signals from *Trichoderma* fungi stimulate iron uptake responses in roots resulting in priming of jasmonic acid-dependent defences in shoots of *Arabidopsis thaliana* and *Solanum lycopersicum*. *Plant Cell Environ.* **2017**, *40*, 2691–2705. [CrossRef]
75. Guerra, C.; Anderson, P. The effect of iron and boron amendments on infection of bean by *Fusarium solani*. *Phytopathology* **1985**, *75*, 989–991. [CrossRef]
76. Navarro-Velasco, G.Y.; Prados-Rosales, R.C.; Ortiz-Urquiza, A.; Quesada-Moraga, E.; Di Pietro, A. *Galleria mellonella* as model host for the trans-kingdom pathogen *Fusarium oxysporum*. *Fungal Genet. Biol.* **2011**, *48*, 1124–1129. [CrossRef]
77. Inal, A.; Gunes, A.; Zhang, F.; Cakmak, I. Peanut/maize intercropping induced changes in rhizosphere and nutrient concentrations in shoots. *Plant Physiol. Biochem.* **2007**, *45*, 350–356. [CrossRef]
78. Reichman, S.M.; Parker, D.R. Critical evaluation of three indirect assays for quantifying phytosiderophores released by the roots of Poaceae. *Eur. J. Soil Sci.* **2007**, *58*, 844–853. [CrossRef]
79. Prieto, P.; Moore, G.; Shaw, P. Fluorescence in situ hybridization on vibratome sections of plant tissues. *Nat. Protoc.* **2007**, *2*, 1831–1838. [CrossRef]

plants

Review

Iron in the Symbiosis of Plants and Microorganisms

Yi Liu [1,†], Zimo Xiong [1,†], Weifeng Wu [1], Hong-Qing Ling [2,3] and Danyu Kong [1,*]

[1] Lushan Botanical Garden, Chinese Academy of Sciences, Jiujiang 332900, China; liuy@lsbg.cn (Y.L.)
[2] Hainan Yazhou Bay Seed Laboratory, Sanya 572024, China; hqling@genetics.ac.cn
[3] State Key Laboratory of Plant Cell and Chromosome Engineering, Institute of Genetics and Developmental Biology, Chinese Academy of Sciences, Beijing 100101, China
* Correspondence: kongdy@lsbg.cn
† These authors contributed equally to this work.

Abstract: Iron is an essential element for most organisms. Both plants and microorganisms have developed different mechanisms for iron uptake, transport and storage. In the symbiosis systems, such as rhizobia–legume symbiosis and arbuscular mycorrhizal (AM) symbiosis, maintaining iron homeostasis to meet the requirements for the interaction between the host plants and the symbiotic microbes is a new challenge. This intriguing topic has drawn the attention of many botanists and microbiologists, and many discoveries have been achieved so far. In this review, we discuss the current progress on iron uptake and transport in the nodules and iron homeostasis in rhizobia–legume symbiosis. The discoveries with regard to iron uptake in AM fungi, iron uptake regulation in AM plants and interactions between iron and other nutrient elements during AM symbiosis are also summarized. At the end of this review, we propose prospects for future studies in this fascinating research area.

Keywords: iron; plant symbiosis; iron uptake; iron homeostasis; rhizobium; mycorrhiza

Citation: Liu, Y.; Xiong, Z.; Wu, W.; Ling, H.-Q.; Kong, D. Iron in the Symbiosis of Plants and Microorganisms. *Plants* **2023**, *12*, 1958. https://doi.org/10.3390/plants12101958

Academic Editor: Ferenc Fodor

Received: 10 April 2023
Revised: 8 May 2023
Accepted: 8 May 2023
Published: 11 May 2023

1. Introduction

Iron, as one of the most abundant elements on earth, is an essential element for most organisms since it functions as an indispensable co-factor of many enzymes in various crucial metabolic processes [1]. In plants, iron deficiency results in reduced chlorophyll synthesis and photosynthesis, and causes chlorosis and dramatic growth defects. To obtain sufficient iron from soil, plants evolved different strategies for effective iron uptake and homeostasis. The first one is the reduction strategy (Strategy I), which is widely used in all non-graminaceous plants. The second one is the chelation strategy (Strategy II), which is applied by graminaceous plants [2]. Compared with plants, microorganisms engage high-affinity and low-affinity uptake systems for iron uptake [3,4]. The high-affinity uptake pathways include the siderophore-mediated iron uptake pathway and the reductive iron assimilation (RIA) pathway [4]. The low-affinity uptake pathways include the iron-containing protein (e.g., heme, ferredoxin) uptake pathway and the ferrous iron uptake pathway. In general, the high-affinity uptake pathways are adopted by microorganisms when limited iron is available. In contrast, when iron is sufficient, the low-affinity pathways are applied by microorganisms.

Plants and microorganisms in the rhizosphere recruit their own ways to acquire iron from soil until an infection or a symbiosis event takes place between them. In the competition of host plants with pathogens, iron also plays an important role in restricting pathogen growth either by the overaccumulation of iron at the pathogen attack site to induce ROS burst, which leads to the infected cell's death [5,6], or by withholding iron out of the vicinity of the infection site [2,7].

Symbiosis between a plant and microorganism is a very common phenomenon in nature. In the symbiont, both the plant and microorganism can obtain nutrients from each other, supporting better growth and development. For example, leguminous plants obtain ammonia from symbiotic rhizobia and bacteria obtain carbon compounds from host plants. Similarly, AM fungi supply nitrogen and phosphate to host plants and host plants provide lipids to mutualistic AM fungi [8,9]. Therefore, symbiosis is very important for both mutualistic microorganisms and host plants. In the symbiosis, such as legume–rhizobium and plant–arbuscular mycorrhizal (AM) fungi, iron is coordinated as an important micronutrient for both plants and symbiotic microbes. For legume–rhizobium symbiosis, iron is an essential co-factor for the nitrogenase and the enzymes of the bacterial respiration. A large amount of iron has to be transported from the roots to the bacteroids to support symbiotic nitrogen fixation (SNF). In the last thirty years, many studies focused on this process have been conducted. For the plant–AM fungi symbiosis, recent studies revealed that AM symbiosis can influence the iron uptake of plants. However, the iron transport mechanisms between the plants and AM fungi are still unknown. In this review, we introduce the iron transportation mechanism in the symbiosis and interaction between the plants and symbiotic microbes in terms of the iron uptake, transport and homeostasis.

2. Iron Transport in Rhizobia–Legume Symbiosis

Legumes play a crucial role in the nitrogen cycle in agricultural and natural environments since they form symbioses with nitrogen-fixing soil bacteria (rhizobia) that enable the plants to utilize atmospheric nitrogen. Starting with the infection of the legume roots by rhizobia, the symbiotic process eventually forms a new organ, root nodule, where the symbiotic nitrogen fixation occurs. The mature nodule possesses a central infection zone, containing infected and uninfected cells, surrounded by layers of cells termed the cortex [10,11]. Metabolites are transported to the nodule through the vasculature, which terminates in the cortex [10,11]. Nodules are classified as two distinct types, determinate and indeterminate. Determinate nodules, such as those in *Glycine max* (soybean) and *Lotus japonicus*, are spherical without a meristem and embed the infected region in the center surrounded by the cortex layer on the outside [10,11]. Indeterminate nodules, such as those of *Medicago truncatula*, *Pisum sativum* (pea) and *Vicia faba* (broad bean), are elongated or branched shape because they have persistent meristems [11,12]. A mature indeterminate nodule can be divided into at least four zones: zone I is the meristematic region that drives nodule growth; zone II is where rhizobia are released from the infection thread and differentiate into bacteroids; zone III is the site of nitrogen fixation; and zone IV is the senescence zone, where bacteroids are degraded and nutrients are recycled [13]. Once the rhizobia attach to the root hair, they are transported to the cortical cell through an infection thread, and then release into the root's cortical cells surrounded by the symbiosome membrane (SM), which is a plant-derived membrane [14].

Iron is a crucial micronutrient for nodule development and symbiotic nitrogen fixation (SNF). Symbiotically grown legumes have a particularly high requirement for iron, and iron deficiency severely interferes with their growth and the ability to fix atmospheric nitrogen [15,16]. In soybean, nitrogen-fixing nodules contain approximately 44% of the total iron of the plant, compared to 39% in the leaves, 7% in the seeds and 5% in the roots [17]. To achieve enough iron for nodulation and SNF, the plants and rhizobia mutually stimulate their iron uptake mechanisms. For example, *Sinorhizobium meliloti*, a typical rhizobium, secretes some volatile organic compounds (VOCs), which are able to significantly induce the ferric reductase activity and enhance the rhizosphere acidification [18]. On the other hand, a plant peptide NCR247 secreted by *Medicago truncatula* can be transported into the rhizobium cytoplasm and binds to the haem [19], resulting in a decrease in free haem and the release of the activity of haem-inactivated transcription factor Irr in the rhizobium. Active Irr represses the expression of the *rirA* gene, which encodes a transcriptional repressor of iron-uptake genes [19]. Thus, the secreted NCR247 peptide from host plants will induce the expression of rhizobium iron-uptake genes by binding the haem in the rhizobium cytoplasm [19].

2.1. Iron Transport from Root to Nodule

Iron is transported as a form of ferric citrate complex into the xylem of plants [20]. Therefore, iron is also most likely to be imported into the nodule as a ferric citrate complex from the vasculature via the crossing of a number of cell layers to the infected cells [21,22]. Several transporters responsible for iron uptake and transport to nodules have been identified. The citrate transporters, MtMATE67 in *Medicago truncatula* and LjMATE1 in *Lotus japonicus*, which are located in nodules (Figure 1), assist the translocation of iron from the roots to nodules via the transportation of citrate to improve the solubility and availability of iron [23,24]. An iron and manganese transporter MtNRAMP1, a member of the Natural Resistance-Associated Macrophage (NRAMP) protein and located on the plasma membrane of nodules, is mainly expressed in zone II (Figure 1B) and has lower expression level in zone I, zone III, the nodule cortex and root vasculature (Figure 1B). Nitrogenase activity is reduced in MtNRAMP1 knockout mutant *nramp1-1*. Exogenous iron supply can rescue the nitrogen fixation to normal levels in *nramp1-1* mutant nodules. This indicates that MtNRAMP1 is involved in the iron uptake of the infected cells in nodules [25]. Recently, Yellow Stripe 1-like (YSL) transporters MtYSL3 in *Medicago truncatula* and GmYSL7 in soybean have been demonstrated as being important for iron transport from roots to nodules [26,27]. It was confirmed that GmYSL7 is located in both cortical cells and infected cells, whereas MtYSL3 is only localized in the vascular tissue (Figure 1) [26].

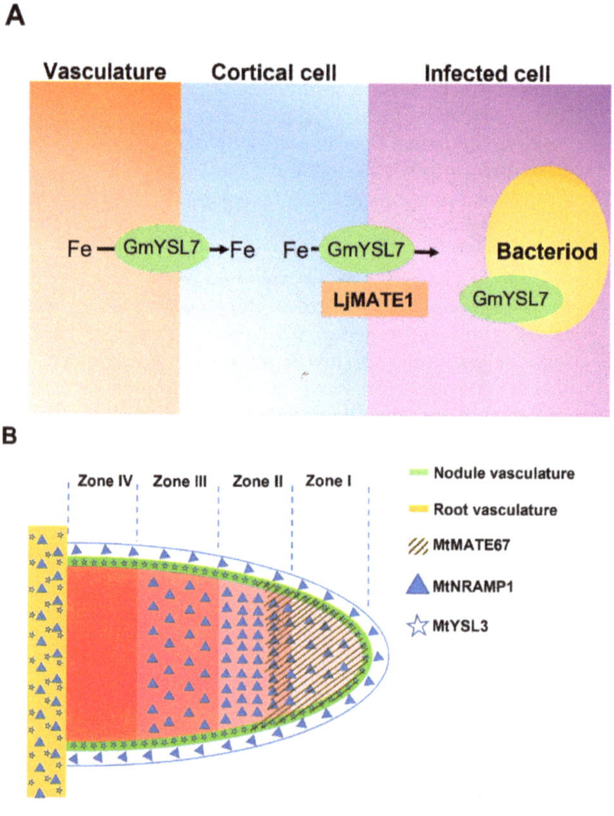

Figure 1. Iron transport from root to nodule. (**A**) Iron transporters in determinate nodule. (**B**) Iron transporters in indeterminate nodule. MtMATE67 is located in zone I and partially in zone II, which next to zone I; MtNRAMP1 is located in root vasculature, cortex, and zone I, II, and III; MtYSL3 is located in both root vasculature and nodule vasculature.

2.2. Iron Transport through the Symbiosome Membrane

It was proved that the symbiosome enables the uptake of both ferric and ferrous iron by studying isolated symbiosome with a given amount of radiolabeled ferric and ferrous iron [28–30]. The uptake of ferrous iron by isolated symbiosome is faster than that of ferric iron [28], indicating that ferrous iron may be the major form transported from the infected cells into the symbiosome. Meanwhile, the transporter for ferrous iron is probably not specific for the acquisition of iron because the transporting efficiency of ferrous iron can be inhibited by copper [28]. The NRAMP family metal transporter GmDMT1 (*Glycine max* Divalent Metal Transporter 1), which is localized on the symbiosome membrane (SM), is able to transport ferrous iron [31], but its transporting direction is still unclear. In the complementary experiment of yeast iron-transport-deficient mutant *fet3fet4*, GmDMT1 was located on the plasm membrane and restored the iron uptake of *fet3fet4* [31]. This indicates that GmDMT1 transports ferrous iron from the extracellular space into the cytoplasm. However, the symbiosome is a vacuole-like structure [32] and the iron uptake into the symbiosome means iron transported out of the cytoplasm. Considering GmDMT1 transports ferrous iron into the cytoplasm in *fet3fet4*, GmDMT1 most likely transports ferrous iron from the symbiosome into the cytoplasm in the infected cells as well (Figure 2) [22,31].

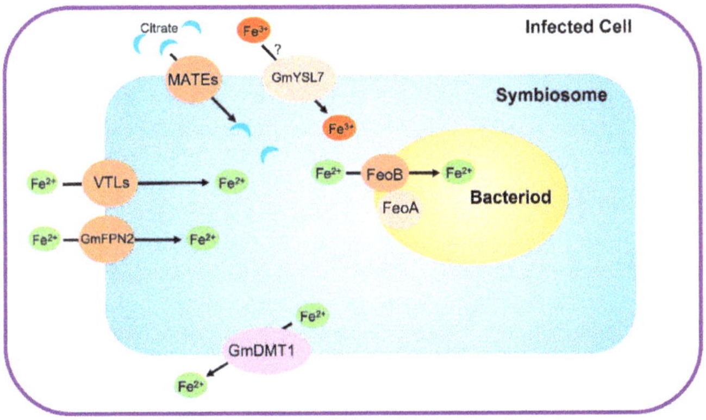

Figure 2. Iron uptake for symbiosome and bacteroid. VTLs and GmFPN2 transport ferrous iron from infected cell cytoplasm into symbiosome, MATEs transport citrate from infected cell into symbiosome, YSLs contribute the transportation of ferric iron from infected cell into symbiosome, FeoAB system transports ferrous iron from symbiosome into bacteroid, GmDMT1 transports ferrous iron from symbiosome to infected cell.

If the symbiosome is considered as an organelle of the infected cell, the transporters that export iron out of the cytoplasm may be involved in iron uptake of the symbiosome. The Vacuolar Iron Transport (VIT) and VIT-like (VLT) proteins are known to transport iron from the cytoplasm into the vacuole [33–35]. Two members of the VIT gene family, *GmVTLa* (known as Nodulin-21) and *GmVTLb*, which are highly expressed in nodules [36,37], have been identified by transcriptome data analysis of soybean [38–40]. Both GmVTL1a and GmVTL1b are located on the SM of the infected cells (Figure 2), and are able to complement the yeast ferrous iron transport mutant Δ*ccc1* [37]. Knockdown of *GmVLT1a* and *1b* causes defected nodule development and reduced iron content in nodules and bacteroids [36]. The GmVTL1a homologues in *Medicago truncatula*, MtVTL4 and MtVTL8, are also expressed in nodules and MtVTL8 is localized on the SM [41]. Both MtVTL4 and MtVTL8 can rescue the yeast Δ*ccc1* mutant, indicating that they are ferrous iron transporters. Expression of *MtVTL8* alone reverted the defect phenotype of nodule development in the 13U mutant, which has a 30 kb deletion spanning *MtVTL4* and *MtVTL8* in the genome, while expression of *MtVTL4* did not [41]. Further analysis of the iron contents in the 13U and *vlt4* mutants revealed that the iron content of the rhizobium in the mutant roots was significantly lower than that of the wild type, while the iron content in the cytoplasm of the infected cells remained similar to that of the wild type. These results indicate that MtVTL4 and MtVTL8 are specifically required for iron delivery into the bacteroids [41]. Another VIT protein of *Lotus japonicus*, LjSEN1, an important regulator of rhizobia differentiation in the nodule [42], was suggested to be an iron transporter in the infected cells as well [22,37]. However, there is still no direct evidence to show whether LjSEN1 is located on the SM.

Another type of ferrous iron transporter mediating iron out of the cytoplasm into the organelles is ferroportins [43]. *Medicago truncatula* nodule-specific gene *Ferroportin 2* (*MtFPN2*) encodes an iron efflux protein located on the SM of the infected cells (Figure 2) and the vascular intracellular compartments [44]. Mutation of *MtFPN2* resulted in a severe reduction in nitrogenase activity, and the mislocalization of iron in the nodules [44]. This suggests that MtFPN2 is indispensable for iron transport into the symbiosome.

As well as ferrous iron, a symbiosome is able to uptake ferric iron [29,30]. It was found that the iron-activated citrate transporter MtMATE67 is located not only on the plasma membrane of nodule cells but also on the symbiosome membrane in *Medicago truncatula* [23] (Figure 1B). The loss function of *MtMATE67* resulted in the accumulation of iron in the apoplasm of nodule cells. MtMATE67 is responsible for citrate efflux from nodule cells into the symbiosome to ensure the solubility and mobility of ferric iron in the apoplasm and further uptake into nodule cells [23]. Based on the fact that MtMATE67 itself does not transport the iron–citrate complex, there must be another system for iron translocation. MtFPN2 or MtVTL8 may be good candidates since these two ferrous transporters are located on the SM [22,44]. Considering that MtFPN2 or MtVTL8 only transport ferrous iron, a ferric reductase is needed to reduce ferric iron to ferrous iron before the iron can be transported by MtFPN2 or MtVTL8. Another possibility is that there is an unknown ferric iron transporter working with MtMATE67 on the SM.

Yellow Stripe 1 (YS1) is induced by iron deficiency and transports the Fe^{3+}-DMA (deoxymugineic acid) complex in maize (*Zea mays*) [45]. Yellow Stripe 1-like (YSL) transporters also play a significant role in plant iron homeostasis. In soybean, GmYSL7 is localized on the plasma membrane of cortical cells in nodules and the SM of the infected cells (Figures 1A and 2) [26,46]. Its iron uptake activity (for both ferric and ferrous) was proved by complementary tests of yeast mutants *fet3fet4* and Δ*ccc1*. Furthermore, it was demonstrated that GmYSL7 as an iron transporter preferentially transports chelated iron (both ferric and ferrous iron) [26].

2.3. *Transporting of Iron into the Bacteroids*

Free rhizobia in soil need to manage iron uptake from an oxidizing environment. Some transporters in free-living rhizobia have been well studied, such as the TonB-ExbBD complex and ABC (ATP-binding cassette-type) transporters, which transfer ferric siderophore and the heme into rhizobia. However, the expression of TonB-dependent receptors, TonB and ABC transporters, are down-regulated in the bacteroid, and neither the rhizobia mutants of heme transporters or the mutants of TonB/ExbB/ExbD have a defect in symbiosis or nitrogen fixation [47–50]. These results suggest that the major way that the bacteroid obtains iron is different from free rhizobia [11]. It is worth noting that an isolated bacteroid exhibited ferrous iron uptake activity [28,30]. A FeoAB system, which transports ferrous iron, has been reported in pathogenic facultative anaerobic bacteria [12,51]. FeoB is a ferrous iron transporter widely distributed among bacteria and archaea, while *feoA* encoding an auxiliary protein which is necessary for ferrous iron uptake [52]. *FeoA* and *FeoB* are located in the same expression operon [52]. The incubation of soybean roots with the *feoA* or *feoB* deletion strains led to small and ineffective nodules with few bacteria and compromised nitrogen fixation activity [53]. A mutant strain *E40K*, which carried a missense mutation within the *feoA* gene, exhibited a diminished iron uptake activity, although it is able to develop nodules with nitrogen fixation activity in soybean [53]. This character of *E40K* makes it possible to evaluate the function of the FeoAB system under the state of symbiosis.

The measurement of $^{55}Fe^{2+}$ uptake by isolated bacteroids illustrated that the *E40K* strain had lower uptake activity than the wild type [53]. These studies indicate that the FeoAB transport system is responsible for ferrous iron uptake into bacteroids.

2.4. Iron Homeostasis in Rhizobia–Legume Symbiosis

As the most important function of nodules, symbiotic nitrogen fixation (SNF) requires a large amount of iron. It was shown that iron deficiency significantly inhibits nodule development and nitrogen fixation [54]. To meet the increased requirement for iron, legumes stimulate the iron-deficiency-induced responses in roots to obtain more iron from soil [55], such as secreting more protons and reductants [56,57], up-regulating the activity of the ferric chelate reductase [58,59], and the accumulation of H-ATPase and IRT1 proteins around the cortex cells of nodules [60]. The bHLH (basic helix–loop–helix) transcription factors play crucial role in the regulation of the iron deficiency response [61–64]. GmbHLH300 is remarkably up-regulated in the nodules of the *Gmysl7* mutant and *GmYSL7OE* plants [26]. Meanwhile, the expression of *GmbHLH300* was induced under both low and high iron conditions, suggesting that it plays a crucial role for iron homeostasis in soybean nodules. Wu et al. also found that GmbHLH300 negatively regulates the expression of *GmYSL7* and *ENOD93*, which are two positive nodulation regulator genes [26].

3. Iron in Arbuscular Mycorrhizal Symbiosis

Arbuscular mycorrhizal (AM) symbiosis is more popular in nature compared with rhizobia–legume symbiosis since 80% of plants growing under natural conditions are associated with mycorrhizae [55,65,66]. AM fungi provide plants with essential mineral nutrients, such as phosphorus (P) and nitrogen (N), and fungi obtain their carbon from the host plants in the form of plant photosynthates and fatty acids in return [67–70]. Additionally, AM fungi can improve the biotic and abiotic stress tolerance in plants. It has been reported that AM symbiosis can both increase and decrease the iron uptake of host plants [55,71]. A meta-analysis of 233 studies supports that there is a positive impact of AM fungi on crop plants' iron nutrition [72]. Studies showed that the impact of AM fungi on iron uptake is dependent on the growth conditions, host plant species and AM fungi species. Using ^{59}Fe as a tracer, Caris et al. found that the iron uptake of *Glomus mosseae* (an AM fungus)-cultured sorghums (Strategy II plant) increased in calcareous iron-deficient soil, but the iron content of peanut (Strategy I plant) had no change [73]. This result suggests that AM fungi increase the iron uptake in Strategy II plant but not Strategy I plant. However, Kabir et al. discovered that AM fungi increase the iron uptake in sunflower (Strategy I plant) when the plants are inoculated with a mixed endomycorrhizal spore including *Glomus intraradices*, *Glomus mosseae*, *Glomus aggregatum* and *Glomus etunicatum* [74]. On the other hand, the AM plants grown at a low pH showed higher iron uptake than those grown at a high pH [75,76]. As well as pH, the temperature influences the iron uptake of the host plant in AM symbiosis. The sorghum inoculated with *Glomus macrocarpum* (a vesicular-arbuscular mycorrhizal fungus) had more than 10-fold iron grown at 25 °C or 30 °C than that grown at 20 °C [77]. Al-Karaki et al. found that water stress promoted the iron uptake in AM-fungi-cultured wheat [78,79]. All of the above studies indicate that AM symbiosis increases the iron uptake of host plants in some certain conditions. Nevertheless, further understanding on how the growth conditions, host plants and AM fungi species affect the iron uptake of the AM plant is still lacking.

3.1. Iron Uptake in AM Fungi

In AM roots, AM fungi develop arbuscules to facilitate nutrient exchange with the host plants. In the soil, AM fungi develop extensive and highly branched external mycelium to absorb nutrients beyond the depletion zone that develops around the roots [80]. In this way, the AM plants can absorb nutrients through both the plant roots and mycorrhiza [80]. Caris et al. found that providing ^{59}Fe to the hyphae resulted in a radioactive signal appearing in the shoots of host plant in both peanut and sorghum [73]. This result indicated that iron was delivered from the AM to the host plant. Kobae et al. performed similar isotopic tracing in maize and observed the same result [81].

Only a few iron transporters in AM fungi have been identified so far. Genome sequencing and transcriptomic analyses of *Rhizophagus irregularis*, an arbuscular mycorrhizal fungus, revealed two potential iron permeases, RiFTR1 and RiFTR2, which are expressed in germinated spores and mycorrhizal roots [82]. RiFTR1 is up-regulated (10-fold) during the symbiotic phase of the fungus, which indicates that RiFTR1 plays an important role in the biotrophic phase of AM fungus [82]. The genes encoding NRAMP family members *RiSMF1*, *RiSMF2*, *RiSMF3.1* and *RiSMF3.2* were characterized in *Rhizophagus irregularis*. In the complementary experiment of yeast *fet3fet4* mutant, only *RiSMF3.2* was proved to be an iron transporter [83].

3.2. *Iron Uptake Regulation in AM Plants*

The research on the iron uptake regulation in AM symbiosis is relatively lacking compared with that of rhizobia–legume symbiosis. Recently, transcriptome analysis in the wheat roots colonized by AM fungus revealed that two iron–phytosiderophore transporters were up-regulated during AM symbiosis [84], indicating that wheat may secrete more siderophores in the symbiosis with AM fungus. On the other hand, in a genome-wide analysis of the nodulin-like gene family in bread wheat, six *VIT* subfamily genes were found to be down-regulated in response to AM inoculation, and five of them were significantly down-regulated only at the fully colonized stage [85]. These results suggest that iron might be involved in the AM symbiosis process.

Some evidences have provided clues on the molecular mechanism of AM fungi positively alleviating iron deficiency for host plants. In Strategy I plants, AM fungi resulted in a significant improvement in iron concentrations in the roots and the shoots of sunflower under iron deficiency [74]. In the presence of AM fungi, the expressions of transport genes *HaIRT1* and *HaNramp1* and the ferric reductase gene *HaFRO1* were up-regulated and the ferric reductase activity was increased as well [74]. Similarly, increased siderophore release was observed in AM fungi symbiosis with *Tagetes patula nana* [86]. In *Medicago sativa* L., the expression of ferric iron reductase gene *MsFRO1* is significantly induced in roots by cultivation with AM fungi under iron-deficient conditions [87].

In Strategy II plants, Prity et al. reported that phytosiderophore (PS) release was increased in the sorghum root cultured with AM fungi, and the expressions of iron uptake related genes *SbDMAS2*, *SbNAS2*, and *SbYS1* were elevated under iron deficiency, compared to the root without AM fungi [88]. Sulfur deficiency had a strong, negative impact on the Strategy II iron acquisition, causing a reduction in iron concentration and induced *ZmNAS1* and *ZmYS1* expression in the roots in non-mycorrhizal maize [89]. However, in mycorrhizal plants, the iron content and the expression of *ZmYS1* remained at a normal level during sulfur depletion, and the expression of *ZmNAS1* was down-regulated [89]. These results imply that the maize with AM symbiosis maintained sufficient iron during sulfur depletion. Chorianopoulou et al. suggested that iron is mainly transported directly to the root from the AM fungi via a special symbiotic iron uptake pathway [89]. This result is consistent with the observations of Caris et al. and Kobae et al. using isotopic tracing which were mentioned before [73,81]. It was also found that the transcript levels of oligopeptide transporter genes *ZmOPT8a* and *ZmOPT8b* in mycorrhizal roots were induced 194- and 62-fold, respectively, than non-mycorrhizal roots [81]. However, the expression levels of other iron-uptake-related genes such as *ZmDMASa*, *ZmYS1*, *ZmIRTa*, *ZmIRTb* and *ZmIRO2a* were similar between non-mycorrhizal roots and mycorrhizal roots [81]. Considering the rice OsOPT1, OsOPT3, and OsOPT4 transport ferric–nicotianamine complex [90], ZmOPT8 is probably involved in iron transport in the mycorrhizal root [81].

3.3. *Interactions of Iron and Other Nutrient Elements during AM Symbiosis*

It was reported that iron interacts with other mineral elements during uptake and homeostasis in plants including AM plants [91]. Phosphorus (Pi) was proven to have a negative effect on the iron uptake of AM plants [92]. Iron uptake decreased under a high Pi condition, whereas it increased under a low Pi supply in AM plants [93]. AM fungi release significant amounts of organic acids to mobilize the phosphorus bounded to iron oxides in soil and increase the Pi acquisition of host plants [94]. There is a possibility that iron is also mobilized by the organic acids under a low Pi supply and are easier to absorb by host plants. The effect of iron on Pi acquisition is barely known. Recently, Pang et al. discovered that AM fungi enhanced Pi uptake when $FePO_4$ was used as Pi sources but not KH_2PO_4 [95]. In addition, co-inoculation of AM fungi and rhizobia enhanced nodulation under the use of $FePO_4$, but not KH_2PO_4, as a Pi source [96].

Zinc is another nutrient element that has been revealed as interacting with iron in AM plants. In sunflower, the uptakes of iron and zinc were both up-regulated under iron deficiency in AM plants [74]. This is consistent with the expression of *HaZIP1*, responsible for Zn uptake, being significantly up-regulated following AM fungi inoculation under Fe deficiency [75]. Ibiang et al. reported that zinc treatment leads to the iron accumulation in roots in AM soybean [96]. However, excess zinc increases the iron content in fruit and decreases iron in mycorrhizal roots in AM tomato [97].

4. Prospects

Although the information with regard to iron in AM symbiosis is limited, it is meaningful to discover that AM symbiosis plays positive roles in the relief of iron deficiency for plants since AM symbiosis is popular in crops suffering from iron deficiency such as sorghum, maize, potato and peanut. This means it is possible to relieve iron deficiency in crops by applying AM fungi in soil. However, the mechanisms involved in this process need to be studied in the future. For instance, AM fungi induce the expression of the iron-uptake-related genes of host plants under iron deficiency. In this process, the signal transport between AM fungi and the host plant needs to be investigated to answer how AM fungi regulate the host's gene expression. In the meantime, studies have shown that plants can obtain iron from AM fungi directly, but how iron is transported from AM fungi into plants is not clear.

In the past decades, researchers have made remarkable progress in understanding the mechanisms of iron transport and distribution in rhizobia–legume symbiosis. However, some questions still remain to be accomplished to illustrate the overall picture for iron homeostasis in rhizobia–legume symbiosis. Firstly, GmYSL7 is located on the SM in the infected cells. However, neither its function for iron transport through the SM, or the transport direction of iron on the SM have been demonstrated (Figure 2). Secondly, the impact of the interaction of the host plants and the symbiosis bacteria on the regulation of iron homeostasis is still not clear. Thirdly, in the high iron environment such as the symbiosome and the bacteroid, the iron efflux and storage mechanisms are crucial for avoiding iron toxicity. It will be meaningful to figure out the detoxification mechanisms. Actually, there are some clues for iron efflux from the symbiosome and the bacteroid. For instance, Wittenberg et al. showed that the bacteroids released a large amount of siderophores-bound iron into the symbiosome space in isolated soybean nodules [98]. GmDMT1 was suggested to transport iron from the symbiosome to the cytoplasm [31]. These results indicate that the iron level is regulated by an unclear system in the symbiosome. Recently, GmbHLH300 has been shown to be involved in the regulation of iron homeostasis in the infected cells. More investigation of the GmbHLH300 network may reveal unknown details for iron detoxification in the symbiosome.

Author Contributions: Y.L., conceptualization, writing—original draft, visualization and funding acquisition. Z.X., Conceptualization and writing—original draft. W.W., writing—review and editing. H.-Q.L., writing—review and editing. D.K., conceptualization, writing—review and editing, and funding acquisition. All authors have read and agreed to the published version of the manuscript.

Funding: This work was supported by the National Natural Science Foundation of China (32260768), Jiangxi Provincial Key R&D Program (20223BBG71003) and Talents Program of Jiangxi Province (jxsq2020101088) and (jxsq2020101020).

Data Availability Statement: Not applicable.

Conflicts of Interest: The authors declare no conflict of interest.

References

1. Johnson, L. Iron and siderophores in fungal-host interactions. *Mycol. Res.* **2008**, *112 Pt 2*, 170–183. [CrossRef] [PubMed]
2. Liu, Y.; Kong, D.; Wu, H.L.; Ling, H.Q. Iron in plant-pathogen interactions. *J. Exp. Bot.* **2021**, *72*, 2114–2124. [CrossRef] [PubMed]
3. Chu, B.C.; Garcia-Herrero, A.; Johanson, T.H.; Krewulak, K.D.; Lau, C.K.; Peacock, R.S.; Slavinskaya, Z.; Vogel, H.J. Siderophore uptake in bacteria and the battle for iron with the host; a bird's eye view. *Biometals* **2010**, *23*, 601–611. [CrossRef]
4. Haas, H. Fungal siderophore metabolism with a focus on Aspergillus fumigatus. *Nat. Prod. Rep.* **2014**, *31*, 1266–1276. [CrossRef] [PubMed]
5. Greenshields, D.L.; Liu, G.; Wei, Y. Roles of iron in plant defence and fungal virulence. *Plant Signal. Behav.* **2007**, *2*, 300–302. [CrossRef]
6. Liu, G.; Greenshields, D.L.; Sammynaiken, R.; Hirji, R.N.; Selvaraj, G.; Wei, Y. Targeted alterations in iron homeostasis underlie plant defense responses. *J. Cell Sci.* **2007**, *120 Pt 4*, 596–605. [CrossRef] [PubMed]
7. Aznar, A.; Chen, N.W.; Thomine, S.; Dellagi, A. Immunity to plant pathogens and iron homeostasis. *Plant Sci.* **2015**, *240*, 90–97. [CrossRef]
8. Jiang, Y.; Wang, W.; Xie, Q.; Liu, N.; Liu, L.; Wang, D.; Zhang, X.; Yang, C.; Chen, X.; Tang, D.; et al. Plants transfer lipids to sustain colonization by mutualistic mycorrhizal and parasitic fungi. *Science* **2017**, *356*, 1172–1175. [CrossRef]
9. Luginbuehl, L.H.; Menard, G.N.; Kurup, S.; Van Erp, H.; Radhakrishnan, G.V.; Breakspear, A.; Oldroyd, G.E.D.; Eastmond, P.J. Fatty acids in arbuscular mycorrhizal fungi are synthesized by the host plant. *Science* **2017**, *356*, 1175–1178. [CrossRef] [PubMed]
10. Udvardi, M.K.; Day, D.A. Metabolite transport across symbiotic membranes of legume nodules. *Annu. Rev. Plant Physiol. Plant Mol. Biol.* **1997**, *48*, 493–523. [CrossRef]
11. Brear, E.M.; Day, D.A.; Smith, P.M. Iron: An essential micronutrient for the legume-rhizobium symbiosis. *Front. Plant Sci.* **2013**, *4*, 359. [CrossRef]
12. Abreu, I.; Mihelj, P.; Raimunda, D. Transition metal transporters in rhizobia: Tuning the inorganic micronutrient requirements to different living styles. *Metallomics* **2019**, *11*, 735–755. [CrossRef]
13. Vasse, J.; de Billy, F.; Camut, S.; Truchet, G. Correlation between ultrastructural differentiation of bacteroids and nitrogen fixation in alfalfa nodules. *J. Bacteriol.* **1990**, *172*, 4295–4306. [CrossRef] [PubMed]
14. Whitehead, L.F.; Day, D.A. The peribacteroid membrane. *Physiol. Plant.* **1997**, *100*, 30–44. [CrossRef]
15. O'Hara, G.W.; Dilworth, M.J.; Boonkerd, N.; Parkpian, P. Iron-deficiency specifically limits nodule development in peanut inoculated with *Bradyrhizobium* sp. *New Phytol.* **1988**, *108*, 51–57. [CrossRef]
16. Tang, C.; Robson, A.D.; Dilworth, M.J. The role of iron in nodulation and nitrogen fixation in *Lupinus angustifolius* L. *New Phytol.* **1990**, *114*, 173–182. [CrossRef]
17. Burton, J.W.; Harlow, C.; Theil, E.C. Evidence for reutilization of nodule iron in soybean seed development. *J. Plant Nutr.* **1998**, *21*, 913–927. [CrossRef]
18. Orozco-Mosqueda Mdel, C.; Macias-Rodriguez, L.I.; Santoyo, G.; Farias-Rodriguez, R.; Valencia-Cantero, E. Medicago truncatula increases its iron-uptake mechanisms in response to volatile organic compounds produced by *Sinorhizobium meliloti*. *Folia Microbiol.* **2013**, *58*, 579–585. [CrossRef]
19. Sankari, S.; Babu, V.M.P.; Bian, K.; Alhhazmi, A.; Andorfer, M.C.; Avalos, D.M.; Smith, T.A.; Yoon, K.; Drennan, C.L.; Yaffe, M.B.; et al. A haem-sequestering plant peptide promotes iron uptake in symbiotic bacteria. *Nat. Microbiol.* **2022**, *7*, 1453–1465. [CrossRef]
20. Cline, G.R.; Powell, P.E.; Szaniszlo, P.J.; Reid, C.P.P. Comparison of the Abilities of Hydroxamic, Synthetic, and Other Natural Organic Acids to Chelate Iron and Other Ions in Nutrient Solution. *Soil Sci. Soc. Am. J.* **1982**, *46*, 1158–1164. [CrossRef]
21. Guinel, F.C. Getting around the legume nodule: I. The structure of the peripheral zone in four nodule types. *Botany* **2009**, *87*, 1117–1138. [CrossRef]
22. Day, D.A.; Smith, P.M.C. Iron Transport across Symbiotic Membranes of Nitrogen-Fixing Legumes. *Int. J. Mol. Sci.* **2021**, *22*, 432. [CrossRef] [PubMed]
23. Kryvoruchko, I.S.; Routray, P.; Sinharoy, S.; Torres-Jerez, I.; Tejada-Jiménez, M.; Finney, L.A.; Nakashima, J.; Pislariu, C.I.; Benedito, V.A.; González-Guerrero, M.; et al. An Iron-Activated Citrate Transporter, MtMATE67, Is Required for Symbiotic Nitrogen Fixation. *Plant Physiol.* **2018**, *176*, 2315–2329. [CrossRef]
24. Takanashi, K.; Yokosho, K.; Saeki, K.; Sugiyama, A.; Sato, S.; Tabata, S.; Ma, J.F.; Yazaki, K. LjMATE1: A citrate transporter responsible for iron supply to the nodule infection zone of *Lotus japonicus*. *Plant Cell Physiol.* **2013**, *54*, 585–594. [CrossRef]
25. Tejada-Jimenez, M.; Castro-Rodriguez, R.; Kryvoruchko, I.; Lucas, M.M.; Udvardi, M.; Imperial, J.; Gonzalez-Guerrero, M. Medicago truncatula natural resistance-associated macrophage Protein1 is required for iron uptake by rhizobia-infected nodule cells. *Plant Physiol.* **2015**, *168*, 258–272. [CrossRef] [PubMed]

26. Wu, X.; Wang, Y.; Ni, Q.; Li, H.; Wu, X.; Yuan, Z.; Xiao, R.; Ren, Z.; Lu, J.; Yun, J.; et al. GmYSL7 controls iron uptake, allocation, and cellular response of nodules in soybean. *J. Integr. Plant Biol.* **2022**, *65*, 167–187. [CrossRef] [PubMed]
27. Castro-Rodriguez, R.; Abreu, I.; Reguera, M.; Novoa-Aponte, L.; Mijovilovich, A.; Escudero, V.; Jimenez-Pastor, F.J.; Abadia, J.; Wen, J.; Mysore, K.S.; et al. The *Medicago truncatula* Yellow Stripe1-Like3 gene is involved in vascular delivery of transition metals to root nodules. *J. Exp. Bot.* **2020**, *71*, 7257–7269. [CrossRef] [PubMed]
28. Moreau, S.; Day, D.A.; Puppo, A. Ferrous iron is transported across the peribacteroid membrane of soybean nodules. *Planta* **1998**, *207*, 83–87. [CrossRef]
29. Moreau, S.; Meyer, J.M.; Puppo, A. Uptake of iron by symbiosomes and bacteroids from soybean nodules. *FEBS Lett.* **1995**, *361*, 225–228. [CrossRef]
30. LeVier, K.; Day, D.A.; Guerinot, M.L. Iron Uptake by Symbiosomes from Soybean Root Nodules. *Plant Physiol.* **1996**, *111*, 893–900. [CrossRef]
31. Kaiser, B.N.; Moreau, S.; Castelli, J.; Thomson, R.; Lambert, A.; Bogliolo, S.; Puppo, A.; Day, D.A. The soybean NRAMP homologue, GmDMT1, is a symbiotic divalent metal transporter capable of ferrous iron transport. *Plant J.* **2003**, *35*, 295–304. [CrossRef] [PubMed]
32. Mellor, R.B. Bacteroids in the Rhizobium-Legume Symbiosis Inhabit a Plant Internal Lytic Compartment: Implications for other Microbial Endosymbioses. *J. Exp. Bot.* **1989**, *40*, 831–839. [CrossRef]
33. Kim, S.A.; Punshon, T.; Lanzirotti, A.; Li, L.; Alonso, J.M.; Ecker, J.R.; Kaplan, J.; Guerinot, M.L. Localization of iron in Arabidopsis seed requires the vacuolar membrane transporter VIT1. *Science* **2006**, *314*, 1295–1298. [CrossRef]
34. Gollhofer, J.; Timofeev, R.; Lan, P.; Schmidt, W.; Buckhout, T.J. Vacuolar-Iron-Transporter1-Like proteins mediate iron homeostasis in Arabidopsis. *PLoS ONE* **2014**, *9*, e110468. [CrossRef] [PubMed]
35. Connorton, J.M.; Jones, E.R.; Rodríguez-Ramiro, I.; Fairweather-Tait, S.; Uauy, C.; Balk, J. Wheat Vacuolar Iron Transporter TaVIT2 Transports Fe and Mn and Is Effective for Biofortification. *Plant Physiol.* **2017**, *174*, 2434–2444. [CrossRef]
36. Liu, S.; Liao, L.L.; Nie, M.M.; Peng, W.T.; Zhang, M.S.; Lei, J.N.; Zhong, Y.J.; Liao, H.; Chen, Z.C. A VIT-like transporter facilitates iron transport into nodule symbiosomes for nitrogen fixation in soybean. *New Phytol.* **2020**, *226*, 1413–1428. [CrossRef]
37. Brear, E.M.; Bedon, F.; Gavrin, A.; Kryvoruchko, I.S.; Torres-Jerez, I.; Udvardi, M.K.; Day, D.A.; Smith, P.M.C. GmVTL1a is an iron transporter on the symbiosome membrane of soybean with an important role in nitrogen fixation. *New Phytol.* **2020**, *228*, 667–681. [CrossRef]
38. Libault, M.; Farmer, A.; Joshi, T.; Takahashi, K.; Langley, R.J.; Franklin, L.D.; He, J.; Xu, D.; May, G.; Stacey, G. An integrated transcriptome atlas of the crop model Glycine max, and its use in comparative analyses in plants. *Plant J.* **2010**, *63*, 86–99. [CrossRef]
39. Severin, A.J.; Woody, J.L.; Bolon, Y.T.; Joseph, B.; Diers, B.W.; Farmer, A.D.; Muehlbauer, G.J.; Nelson, R.T.; Grant, D.; Specht, J.E.; et al. RNA-Seq Atlas of Glycine max: A guide to the soybean transcriptome. *BMC Plant Biol.* **2010**, *10*, 160. [CrossRef]
40. Cao, J. Molecular Evolution of the Vacuolar Iron Transporter (VIT) Family Genes in 14 Plant Species. *Genes* **2019**, *10*, 144. [CrossRef]
41. Walton, J.H.; Kontra-Kovats, G.; Green, R.T.; Domonkos, A.; Horvath, B.; Brear, E.M.; Franceschetti, M.; Kalo, P.; Balk, J. The Medicago truncatula Vacuolar iron Transporter-Like proteins VTL4 and VTL8 deliver iron to symbiotic bacteria at different stages of the infection process. *New Phytol.* **2020**, *228*, 651–666. [CrossRef]
42. Suganuma, N.; Nakamura, Y.; Yamamoto, M.; Ohta, T.; Koiwa, H.; Akao, S.; Kawaguchi, M. The Lotus japonicus Sen1 gene controls rhizobial differentiation into nitrogen-fixing bacteroids in nodules. *Mol. Genet. Genom.* **2003**, *269*, 312–320. [CrossRef] [PubMed]
43. Drakesmith, H.; Nemeth, E.; Ganz, T. Ironing out Ferroportin. *Cell Metab.* **2015**, *22*, 777–787. [CrossRef] [PubMed]
44. Escudero, V.; Abreu, I.; Tejada-Jiménez, M.; Rosa-Núñez, E.; Quintana, J.; Prieto, R.I.; Larue, C.; Wen, J.; Villanova, J.; Mysore, K.S.; et al. Medicago truncatula Ferroportin2 mediates iron import into nodule symbiosomes. *New Phytol.* **2020**, *228*, 194–209. [CrossRef]
45. Curie, C.; Panaviene, Z.; Loulergue, C.; Dellaporta, S.L.; Briat, J.F.; Walker, E.L. Maize yellow stripe1 encodes a membrane protein directly involved in Fe(III) uptake. *Nature* **2001**, *409*, 346–349. [CrossRef]
46. Gavrin, A.; Loughlin, P.C.; Brear, E.; Griffith, O.W.; Bedon, F.; Suter Grotemeyer, M.; Escudero, V.; Reguera, M.; Qu, Y.; Mohd-Noor, S.N.; et al. Soybean Yellow Stripe-like 7 is a symbiosome membrane peptide transporter important for nitrogen fixation. *Plant Physiol.* **2021**, *186*, 581–598. [CrossRef]
47. Wexler, M.; Todd, J.D.; Kolade, O.; Bellini, D.; Hemmings, A.M.; Sawers, G.; Johnston, A.W.B. Fur is not the global regulator of iron uptake genes in Rhizobium leguminosarum. *Microbiology* **2003**, *149 Pt 5*, 1357–1365. [CrossRef]
48. Wexler, M.; Yeoman, K.H.; Stevens, J.B.; de Luca, N.G.; Sawers, G.; Johnston, A.W. The Rhizobium leguminosarum tonB gene is required for the uptake of siderophore and haem as sources of iron. *Mol. Microbiol.* **2001**, *41*, 801–816. [CrossRef]
49. Lynch, D.; O'Brien, J.; Welch, T.; Clarke, P.; Cuív, P.O.; Crosa, J.H.; O'Connell, M. Genetic organization of the region encoding regulation, biosynthesis, and transport of rhizobactin 1021, a siderophore produced by *Sinorhizobium meliloti*. *J. Bacteriol.* **2001**, *183*, 2576–2585. [CrossRef] [PubMed]
50. Nienaber, A.; Hennecke, H.; Fischer, H.M. Discovery of a haem uptake system in the soil bacterium *Bradyrhizobium japonicum*. *Mol. Microbiol.* **2001**, *41*, 787–800. [CrossRef] [PubMed]

51. Sestok, A.E.; Linkous, R.O.; Smith, A.T. Toward a mechanistic understanding of Feo-mediated ferrous iron uptake. *Metallomics* **2018**, *10*, 887–898. [CrossRef]
52. Cartron, M.L.; Maddocks, S.; Gillingham, P.; Craven, C.J.; Andrews, S.C. Feo—Transport of ferrous iron into bacteria. *Biometals* **2006**, *19*, 143–157. [CrossRef]
53. Sankari, S.; O'Brian, M.R. The Bradyrhizobium japonicum Ferrous Iron Transporter FeoAB Is Required for Ferric Iron Utilization in Free Living Aerobic Cells and for Symbiosis. *J. Biol. Chem.* **2016**, *291*, 15653–15662. [CrossRef] [PubMed]
54. Chu, Q.; Sha, Z.; Maruyama, H.; Yang, L.; Pan, G.; Xue, L.; Watanabe, T. Metabolic reprogramming in nodules, roots, and leaves of symbiotic soybean in response to iron deficiency. *Plant Cell Environ.* **2019**, *42*, 3027–3043. [CrossRef]
55. Jin, C.W.; Ye, Y.Q.; Zheng, S.J. An underground tale: Contribution of microbial activity to plant iron acquisition via ecological processes. *Ann. Bot.* **2014**, *113*, 7–18. [CrossRef] [PubMed]
56. Soerensen, K.U.; Terry, R.E.; Jolley, V.D.; Brown, J.C.; Vargas, M.E. The interaction of iron-stress response and root nodules in iron efficient and inefficient soybeans. *J. Plant Nutr.* **1988**, *11*, 853–862. [CrossRef]
57. Terry, R.E.; Hartzook, A.; Jolley, V.D.; Brown, J.C. Interactions of iron nutrition and symbiotic nitrogen fixation in peanuts. *J. Plant Nutr.* **1988**, *11*, 811–820. [CrossRef]
58. Derylo, M.; Skorupska, A. Rhizobial siderophore as an iron source for clover. *Physiol. Plant.* **1992**, *85*, 549–553. [CrossRef]
59. Jin, C.W.; You, G.Y.; He, Y.F.; Tang, C.; Wu, P.; Zheng, S.J. Iron deficiency-induced secretion of phenolics facilitates the reutilization of root apoplastic iron in red clover. *Plant Physiol.* **2007**, *144*, 278–285. [CrossRef]
60. Slatni, T.; Dell'Orto, M.; Ben Salah, I.; Vigani, G.; Smaoui, A.; Gouia, H.; Zocchi, G.; Abdelly, C. Immunolocalization of H(+)-ATPase and IRT1 enzymes in N(2)-fixing common bean nodules subjected to iron deficiency. *J. Plant Physiol.* **2012**, *169*, 242–248. [CrossRef]
61. Ling, H.Q.; Bauer, P.; Bereczky, Z.; Keller, B.; Ganal, M. The tomato fer gene encoding a bHLH protein controls iron-uptake responses in roots. *Proc. Natl. Acad. Sci. USA* **2002**, *99*, 13938–13943. [CrossRef]
62. Yuan, Y.X.; Zhang, J.; Wang, D.W.; Ling, H.Q. AtbHLH29 of Arabidopsis thaliana is a functional ortholog of tomato FER involved in controlling iron acquisition in strategy I plants. *Cell Res.* **2005**, *15*, 613–621. [CrossRef]
63. Du, J.; Huang, Z.; Wang, B.; Sun, H.; Chen, C.; Ling, H.Q.; Wu, H. SlbHLH068 interacts with FER to regulate the iron-deficiency response in tomato. *Ann. Bot.* **2015**, *116*, 23–34. [CrossRef] [PubMed]
64. Yuan, Y.; Wu, H.; Wang, N.; Li, J.; Zhao, W.; Du, J.; Wang, D.; Ling, H.Q. FIT interacts with AtbHLH38 and AtbHLH39 in regulating iron uptake gene expression for iron homeostasis in Arabidopsis. *Cell Res.* **2008**, *18*, 385–397. [CrossRef]
65. Harrison, M.J. Cellular programs for arbuscular mycorrhizal symbiosis. *Curr. Opin. Plant Biol.* **2012**, *15*, 691–698. [CrossRef]
66. Smith, S.E.; Read, D. 1—The symbionts forming arbuscular mycorrhizas. In *Mycorrhizal Symbiosis*, 3rd ed.; Smith, S.E., Read, D., Eds.; Academic Press: London, UK, 2008; pp. 13–41.
67. Ho-Plágaro, T.; García-Garrido, J.M. Molecular Regulation of Arbuscular Mycorrhizal Symbiosis. *Int. J. Mol. Sci.* **2022**, *23*, 5960. [CrossRef] [PubMed]
68. Xie, K.; Ren, Y.; Chen, A.; Yang, C.; Zheng, Q.; Chen, J.; Wang, D.; Li, Y.; Hu, S.; Xu, G. Plant nitrogen nutrition: The roles of arbuscular mycorrhizal fungi. *J. Plant Physiol.* **2022**, *269*, 153591. [CrossRef] [PubMed]
69. Luginbuehl, L.H.; Oldroyd, G.E.D. Understanding the Arbuscule at the Heart of Endomycorrhizal Symbioses in Plants. *Curr. Biol.* **2017**, *27*, R952–R963. [CrossRef] [PubMed]
70. Bago, B.; Pfeffer, P.E.; Abubaker, J.; Jun, J.; Allen, J.W.; Brouillette, J.; Douds, D.D.; Lammers, P.J.; Shachar-Hill, Y. Carbon export from arbuscular mycorrhizal roots involves the translocation of carbohydrate as well as lipid. *Plant Physiol.* **2003**, *131*, 1496–1507. [CrossRef]
71. Clark, R.B.; Zeto, S.K. Mineral acquisition by arbuscular mycorrhizal plants. *J. Plant Nutr.* **2000**, *23*, 867–902. [CrossRef]
72. Lehmann, A.; Rillig, M.C. Arbuscular mycorrhizal contribution to copper, manganese and iron nutrient concentrations in crops—A meta-analysis. *Soil Biol. Biochem.* **2015**, *81*, 147–158. [CrossRef]
73. Caris, C.; Hördt, W.; Hawkins, H.-J.; Römheld, V.; George, E. Studies of iron transport by arbuscular mycorrhizal hyphae from soil to peanut and sorghum plants. *Mycorrhiza* **1998**, *8*, 35–39. [CrossRef]
74. Kabir, A.H.; Debnath, T.; Das, U.; Prity, S.A.; Haque, A.; Rahman, M.M.; Parvez, M.S. Arbuscular mycorrhizal fungi alleviate Fe-deficiency symptoms in sunflower by increasing iron uptake and its availability along with antioxidant defense. *Plant Physiol. Biochem.* **2020**, *150*, 254–262. [CrossRef] [PubMed]
75. Medeiros, C.A.B.; Clark, R.B.; Ellis, J.R. Effects of mes [2(n-morpholino)-ethanesulfonic acid] and ph on mineral nutrient uptake by mycor-rhizal and nonmycorrhizal maize. *J. Plant Nutr.* **1993**, *16*, 2255–2272. [CrossRef]
76. Medeiros, C.A.B.; Clark, R.B.; Ellis, J.R. Growth and nutrient uptake of sorghum cultivated with vesicular-arbuscular mycorrhiza isolates at varying pH. *Mycorrhiza* **1994**, *4*, 185–191. [CrossRef]
77. Raju, P.S.; Clark, R.B.; Ellis, J.R.; Maranville, J.W. Effects of species of VA-mycorrhizal fungi on growth and mineral uptake of sorghum at different temperatures. *Plant Soil* **1990**, *121*, 165–170. [CrossRef]
78. Al-Karaki, G.N.; Clark, R.B. Growth, mineral acquisition, and water use by mycorrhizal wheat grown under water stress. *J. Plant Nutr.* **1998**, *21*, 263–276. [CrossRef]
79. Al-Karaki, G.N.; Al-Raddad, A.; Clark, R.B. Water stress and mycorrhizal isolate effects on growth and nutrient acquisition of wheat. *J. Plant Nutr.* **1998**, *21*, 891–902. [CrossRef]

80. Ferrol, N.; Tamayo, E.; Vargas, P. The heavy metal paradox in arbuscular mycorrhizas: From mechanisms to biotechnological applications. *J. Exp. Bot.* **2016**, *67*, 6253–6265. [CrossRef]
81. Kobae, Y.; Tomioka, R.; Tanoi, K.; Kobayashi, N.I.; Ohmori, Y.; Nishida, S.; Fujiwara, T. Selective induction of putative iron transporters, OPT8a and OPT8b, in maize by mycorrhizal colonization. *Soil Sci. Plant Nutr.* **2014**, *60*, 843–847. [CrossRef]
82. Tamayo, E.; Gómez-Gallego, T.; Azcón-Aguilar, C.; Ferrol, N. Genome-wide analysis of copper, iron and zinc transporters in the arbuscular mycorrhizal fungus Rhizophagus irregularis. *Front. Plant Sci.* **2014**, *5*, 547. [CrossRef] [PubMed]
83. López-Lorca, V.M.; Molina-Luzón, M.J.; Ferrol, N. Characterization of the NRAMP Gene Family in the Arbuscular Mycorrhizal Fungus Rhizophagus irregularis. *J. Fungi* **2022**, *8*, 592. [CrossRef]
84. Li, M.; Wang, R.; Tian, H.; Gao, Y. Transcriptome responses in wheat roots to colonization by the arbuscular mycorrhizal fungus Rhizophagus irregularis. *Mycorrhiza* **2018**, *28*, 747–759. [CrossRef] [PubMed]
85. Zhang, M.; Zhong, X.; Li, M.; Yang, X.; Abou Elwafa, S.F.; Albaqami, M.; Tian, H. Genome-wide analyses of the Nodulin-like gene family in bread wheat revealed its potential roles during arbuscular mycorrhizal symbiosis. *Int. J. Biol. Macromol.* **2022**, *201*, 424–436. [CrossRef] [PubMed]
86. Haselwandter, K.; Haas, H.; Haninger, G.; Winkelmann, G. Siderophores in plant root tissue: Tagetes patula nana colonized by the arbuscular mycorrhizal fungus *Gigaspora margarita*. *Biometals* **2020**, *33*, 137–146. [CrossRef]
87. Rahman, M.A.; Parvin, M.; Das, U.; Ela, E.J.; Lee, S.H.; Lee, K.W.; Kabir, A.H. Arbuscular Mycorrhizal Symbiosis Mitigates Iron (Fe)-Deficiency Retardation in Alfalfa (*Medicago sativa* L.) through the Enhancement of Fe Accumulation and Sulfur-Assisted Antioxidant Defense. *Int. J. Mol. Sci.* **2020**, *21*, 2219. [CrossRef]
88. Prity, S.A.; Sajib, S.A.; Das, U.; Rahman, M.M.; Haider, S.A.; Kabir, A.H. Arbuscular mycorrhizal fungi mitigate Fe deficiency symptoms in sorghum through phytosiderophore-mediated Fe mobilization and restoration of redox status. *Protoplasma* **2020**, *257*, 1373–1385. [CrossRef]
89. Chorianopoulou, S.N.; Saridis, Y.I.; Dimou, M.; Katinakis, P.; Bouranis, D.L. Arbuscular mycorrhizal symbiosis alters the expression patterns of three key iron homeostasis genes, ZmNAS1, ZmNAS3, and ZmYS1, in S deprived maize plants. *Front. Plant Sci.* **2015**, *6*, 257. [CrossRef]
90. Vasconcelos, M.W.; Li, G.W.; Lubkowitz, M.A.; Grusak, M.A. Characterization of the PT Clade of Oligopeptide Transporters in Rice. *Plant Genome* **2008**, *1*, 77–88. [CrossRef]
91. Xie, X.; Hu, W.; Fan, X.; Chen, H.; Tang, M. Interactions Between Phosphorus, Zinc, and Iron Homeostasis in Nonmycorrhizal and Mycorrhizal Plants. *Front. Plant Sci.* **2019**, *10*, 1172. [CrossRef]
92. Azcón, R.; Ambrosano, E.; Charest, C. Nutrient acquisition in mycorrhizal lettuce plants under different phosphorus and nitrogen concentration. *Plant Sci.* **2003**, *165*, 1137–1145. [CrossRef]
93. Hoseinzade, H.; Ardakani, M.R.; Shahdi, A.; Rahmani, H.A.; Noormohammadi, G.; Miransari, M. Rice (*Oryza sativa* L.) nutrient management using mycorrhizal fungi and endophytic *Herbaspirillum seropedicae*. *J. Integr. Agric.* **2016**, *15*, 1385–1394. [CrossRef]
94. Andrino, A.; Guggenberger, G.; Kernchen, S.; Mikutta, R.; Sauheitl, L.; Boy, J. Production of Organic Acids by Arbuscular Mycorrhizal Fungi and Their Contribution in the Mobilization of Phosphorus Bound to Iron Oxides. *Front. Plant Sci.* **2021**, *12*, 661842. [CrossRef] [PubMed]
95. Pang, J.; Ryan, M.H.; Wen, Z.; Lambers, H.; Liu, Y.; Zhang, Y.; Tueux, G.; Jenkins, S.; Mickan, B.; Wong, W.S.; et al. Enhanced nodulation and phosphorus acquisition from sparingly-soluble iron phosphate upon treatment with arbuscular mycorrhizal fungi in chickpea. *Physiol. Plant.* **2023**, *175*, e13873. [CrossRef]
96. Ibiang, Y.B.; Mitsumoto, H.; Sakamoto, K. Bradyrhizobia and arbuscular mycorrhizal fungi modulate manganese, iron, phosphorus, and polyphenols in soybean (*Glycine max* (L.) Merr.) under excess zinc. *Environ. Exp. Bot.* **2017**, *137*, 1–13. [CrossRef]
97. Ibiang, Y.B.; Innami, H.; Sakamoto, K. Effect of excess zinc and arbuscular mycorrhizal fungus on bioproduction and trace element nutrition of Tomato (*Solanum lycopersicum* L. cv. Micro-Tom). *Soil Sci. Plant Nutr.* **2018**, *64*, 342–351. [CrossRef]
98. Wittenberg, J.B.; Wittenberg, B.A.; Day, D.A.; Udvardi, M.K.; Appleby, C.A. Siderophore-bound iron in the peribacteriod space of soybean root nodules. *Plant Soil* **1996**, *178*, 161–169. [CrossRef]

plants

MDPI

Review

Understanding the Mechanisms of Fe Deficiency in the Rhizosphere to Promote Plant Resilience

Zoltán Molnár [1,*], Wogene Solomon [1], Lamnganbi Mutum [1] and Tibor Janda [2]

[1] Department of Plant Sciences, Albert Kázmér Faculty of Mosonmagyaróvár, Széchenyi István University, H-9200 Mosonmagyaróvár, Hungary

[2] Agricultural Institute, Centre for Agricultural Research, H-2462 Martonvásár, Hungary

* Correspondence: molnar.zoltan@sze.hu

Abstract: One of the most significant constraints on agricultural productivity is the low availability of iron (Fe) in soil, which is directly related to biological, physical, and chemical activities in the rhizosphere. The rhizosphere has a high iron requirement due to plant absorption and microorganism density. Plant roots and microbes in the rhizosphere play a significant role in promoting plant iron (Fe) uptake, which impacts plant development and physiology by influencing nutritional, biochemical, and soil components. The concentration of iron accessible to these live organisms in most cultivated soil is quite low due to its solubility being limited by stable oxyhydroxide, hydroxide, and oxides. The dissolution and solubility rates of iron are also significantly affected by soil pH, microbial population, organic matter content, redox processes, and particle size of the soil. In Fe-limiting situations, plants and soil microbes have used active strategies such as acidification, chelation, and reduction, which have an important role to play in enhancing soil iron availability to plants. In response to iron deficiency, plant and soil organisms produce organic (carbohydrates, amino acids, organic acids, phytosiderophores, microbial siderophores, and phenolics) and inorganic (protons) chemicals in the rhizosphere to improve the solubility of poorly accessible Fe pools. The investigation of iron-mediated associations among plants and microorganisms influences plant development and health, providing a distinctive prospect to further our understanding of rhizosphere ecology and iron dynamics. This review clarifies current knowledge of the intricate dynamics of iron with the end goal of presenting an overview of the rhizosphere mechanisms that are involved in the uptake of iron by plants and microorganisms.

Keywords: rhizosphere; iron deficiency; iron acquisition; microorganisms; interaction

Citation: Molnár, Z.; Solomon, W.; Mutum, L.; Janda, T. Understanding the Mechanisms of Fe Deficiency in the Rhizosphere to Promote Plant Resilience. *Plants* **2023**, *12*, 1945. https://doi.org/10.3390/plants12101945

Academic Editor: Ferenc Fodor

Received: 14 April 2023
Revised: 4 May 2023
Accepted: 5 May 2023
Published: 10 May 2023

1. Introduction

Iron (Fe), the fourth most abundant and necessary micronutrient for the growth of plants and other organisms, is insoluble in neutral and alkaline soils, making it unavailable to plants. Fe is considered a vital element in the plant system for controlling life-sustaining processes, such as respiration, nitrogen fixation, photosynthesis, assimilation, the synthesis and repair of nucleotides, metal homeostasis, hormonal regulation, and chlorophyll production due to its redox-active nature under biological circumstances [1–5]. Fe can exist in two different oxidation states (Fe^{2+} and Fe^{3+}), and it can switch between them by receiving and giving away electrons. In crucial metabolic pathways, such as respiration and photosynthesis, which are needed for plants to make energy, iron plays a key role in enzyme reactions requiring electron transfer [2,6].

Although total Fe is a highly plentiful element in the soil, its accessibility to plants is generally quite low [7]. Thus, insufficiency of iron is one of the most significant limiting variables that influences crop yields, the quality of food, and human nutrition. Inadequate iron absorption results in interveinal chlorosis, stunted growth, reduced nutritional value, and diminished plant yield. Iron deficiency-induced anemia, one of the world's most common nutritional disorders, requires adequate iron levels in food crops [8]. According to reports, one third of the world's farmed lands suffer from Fe deficiency, resulting in a considerable annual drop in agricultural productivity, especially in calcareous soils [9,10].

Fe deficiency causes major changes in a plant's physiology and metabolism, slowing plant growth, impacting nutritional quality, and lowering yield [11], which, in turn, affect the health of people through the food chain, especially those with diets high in plant-based foods.

Plants must boost soil iron's mobility to overcome its limited availability. In response to Fe deficiency, plants have evolved sophisticated systems to maintain cellular Fe homeostasis by modifying their physiology, morphology, metabolism, and gene expression to enhance Fe availability [12,13]. Basically, these methods involve: (i) acidification, which is facilitated by the secretion of organic acids or protons; (ii) chelation of Fe^{3+} by ligands, which may include siderophores that have an extremely strong affinity for Fe^{3+}; and (iii) the reduction of Fe^{3+} to Fe^{2+} through the action of reductases and reducing substances [14–16].

Most of the physical, chemical, and biological activities in soil are linked to the geochemistry of iron (Fe) [17] and, consequently, to how much iron is available to the microorganisms and plants that grow in the soil. Iron is mostly found in the rhizosphere as Fe^{3+}, which is inaccessible to plants. The rhizosphere is a thin, dynamic zone with substantial abiotic and biotic interactions between soil microorganisms and plant roots [18]. Plant metabolism strongly influences the rhizospheric environment through the release of 5–21% of photosynthetic material by root exudates [19]. Rhizosphere activities and the rhizosphere's impact on plants are mostly controlled by the release of a complex combination of low and high molecular weight compounds from roots, such as carbohydrates, amino acids, organic acids, protons, phytosiderophores, enzymes, and phenolics [20]. Several microorganisms that interact with plants release siderophores in response to an iron deficiency [21]. These substances can modify the physical, biological, and chemical properties of the soil near the roots.

To improve the adaptability of economically significant plants to specific environmental and soil chemical conditions, it is necessary to comprehend the processes that govern the expression of Fe-deficiency responses in plants. Fe insufficiency is a common issue for various crops, especially those grown in calcareous soils, and it is one of the most significant factors reducing crop production. Increasing our knowledge on how plants deal with iron stress will help us to produce more stress-resistant crops in the future. This article discusses the potential for plant roots and microbes to be able to help plants absorb more iron in iron-deficient soils. The significant role of microorganisms in plant iron uptake has been highlighted by growing knowledge of the interaction between microbes and plants related to dynamics of iron in the rhizosphere.

2. Dynamics of Iron in the Rhizosphere

The dynamics of iron in the rhizosphere are controlled by a combination of factors, including the impacts of soil qualities, the absorption and activities of plants and microorganisms, as well as the interactions between these factors. There are two distinct ways that plants can obtain Fe under iron deficiency, namely strategy I and strategy II, to obtain iron from the rhizosphere in an efficient manner [22,23]. In strategy I, Fe is mobilized by the reduction system in most non-graminaceous species. The initial phase of this technique is to acidify the rhizosphere through the H^+ translocating P-type ATPase AHA2 [12]; this acidification increases Fe^{3+} solubility. To deal with the effects of Fe-deficiency stress in strategy I, plants often induce ferric chelate reductase and Fe(II) transporter in their root systems, acidify the rhizosphere media, and exude organic substances such as phenolics [24,25].

Strategy II (grass species) extracts iron from the soil via a chelation technique. This mechanism is highly reliant on the release of phytosiderophores (PSs) (such as mugineic acids, avenic acid, nicotinamine, etc.) by the root, which would result in the formation of stable iron-phosphate chelates. The Fe(III)–phytosiderophore complex transporter known as yellow stripe1 is a membrane protein that facilitates iron absorption [26]. Temperature, rhizosphere pH, and the type of electrolyte may have a major impact on strategy II Fe acquisition by changing the timing and concentration of the "window of Fe absorption." [27]. The two strategies rely on molecular mechanisms that are carefully controlled and involve two main components: the root, especially its cell of plasma membrane, and the rhizosphere, which is the soil in closest proximity to the roots [28,29]. Plants have special proteins known as "transporters" that are present in the plasma membranes, and they assist plants in transferring molecules either into or out of the roots as required. Under conditions of severe Fe shortage, phytosiderophores may account for 50–90% of the exudates secreted at the root tip (Fan et al., 1997, as cited in [30]). It is challenging to differentiate between plant species based on their iron uptake mechanism because certain plants, for example, rice or peanut, use both strategies [31,32]. The influence of plants and microorganisms on iron dynamics and status in the rhizosphere, especially on iron solubilization, has been investigated and presented.

2.1. *Status of Fe in the Rhizosphere and Soil*

The rhizosphere is the active zone across a plant root that is home to a diverse population of microorganisms and is impacted by the chemicals produced by plant roots. Rhizosphere processes are the communications between plant roots–soil–microbes that occur and alter continually, impacting things such as nutrient solubility, their movement through the soil, and plant absorption. These systems' primary driving force seems to be tied to processes of root exudation (Figure 1). Root exudates are organic and inorganic chemicals released by plant roots. They include high and low molecular weight substances, such as carbohydrates, proteins, amino acids, organic acids, protons, polypeptides, enzymes, and hormones, in the rhizospheric soil environment [33,34]. Rhizosphere priming effect occurs when plant roots release recently formed photosynthates into the rhizosphere, which speeds up the breakdown of organic materials by saprotrophic soil bacteria and increases plant nutrient availability [35]. Increased root exudates in the soil improve microbial biomass and soil fertility levels. The dynamics of Fe in the rhizosphere can also be affected by organic compounds generated by the degradation of soil organic matter. These soil microorganisms are essential for the nutrient transformation in the soil and crop plant nutrition absorption. Plants may affect soil qualities by modifying the composition of root exudates, allowing them to adapt and survive under severe environments.

Iron is one of the most plentiful elements in the soil, but after it is weathered, Fe(III) and Fe(II) ions can be released through dissolution and oxidation/reduction. However, when hydroxyl ions (OH^-) are present, it almost always forms Fe hydroxides and oxides, which have very poor solubility [14]. The dissolution and solubility rates of pedogenic iron oxides (oxides, oxyhydroxides, and hydroxides) play an important role in regulating iron accessibility. The dissolution and solubility rates of iron soil oxides are also significantly affected by pH, microbial population, organic matter content, redox processes, and particle size of the soil [7,14,15,36,37]. Soil pH is the most important of these parameters since it can decrease Fe availability by as much as 95% for every unit increase in soil pH above neutral [38]. When the pH is lowered, the ferric iron is released from its bond with the oxide, making it easier for the roots of plants to absorb it [25]. Fe is transformed to an insoluble Fe–hydroxyl compound in salty, calcareous, alkaline, and sodic soils, which prevents the element from being taken up by plant roots [36]. Soil organic matter level and its breakdown rate affect Fe accessibility because of the formation of excess bicarbonates and phosphates, which hinder the uptake of Fe [4,39].

One of the most significant constraints on agricultural productivity is the low accessibility of iron (Fe) in the soil, which is directly linked to the biological, physical, and chemical activities taking place in the rhizosphere due to the interactions between the soil, microorganisms, and plants [20]. It is widely known that plant roots can alter the pH of the rhizosphere by releasing protons through the H-ATPase enzyme in epidermal cells [40]. This can also occur during Fe deficiency; thus, the plant's impact on pH can result in the exudation of inorganic metals through the plant roots. Iron deficiency causes soil organisms to emit organic (carbohydrates, amino acids, organic acids, phytosiderophores, phenolics, siderophores, and enzymes) and inorganic (protons) chemicals to improve the solubility of inaccessible Fe pools [20]. Soil pH can be lowered by the plant's secretion of low molecular weight organic acids [40]. Thus, in order for microorganisms to survive and flourish, the rhizosphere is a geographically and temporally uneven habitat with quick changes in potentially harsh conditions, such as cycles of water stress and anaerobiosis.

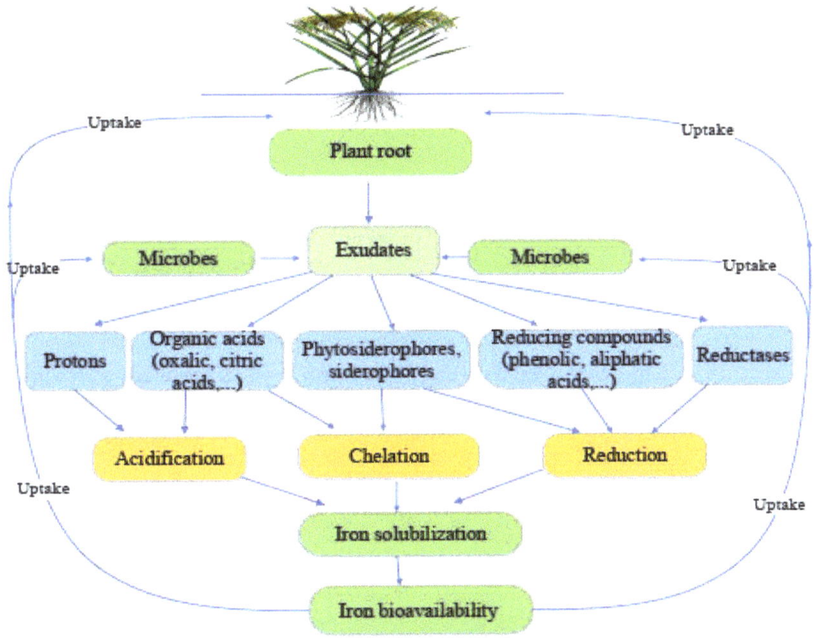

Figure 1. Mechanisms that alter the availability of iron in the rhizosphere. Plants and microbes can improve the bioavailability of iron via (a) acidification—secretion of protons and organic acids, (b) chelation—excretion of complexing molecules with varying affinities for Fe (siderophores, phytosiderophores, carboxylic acids, and phenolics), and (c) reduction—release of substances with reducing characteristics or development of a membrane-bound reductase activity.

It has been extensively documented that metal complexation by humic substances derived from various sources improves plant iron nutrition. The chelation of Fe^{3+} by the organic ligands that comprise the dissolved organic matter has a substantial effect on the solubility of soil iron as well. Reported by [41], more than 95% of the Fe in soil solution is probably complexed or chelated. Depending on the molecular size of humic substances (HS) and solubility, the presence of humified fractions of organic matter in soil sediments and solutions might help provide a reservoir of Fe for plants that exhale metal ligands and supply Fe–HS complexes that are directly utilizable by plant Fe absorption processes [42]. In addition to having iron-chelating qualities, which help enhance iron bioavailability, humic substances also exhibit redox-reactive properties [43].

2.2. Iron Interaction with Plant and Rhizospheric Microorganisms

In the rhizosphere, iron competition is important for microbial and plant–microbe interactions. Competition for Fe occurs among microbes and plants, regarding which has the competitive edge due to their capacity to break down plant-derived chelators and their closeness to the surface of the root. However, plants might avoid direct competition with microbes because the amount and type of exudates they release into the rhizosphere change over time and space [30]. Plant to plant interactions, as well as microbial interactions in non-sterile growth circumstances, can modify the iron status of plants. It is well known that the microorganisms in the soil have a substantial impact on the iron nutrition of plants. The iron content of plants can be significantly increased by intercropping grain and legumes [44]. The intercropping of wheat and chickpeas raised the Fe content in wheat seeds [45], whereas the intercropping of maize and peanuts improved the Fe nutrition of peanuts in a calcareous soil [46]. So, we might postulate that the rhizosphere microbes are responsible for the higher iron absorption with intercropped plants.

Rhizosphere microbes live in an environment where plant activity has a substantial effect on the accessibility of nutrients. In the rhizosphere, a wide variety of biotic interactions take place that might influence the composition and diversity of the microbial population in the soil near the roots. These species' uptake of iron results in complicated interactions, ranging from mutualism to competition [47]. The organization of the microbial community is typically influenced more by biotic interactions in the rhizosphere than by abiotic factors, which are more common in the bulk soil. By excreting rhizodeposits into the rhizosphere, plants provide a fertile and dynamic environment for the microbial populations. The content of iron in solution is further reduced by the iron absorption of these microbes and the host plant. As a result, there is high competition among rhizosphere microbes for iron, encouraging those with the most effective iron absorption strategy [47].

2.3. *Impact of Plants and Microorganisms on the Iron Status*

Plants and microorganisms play important roles in the cycling and availability of iron in the environment. Plant-associated microbes may promote plant development and affect crop output and quality by mobilizing and transporting nutrients [48]. It has been proven that soil microbes play a significant role in promoting plant iron (Fe) absorption in Fe-limiting situations [24,49]. Plant roots and rhizospheric microorganisms release substances such as organic acids, proteins, phenolics, phytosiderophores, and siderophores, which can promote the solubilization of low-availability iron in the soil [14,20,50].

A research report showed that in Chinese cabbage leaves and stalks the concentration of soluble protein, soluble sugar, and vitamin C was significantly decreased under Fe-deficiency stress conditions, whereas the content of cellulose and nitrate was increased [51]. The same study found that Fe-deficiency stress significantly lowered net photosynthetic rate and nitrate reductase activity in the leaves. Iron shortage in the rhizosphere resulted in a 40% rise in root biomass as well as elevated levels of citrate, malate, and phenols in root exudates [52]. The increase in root biomass and elevated levels of these compounds in the root exudates in response to iron deficiency are part of the plant's adaptive response to this micronutrient limitation.

In Fe-limiting situations, plants and soil microorganisms have used active strategies to enhance soil iron availability, which plays a key role in promoting iron absorption. In the rhizosphere, iron oxides are more easily soluble and dissolvable due to processes such as acidification, chelation, and reduction (Figure 1).

2.3.1. Acidification

Plants that experience a shortage of iron may adopt various methods to enhance their absorption of iron. In strategy I plants, such as Arabidopsis, this involves the release of protons by plasma membrane (PM)-localized H^+-ATPases (AHAs) to increase acidity in the rhizosphere, which aids in the solubilization of Fe^{3+} [53]. These Fe^{3+} complexes are then reduced to Fe^{2+} and taken up by plants. The process of rhizosphere acidification occurs when H^+-ATPases, which are bound to the plasma membrane, expel protons from the symplastic area into the rhizosphere, which plays a critical role in nutrient acquisition by plants [5,54]. The FER-like iron deficiency-induced transcription factor (FIT) is crucial for the activation of genes related to iron acquisition and uptake in the root cluster. Its expression is induced in response to iron deficiency and plays a key role in regulating the upregulation of these genes [16,55,56]. FIT is upregulated under low-iron conditions and activates the expression of downstream genes that encode transporters and enzymes involved in iron acquisition, such as the H^+ -ATPases responsible for rhizosphere acidification and enzymes responsible for iron reduction. This mechanism improves the absorption of iron by enhancing the solubility of iron-containing substances in the soil, creating favorable conditions for iron reduction, and generating a proton motive force that aids in the uptake of irons [15,57].

Acidification helps iron become more soluble in the rhizosphere, which is beneficial to plant health. In the rhizosphere, the roots of plants also release organic acids, such as citric acid, malic acid, and oxalic acid, into the soil. These organic acids can chelate cations, including iron (Fe), making them more available to the plant. Acidification occurs as a result of the secretion of protons and organic acids by microorganisms and plants, resulting in proton concentrations in the rhizosphere that are up to 100-fold greater than in bulk soil [14]. Protons are produced through microbial processes such as nitrification [58]. Plants and microbes release protons into the rhizosphere, which makes it more acidic [59,60].

An insufficiency of iron causes a variety of reactions in soil organisms, releasing inorganic (protons) and organic (carbohydrates, amino acids, organic acids, siderophores, phytosiderophores, enzymes, and phenolics) substances that enhance the solubility of low-accessibility Fe pools. Nevertheless, due to the high microbial activity at the soil–root interface, rhizospheric organic compounds (ROCs) have short half-lives, which may restrict their impact on Fe mobility and acquisition [20]. Overall, the acidification of the rhizosphere caused by proton and organic acid release from plant roots and microbes can increase the availability of iron for the plant and promote optimal growth and development.

Despite the advantages it offers, the degree and adaptability of rhizosphere acidification vary significantly among different plant species. In response to iron deficiency, woody and herbaceous plants such as cork oak (*Quercus suber*), M. plum (*Prunus cerasifera E.*), and cucumber (*Cucumis sativus*) can increase proton extrusion to acidify their rhizosphere [57,61,62]. Similarly, when tomato roots are subjected to Fe deficiency, an increased proton extrusion was detected, and Fe-deficient roots exhibit a greater number of proteins that react with monoclonal antibodies targeting a P-type proton-ATPase from maize compared to Fe-sufficient roots [63]. In contrast, peach–almond hybrids (*Prunus amygdalus B. x Prunus persica B.*), *Vaccinium arboreum* (VA), Southern highbush blueberry (SHB), and wild apple (*Malus baccata*) lack the ability to acidify their rhizosphere through proton extrusion in response to iron deficiency [62,64,65]. Moreover, certain species, such as grapevine (*Vitis vinifera*), display diversity in their responses within the same species, for instance, "Cabernet Sauvignon" can acidify its rhizosphere, while "Balta" cannot [66,67].

2.3.2. Chelation

Chelation, which involves a strategy II-based mechanism, is well known in gramineous plants. Grasses have the ability to dissolve and absorb Fe from the relatively insoluble inorganic Fe(III) by secreting chelators, which are non-protein amino acids such as mugineic acid and avenic acid. The roots release chelators called phytosiderophores (PSs) that bind with Fe(III) in the rhizosphere and create a complex called Fe(III)–PS, which is transported into the plant via specific plasmalemma transporter proteins [13,68,69]. The release of MAs from plants increases considerably under Fe-deficient conditions, and their capacity to withstand such conditions is closely linked to the quantity of MA that they produce and release. This results in the formation of Fe(III)–PS complexes, which can then be absorbed by the plant roots. The Fe(III)–PS complex is transported into the root cells through the transporters known as yellow stripe (YS) to yellow stripe-like (YSL) after binding to them [70]. Some crops, such as sorghum, rice, and maize, produce only low amounts of deoxymugineic acid (DMA), making them highly vulnerable to Fe deficiency [23]. On the other hand, barley secretes various types of MAs, including mugineic acid, 3-hydroxymugineic acid (HMA), and 3-epi-hydroxymugineic acid (epi-HMA), in relatively high amounts, making it more resistant to low Fe availability [71].

Microorganisms secrete low molecular weight biomolecules that function as chelating agents for metal ions. These molecules are called microbial siderophores and have a greater affinity for Fe^{3+} than phtosiderophores such as mugineic acids, as reported in studies [72]. A significant population of siderophore-producing bacteria is typically found in the rhizosphere, and their secretions enhance the movement and accessibility of metal ions, thereby promoting phytoremediation [73]. When it comes to iron, the existence of microbes that produce siderophores within the rhizosphere has been shown to improve morphological responses. The process by which siderophores mobilize Fe is not yet fully understood. *Cupriavidus necator*, a bacterium that degrades pollutants, has been found to produce a siderophore called cuprabactin, which is utilized by the bacteria to overcome Fe deficiency [74]. This study could be used in the future to introduce microbes to the rhizosphere as a way to help plants acquire more iron. In reaction to an iron deficiency, almost all microbes make siderophores, which are chelating agents for iron. Siderophore-mediated iron absorption is important in regulating the capacity of various microbes to colonize plant roots and promotes microbial associations in the rhizosphere [75]. Siderophores, which are produced by graminaceous plants and by microorganisms, are among those that have a role in the active iron-uptake strategies.

When iron deficiency occurs, the roots of plants begin to secrete phytosiderophores, while citrate has been shown to be an important chelator of iron [76,77]. Iron is also made more soluble in the rhizosphere by its chelation with siderophores and organic acids, which remove ferric iron from insoluble forms and make it accessible to plants (Figure 1) [24,78]. Most aerobic microorganisms make small molecules that have a strong attraction to ferric iron. These molecules, called siderophores, help plants obtain the iron nutrient in iron stress conditions. In addition to being varied in size and chemical makeup, microbial siderophores exhibit a high but variable affinity for ferric iron (Fe^{3+}) [79,80]. Hence, microbial siderophores improve the solubility of iron in the soil, which can benefit plant growth and development.

The research suggested that red clover's iron-deficiency stress may modify the structure of siderophore-emitting microorganisms in the rhizosphere, possibly because of the root's phenolic secretion, which may increase soil Fe solubility and plant Fe nutrition [24]. Additionally, in rhizosphere soil, a higher concentration of phenolics and number of microorganisms that released siderophores were found in the Fe-deficient treatment than in the Fe-sufficient treatment [24]. However, data have suggested that in a Fe-limited situation, the responses caused by Fe deficiency are not enough for plants to avoid Fe deficiency. According to [81], sunflowers and maize grown in sterile soil grew slowly and had less iron in their tissues than plants grown in non-sterile soil. Similar results were seen for red clover (*Trifolium pratense*) and rape (*Brassica napus*), which grew substantially more slowly in sterile soil and absorbed less iron [82,83]. Furthermore, it was discovered that the barley plant Fe nutritional status affected the microbial population in the rhizosphere, and it was hypothesized that this was affected by variations in root exudates [84]. Hence, it would seem that soil microbial activity is important in promoting plant absorption of iron. Even though some new information has been provided, the precise processes behind the positive influence of microbial siderophores on plant nutrition are still unknown [85].

2.3.3. Reduction

Regarding Fe deficiency, many studies have investigated root responses to this stress condition, showing that most dicot and some monocot species increase NADPH-dependent reductase activity and ATPase-driven proton efflux pumps to solubilize inorganic Fe(III) in the rhizosphere, which enhances Fe(III) reduction to Fe(II) [23,86]. The reduction of Fe(III) in ferric-specific chelates refers to the process of converting Fe(III) ions that are bound to chelating agents or ligands into Fe(II) ions. Reduction of Fe(III) in ferric-specific chelates leads to complex breaking and metal ion release, with the resultant Fe^{2+} being the species of Fe absorbed by the roots [87]. The reduction strategy in the plant Fe uptake process is essential for plant growth and development, as Fe is a critical nutrient required for many important biochemical processes. However, the reduction process is also energy-intensive and can be inhibited by various factors, such as high pH, high levels of calcium, or low levels oxygen [7,15]. As a result, plants have evolved various strategies to cope with Fe deficiency, including the production of various Fe chelators and the modulation of ferric reductase activity. Fe(III) reduction is a necessary and indispensable step in the process of iron uptake by strategy I plant species, since these plants are only capable of absorbing iron in the form of Fe(II). In strategy I, Fe(III) reduction is primarily mediated by reductases, the activities of which are enhanced in response to iron deprivation [14]. H^+ extrusion into the rhizosphere stimulates its reductase activity in the plasma membrane, and in calcareous soils, high levels of HCO_3^- inhibit it. The reduction strategy in the plant Fe uptake process is based on the ability of the plant to release reducing agents that can convert Fe(III) to Fe(II) in the rhizosphere.

The reduction of Fe(III) in the rhizosphere is mediated by two types of compounds secreted by the plant roots: phenolic compounds and organic acids. Phenolic compounds, such as flavins, coumarins, and catechols, are able to reduce Fe(III) to Fe(II) by donating electrons to the Fe(III) ions. Organic acids, such as citrate, malate, and oxalate, also play a role in Fe(III) reduction by chelating Fe(III) ions and facilitating their reduction by donating protons or electrons. When acidified, membrane-bound ferric reductase oxidase (FRO) such as FRO2 reduces Fe^{3+} to Fe^{2+}. Arabidopsis FRO2 was initially recognized as the enzyme that reduces ferric iron chelates at the interface of the root surface and rhizosphere, functioning as a ferric chelate reductase [88]. FRO2 is the specific enzyme accountable for the reduction of Fe(III) chelates in the plasma membrane, which occurs in response to iron deficiency in Arabidopsis roots [89]. The expression of FRO genes is not restricted to the root plasma membrane which suggests that FRO genes play a role in reduction-based Fe transport in other plant organs as well [13,90]. Plants with higher FRO expression exhibit resistance to growth under low iron conditions [91]. Once Fe(III) is reduced to Fe(II) in the rhizosphere, it can be taken up by the root cells through a family of integral membrane proteins called iron-regulated transporter (IRT1) [23,92]. IRT1 is a predominant Fe transporter in *Arabidopsis*, belonging to the ZRT IRT-like protein (ZIP) family. Both IRT1 and FRO2 have specific expression patterns in the root epidermis, where they play crucial roles in the uptake of iron from the soil and are essential for plant growth [93]. The expression of IRT1 is quickly increased upon Fe deficiency, which is likely influenced by signals from both the roots and shoots [94]. These proteins are responsible for the transport of Fe(II) across the plasma membrane into the root cells, where it can be further transported to other parts of the plant.

3. Conclusions

Even though iron is one of the most abundant metals on Earth, its bioavailability is considerably decreased due to the low solubility of Fe(III) oxyhydroxide particles, which are the most common form in neutral pH to alkaline soil and under oxygenated conditions. Due to the enormous demand for Fe(III) in the rhizosphere and its poor accessibility in soils, there is intense competition among living organisms for this nutrient. Iron deficiency in the rhizosphere can have a significant negative impact on plant growth and productivity. However, plants have evolved several mechanisms to increase their resilience to iron deficiency and maintain their growth and development, even in low-iron environments. The mechanisms that plants have developed are active techniques for acquiring iron that rely on acidification and reduction of Fe(III) (strategy I) as well as the production of phytosiderophores (strategy II), enabling plants to adjust more effectively to environments with low levels of iron. Plants emit a large portion of their photosynthates as rhizodeposits, which increase microbial population and activity. The roots of plants can influence the rhizosphere microbiome, by forming distinct chemical niches in the soil through the production of phytochemicals (i.e., root exudates), which is dependent on several factors, such as soil characteristics, plant genotype, climatic conditions, and plant nutritional status. During iron deficiency, both plant roots and microorganisms in the soil can release various compounds, such as organic acids, phenolics, siderophores, reductants, and enzymes, into the rhizosphere. By releasing these compounds, plants can mobilize and acquire more iron, ultimately promoting growth and survival in iron-limited environments. As a result, better understanding these dynamics is a critical issue for increasing plant iron nutrition and productivity in sustainable agriculture.

Author Contributions: Z.M. and W.S. designed, wrote, and approved the contents, L.M. designed the content and edited, and T.J. approved and edited the manuscript. All authors have read and agreed to the published version of the manuscript.

Funding: This work was funded by a grant from the National Research, Development, and Innovation Office (grant No. K142899).

Data Availability Statement: Not applicable.

Conflicts of Interest: The authors declare no conflict of interest.

References

1. Tripathi, D.K.; Singh, S.; Gaur, S.; Singh, S.; Yadav, V.; Liu, S.; Singh, V.P.; Sharma, S.; Srivastava, P.; Prasad, S.M.; et al. Acquisition and homeostasis of iron in higher plants and their probable role in abiotic stress tolerance. *Front. Environ. Sci.* **2018**, *5*, 86. [CrossRef]
2. Balk, J.; Schaedler, T.A. Iron cofactor assembly in plants. *Annu. Rev. Plant Biol.* **2014**, *65*, 125–153. [CrossRef] [PubMed]
3. Rotaru, V.; Sinclair, T.R. Interactive influence of phosphorus and iron on nitrogen fixation by soybean. *Environ. Exp. Bot.* **2009**, *66*, 94–99. [CrossRef]
4. Mahender, A.; Swamy, B.P.M.; Anandan, A.; Ali, J. Tolerance of iron-deficient and -toxic soil conditions in rice. *Plants* **2019**, *8*, 31. [CrossRef]
5. Kim, S.A.; Guerinot, M.L. Mining iron: Iron uptake and transport in plants. *FEBS Lett.* **2007**, *581*, 2273–2280. [CrossRef] [PubMed]
6. Kanwar, P.; Baby, D.; Bauer, P. Interconnection of iron and osmotic stress signalling in plants: Is fit a regulatory hub to cross-connect abscisic acid responses? *Plant Biol.* **2021**, *23* (Suppl. S1), 31–38. [CrossRef]
7. Colombo, C.; Palumbo, G.; He, J.-Z.; Pinton, R.; Cesco, S. Review on iron availability in soil: Interaction of fe minerals, plants, and microbes. *J. Soils Sediments* **2014**, *14*, 538–548. [CrossRef]
8. Schmidt, W.; Thomine, S.; Buckhout, T.J. Editorial: Iron nutrition and interactions in plants. *Front. Plant Sci.* **2019**, *10*, 1670. [CrossRef] [PubMed]
9. Zuo, Y.; Zhang, F. Soil and crop management strategies to prevent iron deficiency in crops. *Plant Soil* **2011**, *339*, 83–95. [CrossRef]
10. Radzki, W.; Gutierrez Mañero, F.J.; Algar, E.; Lucas García, J.A.; García-Villaraco, A.; Ramos Solano, B. Bacterial siderophores efficiently provide iron to iron-starved tomato plants in hydroponics culture. *Antonie Leeuwenhoek* **2013**, *104*, 321–330. [CrossRef] [PubMed]
11. Briat, J.F.; Dubos, C.; Gaymard, F. Iron nutrition, biomass production, and plant product quality. *Trends Plant Sci.* **2015**, *20*, 33–40. [CrossRef] [PubMed]
12. Ivanov, R.; Brumbarova, T.; Bauer, P. Fitting into the harsh reality: Regulation of iron-deficiency responses in dicotyledonous plants. *Mol. Plant* **2012**, *5*, 27–42. [CrossRef] [PubMed]
13. Kobayashi, T.; Nishizawa, N.K. Iron uptake, translocation, and regulation in higher plants. *Annu. Rev. Plant Biol.* **2012**, *63*, 131–152. [CrossRef] [PubMed]
14. Robin, A.; Vansuyt, G.; Hinsinger, P.; Meyer, J.M.; Briat, J.F.; Lemanceau, P. Chapter 4 iron dynamics in the rhizosphere: Consequences for plant health and nutrition. In *Advances in Agronomy*; Academic Press: Cambridge, MA, USA, 2008; pp. 183–225. [CrossRef]
15. Lemanceau, P.; Bauer, P.; Kraemer, S.; Briat, J.-F. Iron dynamics in the rhizosphere as a case study for analyzing interactions between soils, plants and microbes. *Plant Soil* **2009**, *321*, 513–535. [CrossRef]
16. Naranjo-Arcos, M.A.; Maurer, F.; Meiser, J.; Pateyron, S.; Fink-Straube, C.; Bauer, P. Dissection of iron signaling and iron accumulation by overexpression of subgroup ib bhlh039 protein. *Sci. Rep.* **2017**, *7*, 10911. [CrossRef] [PubMed]
17. Carrillo-González, R.; Šimůnek, J.; Sauvé, S.; Adriano, D. Mechanisms and pathways of trace element mobility in soils. In *Advances in Agronomy*; Academic Press: Cambridge, MA, USA, 2006; pp. 111–178. [CrossRef]
18. Mueller, C.W.; Carminati, A.; Kaiser, C.; Subke, J.-A.; Gutjahr, C. Rhizosphere functioning and structural development as complex interplay between plants, microorganisms and soil minerals. *Front. Environ. Sci.* **2019**, *7*, 130. [CrossRef]
19. Badri, D.V.; Chaparro, J.M.; Zhang, R.; Shen, Q.; Vivanco, J.M. Application of natural blends of phytochemicals derived from the root exudates of Arabidopsis to the soil reveal that phenolic-related compounds predominantly modulate the soil microbiome. *J. Biol. Chem.* **2013**, *288*, 4502–4512. [CrossRef] [PubMed]
20. Mimmo, T.; Del Buono, D.; Terzano, R.; Tomasi, N.; Vigani, G.; Crecchio, C.; Pinton, R.; Zocchi, G.; Cesco, S. Rhizospheric organic compounds in the soil-microorganism-plant system: Their role in iron availability. *Eur. J. Soil Sci.* **2014**, *65*, 629–642. [CrossRef]
21. Winkelmann, G.s. Ecology of siderophores with special reference to the fungi. *Biometals* **2007**, *20*, 379–392. [CrossRef]
22. Römheld, V.; Marschner, H. Evidence for a specific uptake system for iron phytosiderophores in roots of grasses. *Plant Physiol.* **1986**, *80*, 175–180. [CrossRef] [PubMed]

23. Dey, S.; Regon, P.; Kar, S.; Panda, S.K. Chelators of iron and their role in plant's iron management. *Physiol. Mol. Biol. Plants* **2020**, *26*, 1541–1549. [CrossRef] [PubMed]
24. Jin, C.W.; Li, G.X.; Yu, X.H.; Zheng, S.J. Plant fe status affects the composition of siderophore-secreting microbes in the rhizosphere. *Ann. Bot.* **2010**, *105*, 835–841. [CrossRef] [PubMed]
25. Morrissey, J.; Guerinot, M.L. Iron uptake and transport in plants: The good, the bad, and the ionome. *Chem. Rev.* **2009**, *109*, 4553–4567. [CrossRef] [PubMed]
26. Curie, C.; Cassin, G.; Couch, D.; Divol, F.; Higuchi, K.; Le Jean, M.; Misson, J.; Schikora, A.; Czernic, P.; Mari, S. Metal movement within the plant: Contribution of nicotianamine and yellow stripe 1-like transporters. *Ann. Bot.* **2008**, *103*, 1–11. [CrossRef] [PubMed]
27. Walter, M.; Kraemer, S.M.; Schenkeveld, W.D.C. The effect of ph, electrolytes and temperature on the rhizosphere geochemistry of phytosiderophores. *Plant Soil* **2017**, *418*, 5–23. [CrossRef]
28. Marzorati, F.; Midali, A.; Morandini, P.; Murgia, I. Good or bad? The double face of iron in plants. *Front. Young Minds* **2022**, *10*, 718162. [CrossRef]
29. Jeong, J.; Connolly, E.L. Iron uptake mechanisms in plants: Functions of the fro family of ferric reductases. *Plant Sci.* **2009**, *176*, 709–714. [CrossRef]
30. Marschner, P.; Crowley, D.; Rengel, Z. Rhizosphere interactions between microorganisms and plants govern iron and phosphorus acquisition along the root axis–model and research methods. *Soil Biol. Biochem.* **2011**, *43*, 883–894. [CrossRef]
31. Xiong, H.; Kakei, Y.; Kobayashi, T.; Guo, X.; Nakazono, M.; Takahashi, H.; Nakanishi, H.; Shen, H.; Zhang, F.; Nishizawa, N.K.; et al. Molecular evidence for phytosiderophore-induced improvement of iron nutrition of peanut intercropped with maize in calcareous soil. *Plant Cell Environ.* **2013**, *36*, 1888–1902. [CrossRef] [PubMed]
32. Ishimaru, Y.; Suzuki, M.; Tsukamoto, T.; Suzuki, K.; Nakazono, M.; Kobayashi, T.; Wada, Y.; Watanabe, S.; Matsuhashi, S.; Takahashi, M.; et al. Rice plants take up iron as an Fe^{3+}-phytosiderophore and as Fe^{2+}. *Plant J.* **2006**, *45*, 335–346. [CrossRef] [PubMed]
33. Upadhyay, S.K.; Srivastava, A.K.; Rajput, V.D.; Chauhan, P.K.; Bhojiya, A.A.; Jain, D.; Chaubey, G.; Dwivedi, P.; Sharma, B.; Minkina, T. Root exudates: Mechanistic insight of plant growth promoting rhizobacteria for sustainable crop production. *Front. Microbiol.* **2022**, *13*, 916488. [CrossRef] [PubMed]
34. Mimmo, T.; Pii, Y.; Valentinuzzi, F.; Astolfi, S.; Lehto, N.; Robinson, B.; Brunetto, G.; Terzano, R.; Cesco, S. *Nutrient Availability in the Rhizosphere: A Review*; International Society for Horticultural Science (ISHS): Leuven, Belgium, 2018. [CrossRef]
35. Gorka, S.; Dietrich, M.; Mayerhofer, W.; Gabriel, R.; Wiesenbauer, J.; Martin, V.; Zheng, Q.; Imai, B.; Prommer, J.; Weidinger, M.; et al. Rapid transfer of plant photosynthates to soil bacteria via ectomycorrhizal hyphae and its interaction with nitrogen availability. *Front. Microbiol.* **2019**, *10*, 168. [CrossRef] [PubMed]
36. Mengel, K.; Kirkby, E.A.; Kosegarten, H.; Appel, T. Iron. In *Principles of Plant Nutrition*; Mengel, K., Kirkby, E.A., Kosegarten, H., Appel, T., Eds.; Springer: Dordrecht, The Netherlands, 2001; pp. 553–571. [CrossRef]
37. Johnson, D.B.; Kanao, T.; Hedrich, S. Redox transformations of iron at extremely low ph: Fundamental and applied aspects. *Front. Microbiol.* **2012**, *3*, 96. [CrossRef] [PubMed]
38. Rengel, Z.s. Availability of mn, zn and fe in the rhizosphere. *J. Soil Sci. Plant Nutr.* **2015**, *15*, 397–409. [CrossRef]
39. Onaga, G.; Edema, R.; Aseas, G. Tolerance of rice germplasm to iron toxicity stress and the relationship between tolerance, Fe^{2+}, P and K content in the leaves and roots. *Arch. Agron. Soil Sci.* **2013**, *59*, 213–229. [CrossRef]
40. Hawkes, C.V.; DeAngelis, K.M.; Firestone, M.K. Chapter 1—Root interactions with soil microbial communities and processes. In *The Rhizosphere*; Cardon, Z.G., Whitbeck, J.L., Eds.; Academic Press: Burlington, VT, USA, 2007; pp. 1–29. [CrossRef]
41. Van Hees, P.a.W.; Lundströms, U.S. Equilibrium models of aluminium and iron complexation with different organic acids in soil solution. *Geoderma* **2000**, *94*, 201–221. [CrossRef]
42. Zanin, L.; Tomasi, N.; Cesco, S.; Varanini, Z.; Pinton, R. Humic substances contribute to plant iron nutrition acting as chelators and biostimulants. *Front. Plant Sci.* **2019**, *10*, 675. [CrossRef]
43. Weber, K.A.; Achenbach, L.A.; Coates, J.D. Microorganisms pumping iron: Anaerobic microbial iron oxidation and reduction. *Nat. Rev. Genet.* **2006**, *4*, 752–764. [CrossRef]
44. Xue, Y.; Xia, H.; Christie, P.; Zhang, Z.; Li, L.; Tang, C. Crop acquisition of phosphorus, iron and zinc from soil in cereal/legume intercropping systems: A critical review. *Ann. Bot.* **2016**, *117*, 363–377. [CrossRef]
45. Gunes, A.; Inal, A.; Adak, M.S.; Alpaslan, M.; Bagci, E.G.; Erol, T.; Pilbeam, D.J. Mineral nutrition of wheat, chickpea and lentil as affected by mixed cropping and soil moisture. *Nutr. Cycl. Agroecosystems* **2007**, *78*, 83–96. [CrossRef]
46. Zuo, Y.; Liu, Y.; Zhang, F.; Christie, P. A study on the improvement iron nutrition of peanut intercropping with maize on nitrogen fixation at early stages of growth of peanut on a calcareous soil. *Soil Sci. Plant Nutr.* **2004**, *50*, 1071–1078. [CrossRef]
47. Lemanceau, P.; Expert, D.; Gaymard, F.; Bakker, P.; Briat, J.F. Chapter 12 role of iron in plant–microbe interactions. In *Advances in Botanical Research*; Academic Press: Cambridge, MA, USA, 2009; pp. 491–549. [CrossRef]
48. Pii, Y.; Borruso, L.; Brusetti, L.; Crecchio, C.; Cesco, S.; Mimmo, T. The interaction between iron nutrition, plant species and soil type shapes the rhizosphere microbiome. *Plant Physiol. Biochem.* **2016**, *99*, 39–48. [CrossRef] [PubMed]
49. Lurthy, T.; Pivato, B.; Lemanceau, P.; Mazurier, S. Importance of the rhizosphere microbiota in iron biofortification of plants. *Front. Plant Sci.* **2021**, *12*, 744445. [CrossRef] [PubMed]

50. Fujii, K. Soil acidification and adaptations of plants and microorganisms in bornean tropical forests. *Ecol. Res.* **2014**, *29*, 371–381. [CrossRef]
51. Wang, Y.; Kang, Y.; Zhong, M.; Zhang, L.; Chai, X.; Jiang, X.; Yang, X. Effects of iron deficiency stress on plant growth and quality in flowering chinese cabbage and its adaptive response. *Agronomy* **2022**, *12*, 875. [CrossRef]
52. M'sehli, W.; Youssfi, S.; Donnini, S.; Dell'orto, M.; De Nisi, P.; Zocchi, G.; Abdelly, C.; Gharsalli, M. Root exudation and rhizosphere acidification by two lines of medicago ciliaris in response to lime-induced iron deficiency. *Plant Soil* **2008**, *312*, 151–162. [CrossRef]
53. Zhang, X.; Zhang, D.; Sun, W.; Wang, T. The adaptive mechanism of plants to iron deficiency via iron uptake, transport, and homeostasis. *Int. J. Mol. Sci.* **2019**, *20*, 2424. [CrossRef] [PubMed]
54. Houmani, H.; Rabhi, M.; Abdelly, C.; Debez, A. Implication of rhizosphere acidification in nutrient uptake by plants: Cases of potassium (K), phosphorus (P), and iron (Fe). In *Crop Production and Global Environmental Issues*; Hakeem, K.R., Ed.; Springer International Publishing: Cham, Switzerland, 2015; pp. 103–122. [CrossRef]
55. Ivanov, R.; Tiedemann, J.; Czihal, A.; Baumlein, H. Transcriptional regulator atet2 is required for the induction of dormancy during late seed development. *J. Plant Physiol.* **2012**, *169*, 501–508. [CrossRef]
56. Schwarz, B.; Bauer, P. Fit, a regulatory hub for iron deficiency and stress signaling in roots, and fit-dependent and -independent gene signatures. *J. Exp. Bot.* **2020**, *71*, 1694–1705. [CrossRef] [PubMed]
57. Dell'Orto, M.; Santi, S.; De Nisi, P.; Cesco, S.; Varanini, Z.; Zocchi, G.; Pinton, R. Development of fe-deficiency responses in cucumber (*Cucumis sativus* L.) roots: Involvement of plasma membrane h+-atpase activity. *J. Exp. Bot.* **2000**, *51*, 695–701. [PubMed]
58. Kuypers, M.M.M.; Marchant, H.K.; Kartal, B. The microbial nitrogen-cycling network. *Nat. Rev. Microbiol.* **2018**, *16*, 263–276. [CrossRef] [PubMed]
59. Norton, J.; Ouyang, Y. Controls and adaptive management of nitrification in agricultural soils. *Front. Microbiol.* **2019**, *10*, 1931. [CrossRef] [PubMed]
60. Hinsinger, P.; Plassard, C.; Tang, C.; Jaillard, B. Origins of root-mediated ph changes in the rhizosphere and their responses to environmental constraints: A review. *Plant Soil* **2003**, *248*, 43–59. [CrossRef]
61. Gogorcena, Y.; Molias, N.; Larbi, A.; Abadia, J.; Abadía, A.s. Characterization of the responses of cork oak (*Quercus suber*) to iron deficiency. *Tree Physiol.* **2001**, *21*, 1335–1340. [CrossRef]
62. Gonzalo, M.J.; Moreno, M.A.; Gogorcena, Y.s. Physiological responses and differential gene expression in *Prunus* rootstocks under iron deficiency conditions. *J. Plant Physiol.* **2011**, *168*, 887–893. [CrossRef]
63. Schmidt, W.; Michalke, W.; Schikora, A. Proton pumping by tomato roots. Effect of fe deficiency and hormones on the activity and distribution of plasma membrane h+-atpase in rhizodermal cells. *Plant Cell Environ.* **2003**, *26*, 361–370. [CrossRef]
64. Wu, T.; Zhang, H.; Wang, Y.; Jia, W.; Xu, X.; Zhang, X.; Han, Z.s. Induction of root Fe(III) reductase activity and proton extrusion by iron deficiency is mediated by auxin-based systemic signaling in *Malys xiaojinensis*. *J. Exp. Bot.* **2012**, *63*, 859–870. [CrossRef]
65. Nunez, G.H.; Olmstead, J.W.; Darnell, R.L. Rhizosphere acidification is not part of the strategy i iron deficiency response of *Vaccinium arboreum* and the southern highbush blueberry. *HortScience* **2015**, *50*, 1064–1069. [CrossRef]
66. Jiménez, S.; Gogorcena, Y.; Hévin, C.; Rombolà, A.D.; Ollat, N.s. Nitrogen nutrition influences some biochemical responses to iron deficiency in tolerant and sensitive genotypes of vitis. *Plant Soil* **2007**, *290*, 343–355. [CrossRef]
67. Ksouri, R.; M'rah, S.; Gharsalli, M.; Lachaal, M.s. Biochemcial responses to true and bicarbonate-induced iron deficiency in grapevine genotypes. *J. Plant Nutr.* **2006**, *29*, 305–315. [CrossRef]
68. Kar, S.; Panda, S.K. Iron homeostasis in rice: Deficit and excess. *Proc. Natl. Acad. Sci. USA India Sect. B Biol. Sci.* **2020**, *90*, 227–235. [CrossRef]
69. Grotz, N.; Guerinot, M.L. Molecular aspects of Cu, Fe and Zn homeostasis in plants. *Biochim. Biophys. Acta (BBA)-Mol. Cell Res.* **2006**, *1763*, 595–608. [CrossRef] [PubMed]
70. Curie, C.; Panaviene, Z.; Loulergue, C.; Dellaporta, S.L.; Briat, J.-F.; Walker, E.L. Maize yellow stripe1 encodes a membrane protein directly involved in Fe(III) uptake. *Nature* **2001**, *409*, 346–349. [CrossRef] [PubMed]
71. Negishi, T.; Nakanishi, H.; Yazaki, J.; Kishimoto, N.; Fujii, F.; Shimbo, K.; Yamamoto, K.; Sakata, K.; Sasaki, T.; Kikuchi, S.; et al. cDNA microarray analysis of gene expression during Fe-deficiency stress in barley suggests that polar transport of vesicles is implicated in phytosiderophore secretion in Fe-deficient barley roots. *Plant J.* **2002**, *30*, 83–94. [CrossRef]
72. Sharma, R.; Bhardwaj, R.; Gautam, V.; Kohli, S.K.; Kaur, P.; Bali, R.S.; Saini, P.; Thukral, A.K.; Arora, S.; Vig, A.P. Microbial siderophores in metal detoxification and therapeutics: Recent prospective and applications. In *Plant Microbiome: Stress Response*; Egamberdieva, D., Ahmad, P., Eds.; Springer: Singapore, 2018; pp. 337–350. [CrossRef]
73. Schalk, I.J.; Hannauer, M.; Braud, A. New roles for bacterial siderophores in metal transport and tolerance. *Environ. Microbiol.* **2011**, *13*, 2844–2854. [CrossRef] [PubMed]
74. Li, C.; Zhu, L.; Pan, D.; Li, S.; Xiao, H.; Zhang, Z.; Shen, X.; Wang, Y.; Long, M. Siderophore-mediated iron acquisition enhances resistance to oxidative and aromatic compound stress in *Cupriavidus necator* jmp134. *Appl. Environ. Microbiol.* **2019**, *85*, e01938-18. [CrossRef] [PubMed]
75. Crowley, D.E. Microbial siderophores in the plant rhizosphere. In *Iron Nutrition in Plants and Rhizospheric Microorganisms*; Barton, L.L., Abadia, J., Eds.; Springer: Dordrecht, The Netherlands, 2006; pp. 169–198. [CrossRef]
76. Yokosho, K.; Yamaji, N.; Ma, J.F. Osfrdl1 expressed in nodes is required for distribution of iron to grains in rice. *J. Exp. Bot.* **2016**, *67*, 5485–5494. [CrossRef]

77. Aznar, A.; Chen, N.W.; Thomine, S.; Dellagi, A. Immunity to plant pathogens and iron homeostasis. *Plant Sci.* **2015**, *240*, 90–97. [CrossRef]
78. Ferret, C.; Sterckeman, T.; Cornu, J.-Y.; Gangloff, S.; Schalk, I.J.; Geoffroy, V.A. Siderophore-promoted dissolution of smectite by fluorescent *Pseudomonas*. *Environ. Microbiol. Rep.* **2014**, *6*, 459–467. [CrossRef]
79. Saha, M.; Sarkar, S.; Sarkar, B.; Sharma, B.K.; Bhattacharjee, S.; Tribedi, P. Microbial siderophores and their potential applications: A review. *Environ. Sci. Pollut. Res.* **2016**, *23*, 3984–3999. [CrossRef]
80. Hider, R.C.; Kong, X. Chemistry and biology of siderophores. *Nat. Prod. Rep.* **2010**, *27*, 637–657. [CrossRef]
81. Masalha, J.; Kosegarten, H.; Elmaci, Ö.; Mengel, K. The central role of microbial activity for iron acquisition in maize and sunflower. *Biol. Fertil. Soils* **2000**, *30*, 433–439. [CrossRef]
82. Rroço, E.; Kosegarten, H.; Harizaj, F.; Imani, J.; Mengel, K. The importance of soil microbial activity for the supply of iron to sorghum and rape. *Eur. J. Agron.* **2003**, *19*, 487–493. [CrossRef]
83. Jin, C.W.; He, Y.F.; Tang, C.X.; Wu, P.; Zheng, S.J. Mechanisms of microbially enhanced fe acquisition in red clover (*Trifolium pratense* L.). *Plant Cell Environ.* **2006**, *29*, 888–897. [CrossRef]
84. Yang, C.-H.; Crowleys, D.E. Rhizosphere microbial community structure in relation to root location and plant iron nutritional status. *Appl. Environ. Microbiol.* **2000**, *66*, 345–351. [CrossRef]
85. González-Guerrero, M.; Escudero, V.; Saéz, Á.; Tejada-Jiménez, M. Transition metal transport in plants and associated endosymbionts: Arbuscular mycorrhizal fungi and rhizobia. *Front. Plant Sci.* **2016**, *7*, 1088. [CrossRef] [PubMed]
86. Hell, R.; Stephan, U.W. Iron uptake, trafficking and homeostasis in plants. *Planta* **2003**, *216*, 541–551. [CrossRef]
87. Schmidt, W. Mechanisms and regulation of reduction-based iron uptake in plants. *New Phytol.* **1999**, *141*, 1–26. [CrossRef]
88. Jain, A.; Wilson, G.T.; Connolly, E.L. The diverse roles of fro family metalloreductases in iron and copper homeostasis. *Front. Plant Sci.* **2014**, *5*, 100. [CrossRef] [PubMed]
89. Robinson, N.J.; Procter, C.M.; Connolly, E.L.; Guerinot, M.L. A ferric-chelate reductase for iron uptake from soils. *Nature* **1999**, *397*, 694–697. [CrossRef]
90. Liang, G.; Zhang, H.; Li, X.; Ai, Q.; Yu, D. Bhlh transcription factor bhlh115 regulates iron homeostasis in *Arabidopsis thaliana*. *J. Exp. Bot.* **2017**, *68*, 1743–1755. [CrossRef] [PubMed]
91. Connolly, E.L.; Campbell, N.H.; Grotz, N.; Prichard, C.L.; Guerinot, M.L. Overexpression of the fro2 ferric chelate reductase confers tolerance to growth on low iron and uncovers posttranscriptional control. *Plant Physiol.* **2003**, *133*, 1102–1110. [CrossRef] [PubMed]
92. Vert, G.; Grotz, N.; Dédaldéchamp, F.; Gaymard, F.; Guerinot, M.L.; Briat, J.-F.; Curie, C. Irt1, an Arabidopsis transporter essential for iron uptake from the soil and for plant growth. *Plant Cell* **2002**, *14*, 1223–1233. [CrossRef] [PubMed]
93. Jeong, J.; Guerinots, M.L. Homing in on iron homeostasis in plants. *Trends Plant Sci.* **2009**, *14*, 280–285. [CrossRef]
94. Gayomba, S.R.; Zhai, Z.; Jung, H.I.; Vatamaniuks, O.K. Local and systemic signaling of iron status and its interactions with homeostasis of other essential elements. *Front. Plant Sci.* **2015**, *6*, 716. [CrossRef] [PubMed]

plants

MDPI

Review

Plant Iron Research in African Countries: Current "Hot Spots", Approaches, and Potentialities

Irene Murgia * and Piero Morandini

Department of Environmental Science and Policy, Università degli Studi di Milano, Via Celoria 10, 20133 Milan, Italy; piero.morandini@unimi.it
* Correspondence: irene.murgia@unimi.it

Abstract: Plant iron (Fe) nutrition and metabolism is a fascinating and challenging research topic; understanding the role of Fe in the life cycle of plants requires knowledge of Fe chemistry and biochemistry and their impact during development. Plant Fe nutritional status is dependent on several factors, including the surrounding biotic and abiotic environments, and influences crop yield and the nutritional quality of edible parts. The relevance of plant Fe research will further increase globally, particularly for Africa, which is expected to reach 2.5 billion people by 2050. The aim of this review is to provide an updated picture of plant Fe research conducted in African countries to favor its dissemination within the scientific community. Three main research hotspots have emerged, and all of them are related to the production of plants of superior quality, i.e., development of Fe-dense crops, development of varieties resilient to Fe toxicity, and alleviation of Fe deficiency, by means of Fe nanoparticles for sustainable agriculture. An intensification of research collaborations between the African research groups and plant Fe groups worldwide would be beneficial for the progression of the identified research topics.

Keywords: Africa; biofortification; crops; Fe deficiency; Fe toxicity; nanoparticles; nutrition; research dissemination; rice; sustainable agriculture

Citation: Murgia, I.; Morandini, P. Plant Iron Research in African Countries: Current "Hot Spots", Approaches, and Potentialities. *Plants* **2024**, *13*, 14. https://doi.org/10.3390/plants13010014

Academic Editors: Ferenc Fodor and Daniela Businelli

Received: 15 November 2023
Revised: 14 December 2023
Accepted: 17 December 2023
Published: 19 December 2023

1. Introduction

Plant iron (Fe) research deals with the multifaceted strategies adopted by plants for Fe uptake from soils of various pHs [1,2], its transport, metabolism, signaling, and distribution from roots to other districts, such as leaves and seeds [3,4], and the biochemistry of Fe-requiring enzymes FeRE [5,6] and Fe complexes [7,8]. Fe homeostasis is influenced by below- and above-ground environments, including detrimental or beneficial living organisms [4,9]. A relevant branch of plant Fe research is devoted to amelioration of bioavailable Fe content in edible parts of staple crops for improving human nutrition and for combating Fe deficiency anemia [10–12]. These research goals require competency in genetics, agronomy, biotechnology, and both human and plant physiology, and they could benefit from the identification of novel hubs in plant nutrition [13]. The various micro- and macronutrients reciprocally influence each other in terms of their homeostasis, as has emerged, for example, between Fe and molybdenum [14] and between Fe and sulfur [15], clearly indicating that "no plant nutrient is an island".

Hence, research of all these aspects of plant Fe metabolism is of worldwide importance, including the African continent. Indeed, African countries can count on incredible plant biodiversity and a variety of environments, and they can contribute to expanding the number of species employed for Fe research. Moreover, several African populations experience severe Fe deficiency anemia (IDA) [16] exacerbated by malaria, which is endemic in various African countries. Fe-rich staples are therefore needed. However, Fe metabolism of some of the species which are nutritionally relevant in Africa (e.g., millets) is not investigated as intensively as other more globally relevant crops.

Unfortunately, plant Fe research prominently conducted by scientists affiliated with African research centers does not always enjoy full international visibility and dissemination. This can be ascribed to several factors, besides research funding, such as participation in international congresses, publications in relevant journals in the field, availability of scholarships, student mobility, rigor of research, etc.

The analysis of all these aspects is beyond the scope of the present work, which instead aims to fill this gap of visibility and dissemination. For that, we focused on research conducted by scientists affiliated with African universities/research institutions and focusing on plant Fe nutrition and metabolism in recent years (2018–2023). Interestingly, three major topics of interests emerged, which are discussed in the following paragraphs.

2. Plant Fe Research Prominently Conducted in Africa, from 2018 to 2023

Various databases of scholarly literature are available; the Scopus database (https://www.scopus.com/home.uri, accessed on 14 September 2023) was used for the present work, as it allows user-friendly searches with multiple exclusions and/or restrictions in various fields. The query in Scopus was conducted by searching all research papers or reviews published in English from 2018 onwards, in which the term "iron" appears in the title. Only "Agricultural and Biological Sciences" and "Biochemistry, Genetics and Molecular Biology" subject areas were considered; all the other subject areas were excluded. To restrict searches to plant science, several keywords (Supplementary Figure S1) as well as various journals (Supplementary Figure S2) were used as criteria for exclusion. The search was then limited to authors affiliated with African countries. The final list of 92 publications was then manually pruned to exclude publications whose main focus was not on plant science (e.g., publications mainly on nutrition were excluded). At the end, 69 publications were retained, affiliated with 26 countries (Supplementary Figure S3). Although not retrieved by this procedure, one more publication [17] was added to the list, as it represents the beginning of a work [18] included among the 69 publications.

About 50% of these 70 publications are focused on rice (various *Oryza* species), maize (*Zea mays*), common bean (*Phaseolus vulgaris*), soybean (*Glycine max*), pearl millet (*Pennisetum glaucum*), finger millet (*Eleusine coracana*), and pea (*Pisum sativum*), which are well-known staple crops in African countries; the remaining publications deal with a further 24 plant species, and a few papers investigate more than one species. The list of all the investigated plant species with the related references is reported in Table 1.

Table 1. List of investigated species in publications on plant Fe science and affiliated with African countries, from 2018 to 2023, as retrieved by the Scopus search described in the text. Columns from left: plant species, references of publications referring to the species, and publications with prominent African affiliations. Species investigated in publications with prominent African affiliations are in bold.

Plant Species	Publications	Publications with African Prominent Authorship
rice (various *Oryza* species)	[17–26]	[17,18,21–26]
maize (*Zea mays*)	[27–34]	[29–34]
common bean (*Phaseolus vulgaris*)	[35–40]	[37–40]
soybean (*Glycine max*)	[27,41–44]	[44]
pearl millet (*Pennisetum glaucum*)	[45–48]	[45–48]
finger millet (*Eleusine coracana*)	[33,49,50]	[33,49,50]

Table 1. *Cont.*

Plant Species	Publications	Publications with African Prominent Authorship
pea **(*Pisum sativum*)**	[51,52]	[51,52]
tomato **(*Solanum lycopersicum*)**	[53–55]	[53–55]
sorghum **(*Sorghum bicolor*)**	[56–58]	[57,58]
lentil (*Lens culinaris*)	[59,60]	
wheat **(various species)**	[61,62]	[61,62]
barrel medic **(*Medicago truncatula*)**	[63,64]	[63,64]
cowpea **(*Vigna unguiculata*)**	[33,65]	[33,65]
broad bean **(*Vicia faba*)**	[66,67]	[66,67]
barley **(*Hordeum vulgare*)**	[68]	[68]
flax (*Linum usitatissimum*)	[69]	
date palm (*Phoenix dactylifera*)	[70]	
grey mangrove (*Avicennia marina*)	[71]	
African wormwood **(*Artemisia afra*)**	[72]	[72]
spinach **(*Spinacia oleracea*)**	[72]	[72]
carrot **(*Dacus carota*)**	[72]	[72]
Chinese mandarine (*Citrus reticulata* Blanco)	[73]	
fenugreek **(*Trigonella foenum-graecum*)**	[74]	[74]
rose-scented geranium **(*Pelargonium graveolens*)**	[75]	[75]
sesame **(*Sesamum indicum*)**	[76]	[76]
durum wheat **(*Triticum durum*)**	[77]	[77]
alfalfa (*Medicago sativa*)	[78]	
Washington navel orange (*Citrus sinensis*)	[79]	
roselle **(*Hibiscus sabdariffa*)**	[80]	[80]
grape (*Vitis vinifera*)	[81]	

Table 1. *Cont.*

Plant Species	Publications	Publications with African Prominent Authorship
Sulla carnosa (*Hedysarum carnosum*)	[82]	
legumes	[83]	[83]
cereals	[84]	[84]

To further restrict our analysis to research prominently conducted in Africa, only 48 publications were retained for further analysis, i.e., those with first and/or corresponding authors affiliated with an African country (Table 1, plant species and references in bold). Hence, 22 papers were excluded, as they either did not report an African affiliation as a prominent authorship [15,19,20,27,28,35,36,41,42,53,56,59,70,71,73,78,81,82,85], or the first author did not have an African affiliation, and there was no clear African research leadership due to the presence of at least three corresponding authors [43,60,69].

Egypt, Tunisia, Nigeria, Ghana, and South Africa are the most represented affiliated countries in these 48 publications (11, 8, 7, 5, and 5 publications each, respectively). Notably, the model plant *Arabidopsis thaliana* is not among the analyzed plants species (Table 1).

A variety of journals with a broad range of impact factors (IFs) and publishers are represented in the 48-publication list, as detailed in Supplementary Table S1. Most important, three "research hotspots" emerged from content analysis of these 48 publications, i.e., (1) Fe deficiency and crops biofortification, (2) sustainable agriculture and fertilization with Fe nanoparticles, (3) Fe toxicity. These hotspots are described in detail in the following paragraphs.

3. Research Hotspot: Fe Deficiency and Crops Biofortification

Fe availability is low in alkaline soils; plants in arid and semiarid regions incur a shortage of Fe availability and hence, a Fe deficiency. Plants activate a complex array of morphological and biochemical responses to counteract Fe deficiency stress [86–88]. The isolation of plant genotypes from different crop species which are more tolerant to Fe deficiency represents a current effort pursued by several plant scientists in Africa. Such effort goes hand in hand with the search for and validation of Fe biofortification approaches. Also, efforts to achieve Fe enrichment in seeds is often associated with efforts to achieve Zn enrichment [29,84]. Surely, some hurdles have been identified in the various approaches, at least for cereals. The variability of the measured Fe content in seeds, as observed in various studies, could be due to the sensitivity of the adopted analytical method to improper post-harvest seed handling and also to data restricted to a single year [84]. The results obtained with rice, wheat, maize, barley, millet, and various legume species are detailed below; it is worth recalling that wild species have higher Fe content but lower yields than cultivated ones [3,84].

3.1. Rice

Oryza glaberrima is an African indigenous low-yielding rice species with various resistance traits, cultivated in West Africa for thousands of years [89,90]. Scientists from the West Africa Rice Development Association (WARDA) succeeded in producing fertile progenies from crosses of this species with *O. sativa* and further backcrosses with *O. sativa,* increasing stability and fertility for better performance in upland cultivation [89]. Later, hybrids from *O. sativa* × *O. glaberrima* crosses were produced for irrigated, lowland cultivation [90]. Some of these hybrids were named NERICAs (New Rice for Africa). Several concerns regarding the rigorous scientific assessments of the claims regarding NERICA rice as the "silver bullet" for Africa's green revolution were raised [91]. Physical properties of NERICA varieties NERICA-1 and NERICA-4, Indica varieties IR-28 and IR-50, and Japonica variety Yumepirika were then compared; NERICA and Indica rice seeds had similar results in these tests [92]. A total of 445 NERICA × *O. sativa* rice lines have been field-tested in two different sites in Liberia during the wet season to identify QTLs for higher tolerance to Fe-toxicity; for that, four traits were scored: days to flowering, plant height, grain yield, and leaf bronzing score [21]. Also, 35 upland rice genotypes were grown in the field, in Ibadan, Nigeria, under Fe sufficiency or deficiency; three varieties tolerant to Fe deficiency were identified (FARO65, IRAT 109, NERICA3) [22].

3.2. Wheat

Foliar treatments with $FeSO_4$ and/or $ZnSO_4$ of winter wheat could alleviate drought symptoms [61], although the molecular mechanisms of interaction between Fe supplementations and such abiotic stress were not investigated.

Various synthetic wheat lines were developed in the past by Japanese groups [93] by crossing the tetraploid wheat cultivar "Langdon" with *Aegilops tauschii* accessions mainly collected from Iran but also from other Asian countries; three lines with high Fe and Zn content in grains were identified; such traits were stable across the seasons [62]. Durum wheat genotypes tolerant to growth in calcareous soil were also selected [77].

3.3. Maize

Low nitrogen levels affect maize cultivation and yield in Sub-Saharan Africa (SSA), as well as the Fe and Zn content of seeds [30]. Several maize hybrids were grown in various experimental conditions (low or optimum N levels), and the most stable genotypes for Fe and Zn content of grains and for yield were identified under low N conditions [31]. Also, hybrid lines grown under low or optimum N conditions or under controlled drought stress were analyzed for genotype x environment interactions [32]. The authors reported a high positive correlation between grain Fe and Zn concentration ($r = 0.97$) and a moderate negative correlation between grain yield and Fe and Zn content ($r = -0.43$ and $r = -0.44$, respectively). The authors concluded that the development of Fe- and Zn-dense maize cultivars with high grain yield is feasible to combat Fe and Zn deficiency in SSA [32]. However, the effect of N treatments on total Fe grain content in maize is less clear, as reported in another study [33].

Efforts for multiple enrichments of maize with Fe, Zn, and also provitamin A have been conducted in Nigeria, with the identification of sets of hybrids which combined high Fe and Zn content (24.45 mg kg^{-1} and 29.24 mg kg^{-1}) and average provitamin A content (7.48 mg kg^{-1}); however, there was a weak but significant negative correlation between Zn and provitamin A, so the hybrids with the highest provitamin A content were also lowest in Zn content [34].

3.4. Barley

Barley (*Hordeum vulgare*) is another staple crop in North Africa and in Ethiopia, and a genome-wide association study (GWAS) of a collection of 496 spring barley genotypes (cultivars, improved lines, landraces) identified several single-nucleotide polymorphisms (SNPs) associated with Fe and Zn content, although several candidate genes are still annotated as "undescribed" [68].

3.5. Millet

Pearl millet is one of the six most important cereal crops for human nutrition in the world, and its relevance is increasing with global climate changes due to its resilience to high temperatures and drought [94]. Moreover, this so-called "nutricereal" has a very high nutritional value with respect to other cereals, and various breeding approaches for pearl millet are currently adopted [94]. Open-pollinated varieties (OPVs) of pearl millet were tested in various locations in West Africa for their growth performance and Fe and Zn grain content, with the identification of stable, high-Fe varieties [45–47]. Interestingly, quantitative genetics were applied to six generations of pearl millet obtained from two sets of parental lines grown at ICRISAT, Sadore, Niger, to model the trait inheritance of Fe grain content [48].

Finger millet also represents an important staple crop in various arid and marginal lands in eastern Africa, where other crops cannot thrive, as in Ethiopia, where finger millet represents one of the major staple foods [49]. Three different genotypes (Diga-01, Urji, Meba) grown in two different regions of Ethiopia and in two different slope configurations for each region were treated with Fe and Zn in a combination of the two micronutrients with the NPKS fertilization [49]. It emerged that, besides genotypes and treatments, the yield is dependent on location and slope, and all these variables should be considered for programs of agronomic biofortification [49].

Fertilization with phosphorus is beneficial to finger millet, causing an increase in calcium, Fe, and Zn content of grains from plants grown in three different locations in Kenya [50]. Again, the optimal level of phosphorus fertilizer is dependent on tested location [50]. This aspect should be considered to avoid the negative effects of unnecessary, supra-optimal concentrations of phosphorus for plants and soils.

3.6. *Legumes*

Legumes show genetic variability in their tolerance to Fe deficiency; their inclusion in crop rotation improves soil nutritional status, thanks to symbiosis with N-fixing bacteria [83]. For these reasons, legumes are good candidates for sustainable agriculture [63,83]. As most Tunisian soils are calcareous, several Tunisian researchers studied Fe deficiency responses in different legume species in order to identify tolerant lines. Twenty genotypes of barrel medic (*Medicago truncatula*), a small annual legume used as forage but also as a model plant, were screened for their physiological response to Fe deficiency during growth. Fe deficiency-tolerant and -sensitive genotypes were identified; tolerance was correlated with greater acidification capacity, modified root architecture, and an induction of Superoxide Dismutase (SOD) activities [63,64]. Similar approaches were applied to investigate genotypic differences in the response to Fe deficiency in pea [51,52] and in common bean [37]. A set of 99 different cultivars and landraces of common bean were tested in different locations in Tanzania for their adaptability and stability [38]. Moreover, SNPs associated with grain Fe and Zn concentration were identified for such species [39]. In another study, Fe and Zn content of broad bean (*Vicia faba*) in Ethiopia was influenced by environmental conditions, especially soil properties [66].

4. Research Hotspot: Sustainable Agriculture and Fe Nanoparticles

Soil fertilization with macro- and microelements is a common agricultural practice which needs to be tailored to the specific crop to limit leakage and soil degradation. The use of nanoparticles (NPs) is a practice explored in recent years; such research has been prompted by the evidence that parts of fertilizers (whether in soil or directly applied to plants) are not used by plants, and they are therefore wasted, with economical as well as ecological consequences. The production of micronutrient NPs would offer a sustainable alternative to traditional fertilizers due to their higher efficiency and reduced contamination. All these aspects, as well as the ways in which they are internalized by plants' leaves and roots, the potential risks associated with their use (human health and food safety), and the need of rigorous assessments, have been recently reviewed [95,96].

A study described the green production of Fe-NPs, starting from extract of African wormwood (*Artemisia afra*) [72]. The authors characterized the physical and chemical properties of the obtained NP, thus demonstrating that they are genuinely formed by Fe oxides of 10–20 nm in diameter. These Fe-NPs influence the germination rate of both spinach and carrot seeds in a concentration-dependent way, and they can therefore act as nano-priming agents [72]. This, in turn, opens the question regarding the mechanisms by which Fe-NPs can promote seed germination in a given concentration range [72]. Another study compared the effects of conventional Fe treatment (ferric sulfate or Fe-EDDHA) with Fe-NPs on the composition in essential oils (EOs) of rose-scented geranium (*Pelargonium graveolens*). This perennial plant is cultivated worldwide, for instance in Egypt, and it is used for cosmetic and perfume production; its EO composition is dependent on nutritional status, and treating plants with Fe-NPs and humic acid boosts EO production in a superior way with respect to the other Fe treatments [75]. Also, the effects of treatment of orange trees (*Citrus sinensis*) grown in Egypt with green Fe-NPs (produced from fresh leaves of *Psidium guiaiava*, commonly known as guava) were evaluated in comparison with other conventional Fe treatments ($FeSO_4$, Fe-EDTA) [79]. The appearance of produced Fe-NPs of needle-like shape at SEM seems different from the green Fe-NPs characterized in [72], and their chemical composition was not investigated; nonetheless, the improvement in most of the analyzed parameters in terms of physical and chemical features of fruits (yields, shelf life, mineral contents of fruits, etc.) would encourage the use of such Fe-NPs with respect to the other treatments for orange cultivation in arid regions [79]. Effects of foliar application of Fe-NPs, $FeSO_4$, and Fe-chelate were also evaluated in broad bean grown in sandy soil, and Fe-NP gave superior results in terms of plant growth parameters and harvest index [67]. The superiority of Fe-NP fertilization with respect to conventional Fe or chelated Fe was also demonstrated in tomato for growth and yield [54].

Fe-NPs were also effective in ameliorating growth parameters in soybean [44]; commercial Fe-NPs (n-Fe_2O_3) were also effective in ameliorating growth parameters in sorghum seedlings, as well as their tolerance to salinity stress [57].

Sesame (*Sesamum indicum*) is produced in many African countries. An Egyptian group [76] studied the effect of treating three different sesame genotypes with a mixture of Fe, Mn, and Zn in nano form; however, the production of such Fe, Zn, and Mn fertilizers in the nano form has not been clearly described by authors.

Although these studies are promising and in line with the worldwide interest in precision agricultural practices, benefits observed from use of Fe-NPs have been not always confirmed, and indeed, in some cases, treatment with classic Fe-EDTA was more beneficial than with Fe-NPs, as described for sorghum plants [58].

Within the above cited results on this African research hotspot, we believe that discussion on health risks for exposed workers and on food safety for consumers due to exposure to Fe-NP via inhalation or via ingestion is still missing. Such risks should not be underestimated, as several studies reported endotheliar disfunction and inflammation, with increased pulmonary, vascular, and cardiac diseases, due to exposure to Fe-NPs of various diameters [97,98].

5. Research Hotspot: Fe Toxicity

Fe toxicity is less frequent than Fe deficiency due to Fe chemistry; indeed, although Fe is abundant in the Earth's crust, it is usually in highly insoluble Fe (III) oxidized form. Still, occasionally, Fe solubility can massively increase due to Fe reduction to Fe (II) form. This increase in Fe (II) availability can occur in anoxic conditions, such as in the so-called "rainfed lowlands on Fe-rich soils", i.e., in Fe-rich soils exposed to prolonged floods and without satisfactory drainage ([23,99], or in acid sulfate soils, acid clay soils, and peat soils [100]. In all these conditions, Fe can be taken up in excess by root cells and exert its toxicity, with a consequent reduction in crop yields. The typical phenotypic hallmark of Fe toxicity is a brownish-red color, known as "bronzing".

Four plant defense mechanisms against Fe toxicity are adopted by plants [19,24]: (1) Fe exclusion, achieved by oxidation of Fe (II) into Fe (III) via oxygen released by roots; (2) Fe retention in metabolically inactive forms in cell vacuoles and within the Fe storage protein ferritin, mainly in plastids; (3) Fe partitioning in older plant tissues; and (4) ROS detoxification to alleviate oxidative load. Rice genotypes are broadly classified as Fe "excluders" or "includers", relative to the adopted strategy [24].

Rice is the only major crop species which can incur Fe toxicity [99]; rice cultivations in several African regions are potentially exposed to Fe-toxicity conditions [23,24].

Praiseworthily, van Oort [23] mapped the Fe-toxicity spots in the African regions for the first time, together with maps of spots for other abiotic stresses (drought, cold, salinity, and sodicity). The various countries with the largest areas potentially at risk of Fe toxicity were identified, with Nigeria being at the highest risk. Also, a list of five countries with the largest surfaces cultivated with rice at risk of Fe toxicity was provided; such values were expressed as "rainfed lowland rice areas on Fe rich soil"/"total rice area", and Togo was at the top of this list [23]. A predominance of West African countries is exposed to Fe toxicity, involving 12% of total rice area, thus representing a relevant threat for rice production. This percentage might become even larger, as further expansion of crop areas, which is expected in Africa for rice as well as for other cereals, could involve wetlands, which are currently neglected [23].

Tolerance mechanisms were investigated in seven different rice genotypes (KA-28, Bahia, Ciherang, IR64, L-43, Tsipala, X265), either tolerant of or sensitive to Fe toxicity, which were grown in the central highlands of Madagascar. Fe uptake rate, growth rate, and Fe partitioning between shoot tissues during vegetative and reproductive growth were measured, together with grain yield. Exclusion mechanisms, even in "excluder" genotypes, were relaxed during reproductive growth, indicating that mechanisms activated by plants to counteract Fe toxicity are dependent on developmental stage [24]. Strikingly, no correlation between grain yield and visual symptoms has been observed, indicating that the selection of tolerant lines based on visual symptoms is a simplistic approach [24]. These findings, together with the environmental dependence of intensity and the dynamics of Fe toxicity, highlight the need to unveil the genetic factors orchestrating responses against Fe toxicity. An RNA-seq analysis was performed to investigate the effects of magnesium (Mg) on Fe toxicity in the central highlands of Madagascar, as well as in hydroponic conditions, thus uncovering various genes potentially affecting the enhancement of tolerance by Mg [25]. Another research group analyzed Fe toxicity-tolerant rice varieties CK801 and Suakoko8, as well as sensitive varieties IR64 and Supa, under hydroponic conditions for their morphological and physiological responses to Fe toxicity [17,18]. In these studies, tolerant lines showed more lateral roots, a better development of aerenchyma, and higher O_2 release. Hundreds of African rice *Oryza glaberrima* accessions were grown in the Plateaux region of Togo and analyzed for growth parameters, together with symptoms of Fe toxicity [26]. Physiological and biochemical responses against Fe toxicity were also studied in 15 accessions of cowpea (*Vigna unguiculata*) [65].

6. Other Research Lines

Analysis of fenugreek (*Trigonella foenum-grecum*) under increasing concentrations of Fe in the form of $FeSO_4$ would encourage the use of such plants in Tunisian phytoremediation programs [74]. Roselle (*Hibiscus sabdaiffa* L.), known as karkadeh, is cultivated in various African regions, and it is a valuable commercial plant, as its sepals are used for human nutrition as well as for beauty cosmetic industries. Roselle plants grown in-field were sprayed with different compounds (Fe-EDTA, arginine, hemin, and their combinations), and the yield and chemical and elemental composition of their sepals were analyzed, showing stimulatory effects on flavonoid pigments [80].

7. Discussion

The responses of crops to Fe deficiency, including various approaches of plant biofortification with Fe, whether via fertilization or via genetic approaches, is a well-represented research topic in Africa; several lines of wheat, maize, rice, and various legumes have been examined to identify those of superior quality. Such research, in Africa, also involves millets; this is promising, since these species can better withstand adverse growth conditions and have the potential to also become an important crop outside Africa. Notably, the Food and Agriculture Organization of the United Nations (FAO) declared 2023 the International Year of Millets (IYM 2023) at https://www.fao.org/millets-2023/en (accessed on 14 September 2023) Hoboken, NJ, USA. The screening for Fe-dense millets is therefore of great importance [94]. Still, the molecular basis of Fe nutrition and homeostasis in millet is not known in the same detail as in other plant species. The international expansion of millet as a model plant in the plant Fe research community could therefore contribute to boosting its diffusion as a "nutricereal".

Sustainable agriculture is becoming a growing concern not only in Africa but worldwide, requiring the adoption of eco-friendly cultivation practices for correct soil management and for the optimization of costs versus crop yield. Several scientists in the African continent are involved in the analysis of the effects of Fe-NP, indicating that this subject is very relevant for such parts of the world.

Several approaches can be followed to synthesize Fe-NP, and one of them is based on a bottom-up approach in which NPs can grow from reaction precursors by using plant extracts rich in organic reductants [101]. Several plant extracts have already been used for such synthesis, and various mechanisms have been proposed for the reactions involved [101]. Still, the establishment of species- and developmental stage-specific protocols for Fe-NP preparation and use in terms of concentration and times of treatments is still needed. These shared protocols should also clearly include the green Fe-NP preparations. A systematic comparison of green Fe-NP versus Fe-NP prepared with other chemical methods would help disentangle the variety of results which are emerging from various research groups; this, in turn, would offer shared avenues of collaborative approaches to this relevant research field among scientists from different countries and with different expertise (Figure 1). Moreover, in this identified research hotspot, we noticed a general lack of risk assessment of Fe-NPs in terms of human health upon exposure to Fe-NPs during handling, but also in terms of food safety. Fe-NP risk assessment is for sure a compulsory step in any future research involving use of nanomaterials in agriculture (Figure 1). Historically, NP emissions from industrial or traffic sources have been associated with health risks; however, the expansion in the use of Fe-NPs for novel agricultural approaches requires a rigorous risk assessment of the nanomaterials within this application field, for human health [97,98] but also for the whole environment [102]. Notably, proposals to modify surfaces of nanomaterials to make them more biocompatible and reduce the inflammatory/oxidative responses of human tissues have been put forward [97]; such hazard reduction strategies have been named "safe by design" [103].

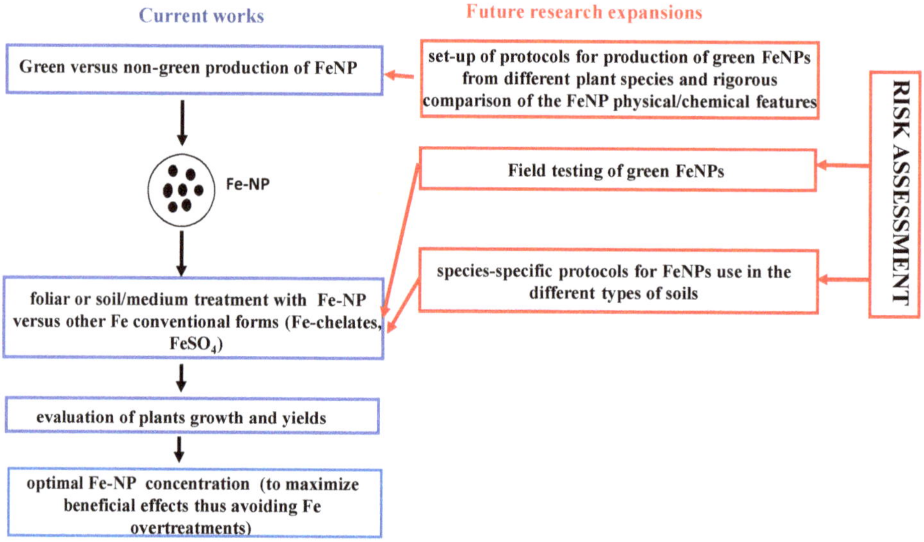

Figure 1. Current works and approaches on Fe nanoparticles (Fe-NPs) in the research conducted in the time span from 2018 to 2023 by prominently African-affiliated scientists (text boxes in blue, on the left) and the future research expansions suggested in the present work (text boxes in red, on the right).

Current works: green and non-green Fe-NPs, produced with or without plant extracts, respectively; soil, growth medium, or foliar treatments with various concentrations of Fe-NP, Fe-chelates, or other conventional Fe treatments (e.g., $FeSO_4$) have been performed on different plant species. To identify the best Fe-NP concentration to avoid overtreatments, plant growth and yield have been assessed. Future research expansions: a higher number of plant species could be tested as starting green extracts for the production of green Fe-NP; the physical/chemical features of the produced Fe-NP should be rigorously assessed. To optimize Fe-NP use in the environment, protocols specifically adapted for optimal treatment of each plant species with green Fe-NP should be established and shared within the scientific community. A rigorous risk assessment is needed: direct effects on human health during Fe-NP treatments and indirect effects on human health in terms of food and soil safety should be considered.

Research on Fe toxicity is also well represented in African countries, as it is driven by the need to ameliorate crop yield and quality of lowland rice. The extent of African soil prone to Fe toxicity and the increasing African population and consequent increasing need of crop production, including rice, make the understanding of Fe toxicity of great importance for millions of people. Some molecular aspects of plant exposure to Fe excess have been analyzed in the past in model and/or crop plants, such as the regulatory mechanisms of Fe sequestration through the iron storage protein ferritin [104–109]. Also, below-ground as well as above-ground adaptive mechanisms of rice to Fe toxicity have been elucidated [19,100]. Still, various aspects of plant response to Fe excess and defense mechanisms against Fe toxicity need to be disentangled. Fe toxicity is indeed often accompanied by deficiencies in other nutrients, such as potassium, magnesium, calcium, and silicon [19]. The impact of such deficiencies in the development of Fe toxicity symptoms during exposure to Fe excess and their homeostatic adjustments from root to seed in tolerant versus Fe-sensitive species might contribute to the identification of markers of tolerance for breeding programs (Figure 2). Most of the gene regulators involved in the responses against Fe toxicity are also unknown [100].

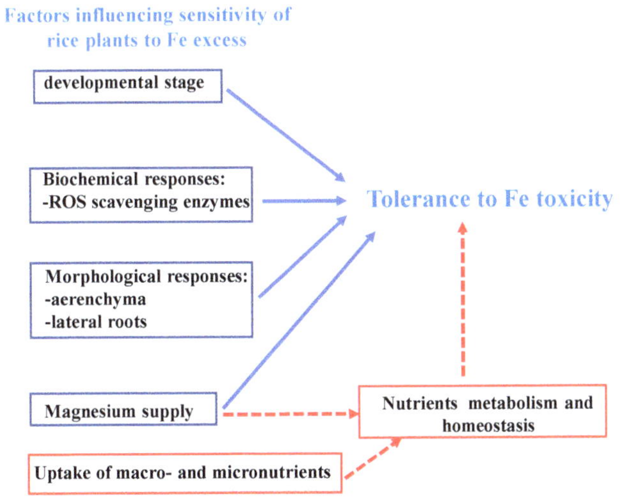

Figure 2. Current findings on tolerance/sensitivity to Fe toxicity in rice in the research conducted in the time span from 2018 to 2023 by prominently African-affiliated scientists (blue boxes) and the research potentialities suggested in the present work (red boxes). The sensitivity of rice plants to Fe excess is dependent on anatomical traits (extent of root aerenchyma and expansion of lateral roots), on the involvement of the ROS scavenging enzymes, and on Mg nutritional status of the plant; also,

responses to Fe excess are dependent on developmental stage. All these features contribute to tolerance/sensitivity of rice plants to Fe toxicity. Possible future research potentialities could involve the analysis of the uptake processes and of the nutritional status of the other essential nutrients under Fe excess, together with the analysis of their homeostatic adjustments.

8. Conclusions and Future Directions

Predictions on climate change and growth of human population make plant Fe science of utmost relevance; more accurate working models on Fe nutrition and metabolism in plants are needed for crops of established economic and nutritional value, as well as for plants with emerging nutritional, economic, and ecological importance. Hence, any relevant advancement in this field should be worthy of consideration, and its dissemination within the scientific community should be favored. The present work therefore turns the spotlight on plant Fe research mainly conducted in Africa, with an overview of the current research situation. Three main research hotspots were identified, and future research expansions were proposed, hopefully fostering initiatives with research groups worldwide.

The 48 research products identified in the time span from 2018 to 2023 are affiliated with 22 countries; hence, more than half of the African countries are either not involved in any research on plant Fe or did not succeed in publishing any Scopus-indexed work. Also, around one fifth of such research products have been published in minor journals without an impact factor. This analysis reveals that an expansion of African plant Fe research, both quantitative and qualitative, is welcome, as it could be beneficial for Africa itself, but also outside this continent. For example, a similar Scopus search on plant Fe research conducted on Southeast Asian territories highlighted a limited number of total publications (22), with Thailand, Indonesia, the Philippines, and Vietnam as affiliated countries; interestingly, half of the publications were related to plant response against Fe toxicity (not shown). Hence, Fe toxicity is an economical and nutritional burden for many low-income countries worldwide. Last, ongoing projects of Fe research in Africa, if unpublished, have not been included in the present review to avoid any bias regarding retrieval, visibility, sponsorship, or funding of unpublished results. We nonetheless hope that the present work can be handled as a useful tool by plant Fe scientists worldwide to come in direct contact with Fe research conducted in Africa, as well as for future collaborations and shared initiatives. Scholarships and shared organizations of scientific sessions at congresses in plant physiology and nutrition might represent starting points for such collaborative initiatives.

Supplementary Materials: The following supporting information can be downloaded at https://www.mdpi.com/article/10.3390/plants13010014/s1. Figure S1: Keywords excluded from Scopus search; Figure S2: Journals excluded from Scopus search; Figure S3: Affiliated African countries in the 48 publications on plant science prominently produced by African research groups in the time span from 2018 to 2023; Table S1: List of journals in which the 48 publications of plant Fe science with African affiliations in a prominent position have been published, from 2018 to 2023. For each journal, impact factor IF and the number of publications published in that journal are given. Lack of IF is indicated as "not available" (n.a.).

Author Contributions: I.M. conceived the work and performed the search; I.M. and P.M. analyzed the literature data; I.M. wrote the manuscript, with contributions by P.M. All authors have read and agreed to the published version of the manuscript.

Funding: This research received no external funding.

Data Availability Statement: The data presented in this study are available in this article and in its Supplementary Materials.

Acknowledgments: We are grateful to Gianpiero Vigani for reading the manuscript draft and for his suggestions.

Conflicts of Interest: The authors declare no conflict of interest.

References

1. Rajniak, J.; Giehl, R.F.; Chang, E.; Murgia, I.; von Wirén, N.; Sattely, E.S. Biosynthesis of redox-active metabolites in response to iron deficiency in plants. *Nat. Chem. Biol.* **2018**, *14*, 442–450. [CrossRef]
2. Liang, G. Iron uptake, signaling, and sensing in plants. *Plant Commun.* **2022**, *3*, 100349. [CrossRef] [PubMed]
3. Murgia, I.; Marzorati, F.; Vigani, G.; Morandini, P. Plant iron nutrition in the long road from soil to seeds. *J. Exp. Bot.* **2022**, *73*, 1809–1824. [CrossRef] [PubMed]
4. Murgia, I.; Midali, A.; Cimini, S.; De Gara, L.; Manasherova, E.; Cohen, H.; Paucelle, A.; Morandini, P. The *Arabidopsis thaliana* Gulono-1,4 γ-lactone oxidase 2 (GULLO2) facilitates iron transport from endosperm into developing embryos and affects seed coat suberization. *Plant Physiol. Biochem.* **2023**, *196*, 712–723. [CrossRef] [PubMed]
5. Vigani, G.; Morandini, P.; Murgia, I. Searching iron sensors in plants by exploring the link among 2′OG-dependent dioxygenases, the iron deficiency response and metabolic adjustments occurring under iron deficiency. *Front. Plant Sci.* **2013**, *4*, 169. [CrossRef] [PubMed]
6. Vigani, G.; Murgia, I. Iron-requiring enzymes in the spotlight of oxygen. *Trends Plant Sci.* **2018**, *23*, 874–882. [CrossRef] [PubMed]
7. Ramirez, L.; Simontacchi, M.; Murgia, I.; Zabaleta, E.; Lamattina, L. Nitric Oxide, Nitrosyl Iron complexes, ferritin and frataxin: A well equipped team to preserve plant iron homeostasis. *Plant Sci.* **2011**, *181*, 582–592. [CrossRef]
8. Sági-Kazár, M.; Solymosi, K.; Solti, Á. Iron in leaves: Chemical forms, signalling, and in-cell distribution. *J. Exp. Bot.* **2022**, *73*, 1717–1734. [CrossRef]
9. Harbort, C.J.; Hashimoto, M.; Inoue, H.; Niu, Y.; Guan, R.; Rombolà, A.D.; Kopriva, S.; Voges, M.J.E.E.E.; Sattely, E.S.; Garrido-Oter, R.; et al. Root-secreted coumarins and the microbiota interact to improve iron nutrition in *Arabidopsis*. *Cell Host Microbe* **2020**, *28*, 825–837. [CrossRef]
10. Murgia, I.; Arosio, P.; Tarantino, D.; Soave, C. Biofortification for combating "hidden hunger" for iron. *Trends Plant Sci.* **2012**, *17*, 47–55. [CrossRef]
11. Connorton, J.M.; Balk, J. Iron biofortification of staple crops: Lessons and challenges in plant genetics. *Plant Cell Physiol.* **2019**, *60*, 1447–1456. [CrossRef] [PubMed]
12. Stangoulis, J.C.R.; Knez, M. Biofortification of major crop plants with iron and zinc—Achievements and future directions. *Plant Soil* **2022**, *474*, 57–76. [CrossRef]
13. Di Silvestre, D.; Vigani, G.; Mauri, P.; Hammadi, S.; Morandini, P.; Murgia, I. Network topological analysis for the identification of novel hubs in plant nutrition. *Front. Plant Sci.* **2021**, *12*, 629013. [CrossRef] [PubMed]
14. Vigani, G.; Di Silvestre, D.; Agresta, A.M.; Donnini, S.; Mauri, P.; Gehl, C.; Bittner, F.; Murgia, I. Molybdenum and iron mutually impact their homeostasis in cucumber (*Cucumis sativus*) plants. *New Phytol.* **2017**, *213*, 1222–1241. [CrossRef] [PubMed]
15. Mendoza-Cózatl, D.G.; Gokul, A.; Carelse, M.F.; Jobe, T.O.; Long, T.A.; Keyster, M. Keep talking: Crosstalk between iron and sulfur networks fine-tunes growth and development to promote survival under iron limitation. *J. Exp. Bot.* **2019**, *70*, 4197–4210. [CrossRef] [PubMed]
16. Mwangi, M.N.; Mzembe, G.; Moya, E.; Verhoef, H. Iron deficiency anaemia in sub-Saharan Africa: A review of current evidence and primary care recommendations for high-risk groups. *Lancet Haematol.* **2021**, *8*, e732–e743. [CrossRef]
17. Onyango, D.A.; Entila, F.; Dida, M.M.; Ismail, A.M.; Drame, K.N. Mechanistic understanding of iron toxicity tolerance in contrasting rice varieties from Africa: 1. Morpho-physiological and biochemical responses. *Funct. Plant Biol.* **2018**, *46*, 93–105. [CrossRef]
18. Onyango, D.A.; Entila, F.; Egdane, J.; Pacleb, M.; Katimbang, M.L.; Dida, M.M.; Ismail, A.M.; Drame, K.N. Mechanistic understanding of iron toxicity tolerance in contrasting rice varieties from Africa: 2. Root oxidation ability and oxidative stress control. *Funct. Plant Biol.* **2020**, *47*, 145–155. [CrossRef]
19. Kirk, G.J.D.; Manwaring, H.R.; Ueda, Y.; Semwal, V.K.; Wissuwa, M. Below-ground plant–soil interactions affecting adaptations of rice to iron toxicity. *Plant Cell Environ.* **2022**, *45*, 705–718. [CrossRef]
20. Ullah, H.; Ahmed, S.F.; Santiago-Arenas, R.; Himanshu, S.K.; Mansour, E.; Chaum, S.; Datta, A. Tolerance mechanism and management concepts of iron toxicity in rice: A critical review. *Adv. Agron.* **2023**, *177*, 215–257.
21. Melandri, G.; Sikirou, M.; Arbelaez, J.D.; Shittu, A.; Semwal, V.K.; Konaté, K.A.; Maji, A.T.; Ngaujah, S.A.; Akintayo, I.; Govindaraj, V.; et al. Multiple small-effect alleles of indica origin enhance high iron-associated stress tolerance in rice under field conditions in West Africa. *Front. Plant Sci.* **2021**, *11*, 604938. [CrossRef]
22. Sakariyawo, O.S.; Oyedeji, O.E.; Soretire, A.A. Effect of iron deficiency on the growth, development and grain yield of some selected upland rice genotypes in the rainforest. *J. Plant Nutr.* **2020**, *43*, 851–863. [CrossRef]
23. Van Oort, P.A.J. Mapping abiotic stresses for rice in Africa: Drought, cold, iron toxicity, salinity and sodicity. *Field Crops Res.* **2018**, *219*, 55–75. [CrossRef] [PubMed]
24. Rajonandraina, T.; Rakotoson, T.; Wissuwa, M.; Ueda, Y.; Razafimbelo, T.; Andriamananjara, A.; Kirk, G.J.D. Mechanisms of genotypic differences in tolerance of iron toxicity in field-grown rice. *Field Crops Res.* **2023**, *298*, 108953. [CrossRef]
25. Rajonandraina, T.; Ueda, Y.; Wissuwa, M.; Kirk, G.J.D.; Rakotoson, T.; Manwaring, H.; Andriamananjara, A.; Razafimbelo, T. Magnesium supply alleviates iron toxicity-induced leaf bronzing in rice through exclusion and tissue-tolerance mechanisms. *Front. Plant Sci.* **2023**, *14*, 1213456. [CrossRef] [PubMed]
26. Mayaba, T.; Sawadogo, N.; Ouédraogo, M.H.; Sawadogo, B.; Aziadekey, M.; Sié, M.; Sawadogo, M. Genetic diversity of African's rice (*Oryza glaberrima* Steud.) accessions cultivated under iron toxicity. *Aust. J. Crop Sci.* **2020**, *14*, 415–421. [CrossRef]

27. Nasar, J.; Wang, G.Y.; Zhou, F.J.; Gitari, H.; Zhou, X.B.; Tabl, K.M.; Hasan, M.E.; Ali, H.; Waqas, M.M.; Ali, I.; et al. Nitrogen fertilization coupled with foliar application of iron and molybdenum improves shade tolerance of soybean under maize-soybean intercropping. *Front. Plant Sci.* **2022**, *13*, 1014640. [CrossRef] [PubMed]
28. Nasar, J.; Wang, G.-Y.; Ahmad, S.; Muhammad, I.; Zeeshan, M.; Gitari, H.; Adnan, M.; Fahad, S.; Khalid, M.H.B.; Zhou, X.-B.; et al. Nitrogen fertilization coupled with iron foliar application improves the photosynthetic characteristics, photosynthetic nitrogen use efficiency, and the related enzymes of maize crops under different planting patterns. *Front. Plant Sci.* **2022**, *13*, 988055. [CrossRef]
29. Akhtar, S.; Osthoff, G.; Mashingaidze, K.; Labuschagne, M. Iron and zinc in maize in the developing world: Deficiency, availability, and breeding. *Crop Sci.* **2018**, *58*, 2200–2213. [CrossRef]
30. Akhtar, S.; Labuschagne, M.; Osthoff, G.; Mashingaidze, K.; Hossain, A. Xenia and deficit nitrogen influence the iron and zinc concentration in the grains of hybrid maize. *Agronomy* **2021**, *11*, 1388. [CrossRef]
31. Akhtar, S.; Mekonnen, T.W.; Osthoff, G.; Mashingaidz, K.; Labuschagne, M. Genotype by environment interaction in grain iron and zinc concentration and yield of maize hybrids under low nitrogen and optimal conditions. *Plants* **2023**, *12*, 1463. [CrossRef] [PubMed]
32. Goredema-Matongera, N.; Ndhlela, T.; van Biljon, A.; Kamutando, C.N.; Cairns, J.E.; Baudron, F.; Labuschagne, M. Genetic variation of zinc and iron concentration in normal, provitamin a and quality protein maize under stress and non-stress conditions. *Plants* **2023**, *12*, 270. [CrossRef] [PubMed]
33. Manzeke-Kangara, M.G.; Mtambanengwe, F.; Watts, M.J.; Broadley, M.R.; Lark, R.M.; Mapfumo, P. Can nitrogen fertilizer management improve grain iron concentration of agro-biofortified crops in Zimbabwe? *Agronomy* **2021**, *11*, 124. [CrossRef]
34. Udo, E.; Abe, A.; Meseka, S.; Mengesha, W.; Menkir, A. Genetic Analysis of zinc, iron and provitamin A content in tropical maize (*Zea mays* L.). *Agronomy* **2023**, *13*, 266. [CrossRef]
35. Saradadevi, R.; Mukankusi, C.; Li, L.; Amongi, W.; Mbiu, J.P.; Raatz, B.; Ariza, D.; Beebe, S.; Varshney, R.K.; Huttner, E.; et al. Multivariate genomic analysis and optimal contributions selection predicts high genetic gains in cooking time, iron, zinc, and grain yield in common beans in East Africa. *Plant Genome* **2021**, *14*, e20156. [CrossRef] [PubMed]
36. Katuuramu, D.N.; Wiesinger, J.A.; Luyima, G.B.; Nkalubo, S.T.; Glahn, R.P.; Cichy, K.A. Investigation of genotype by environment interactions for seed zinc and iron concentration and iron bioavailability in common bean. *Front. Plant Sci.* **2021**, *12*, 670965. [CrossRef] [PubMed]
37. Nsiri, K.; Krouma, A. The key physiological and biochemical traits underlying common bean (*Phaseolus vulgaris* L.) response to iron deficiency, and related interrelationships. *Agronomy* **2023**, *13*, 2148. [CrossRef]
38. Philipo, M.; Ndakidemi, P.A.; Mbega, E.R. Environmental and genotypes influence on seed iron and zinc levels of landraces and improved varieties of common bean (*Phaseolus vulgaris* L.) in Tanzania. *Ecol. Genet. Genom.* **2020**, *15*, 100056. [CrossRef]
39. Gelaw, Y.M.; Eleblu, J.S.Y.; Ofori, K.; Fenta, B.A.; Mukankusi, C.; Offei, S. Genome-wide association study of grain iron and zinc concentration in common bean (*Phaseolus vulgaris*). *Plant Breed.* **2023**, *142*, 357–371. [CrossRef]
40. Salama, Z.A.; El Fouly, M.M. Can copper and zinc in different chemical forms can improve iron deficient in phaseolus plant. *Iraqi J. Agric. Sci.* **2020**, *51*, 278–286.
41. Mira, M.M.; Asmundson, B.; Renault, S.; Hill, R.D.; Stasolla, C. Suppression of the soybean (*Glycine max*) phytoglobin GmPgb1 improves tolerance to iron stress. *Acta Physiol. Plant.* **2021**, *43*, 147. [CrossRef]
42. Haque, A.F.M.M.; Rahman, M.A.; Das, U.; Rahman, M.M.; Elseehy, M.M.; El-Shehawi, A.M.; Parvez, M.S.; Kabir, A.H. Changes in physiological responses and MTP (metal tolerance protein) transcripts in soybean (*Glycine max*) exposed to differential iron availability. *Plant Physiol. Biochem.* **2022**, *179*, 1–9. [CrossRef] [PubMed]
43. Knijnenburg, J.T.N.; Hilty, F.M.; Oelofse, J.; Buitendag, R.; Zimmermann, M.B.; Cakmak, I.; Grobler, A.F. Nano- and Pheroid technologies for development of foliar iron fertilizers and iron biofortification of soybean grown in South Africa. *Chem. Biol. Technol. Agric.* **2018**, *5*, 26. [CrossRef]
44. Ramadan, A.A.E.-M.; El-Bassiouny, H.M.S.; Bakry, B.A.; Abdallah, M.M.S.; El-Enany, M.A.M. Growth, yield and biochemical changes of soybean plant in response to iron and magnesium oxide nanoparticles. *Pak. J. Biol. Sci.* **2020**, *23*, 406–417. [CrossRef]
45. Gangashetty, P.I.; Riyazaddin, M.; Sanogo, M.D.; Inousa, D.; Issoufou, K.A.; Asungre, P.A.; Sy, O.; Govindaraj, M.; Ignatius, A.I. Identification of high-yielding iron-biofortified open-pollinated varieties of pearl millet in West Africa. *Front. Plant Sci.* **2021**, *12*, 688937. [CrossRef] [PubMed]
46. Asungre, P.A.; Akromah, R.; Kena, A.W.; Gangashetty, P. Genotype by environment interaction on grain yield stability and iron and zinc content in OPV of pearl millet in Ghana using the AMMI method. *Int. J. Agron.* **2021**, *2021*, 9656653. [CrossRef]
47. Asungre, P.A.; Akromah, R.; Kena, A.W.; Gangashetty, P. Assessing the adaptability and stability of new pearl millet hybrids for grain yield, grain iron and zinc content in Ghana using AMMI analysis. *J. Crop Sci. Biotechnol.* **2022**, *25*, 501–514. [CrossRef]
48. Gaoh, B.B.S.; Gangashetty, P.I.; Mohammed, R.; Dzidzienyo, D.K.; Tongoona, P. Generation mean analysis of pearl millet [*Pennisetum glaucum* (L.) R. Br.] grain iron and zinc contents and agronomic traits in West Africa. *J. Cereal Sci.* **2020**, *96*, 103066. [CrossRef]
49. Teklu, D.; Gashu, D.; Joy, E.J.M.; Lark, R.M.; Bailey, E.H.; Wilson, L.; Amede, T.; Broadley, M.R. Genotypic response of finger millet to zinc and iron agronomic biofortification, location and slope position towards yield. *Agronomy* **2023**, *13*, 1452. [CrossRef]
50. Wafula, W.N.; Korir, N.K.; Ojulong, H.F.; Siambi, M.; Gweyi-Onyango, J.P. Protein, calcium, zinc, and iron contents of finger millet grain response to varietal differences and phosphorus application in Kenya. *Agronomy* **2018**, *8*, 24. [CrossRef]

51. Krouma, A. Differential response of pea (*Pisum sativum* L.) genotypes to iron deficiency in relation to the growth, rhizosphere acidification and ferric chelate reductase activities. *Aust. J. Crop Sci.* **2021**, *15*, 925–932. [CrossRef]
52. Barhoumi, S.; Ellouzi, H.; Krouma, A. Functional analysis of the genotypic differences in response of pea (*Pisum sativum* L.) to calcareous-induced iron deficiency. *Phyton-Int. J. Exp. Bot.* **2023**, *92*, 521–536. [CrossRef]
53. Coppa, E.; Vigani, G.; Aref, R.; Savatin, D.; Bigini, V.; Hell, R.; Astolfi, S. Differential modulation of Target of Rapamycin activity under single and combined iron and sulfur deficiency in tomato plants. *Plant J.* **2023**, *115*, 127–138. [CrossRef] [PubMed]
54. El-Desouky, H.S.; Islam, K.R.; Bergefurd, B.; Gao, G.; Harker, T.; Abd-El-Dayem, H.; Ismail, F.; Mady, M.; Zewail, R.M.Y. Nano iron fertilization significantly increases tomato yield by increasing plants' vegetable growth and photosynthetic efficiency. *J. Plant Nutr.* **2021**, *44*, 1649–1663. [CrossRef]
55. Olowolaju, E.D.; Okunlola, G.O.; Ayeotan, O.J. Growth, yield and uptake of some nutrients by tomato as affected by iron concentration. *Int. J. Veg. Sci.* **2021**, *27*, 378–387. [CrossRef]
56. Prity, S.A.; El-Shehawi, A.M.; Elseehy, M.M.; Tahura, S.; Kabir, A.H. Early-stage iron deficiency alters physiological processes and iron transporter expression, along with photosynthetic and oxidative damage to sorghum. *Saudi J. Biol. Sci.* **2021**, *28*, 4770–4777. [CrossRef] [PubMed]
57. Maswada, H.F.; Djanaguiraman, M.; Prasad, P.V.V. Seed treatment with nano-iron (III) oxide enhances germination, seeding growth and salinity tolerance of sorghum. *J. Agron. Crop Sci.* **2018**, *204*, 577–587. [CrossRef]
58. Abou-Baker, N.H.; Hussein, M.M.; El-Ashry, S.M. Comparison between nano iron and iron EDTA as foliar fertilizers under salt stress conditions. *Plant Cell Biotechnol. Mol. Biol.* **2020**, *21*, 17–32.
59. Bhattacharya, S.; Das, A.; Banerjee, J.; Mandal, S.N.; Kumar, S.; Gupta, S. Elucidating genetic variability and genotype × environment interactions for grain iron and zinc content among diverse genotypes of lentils (*Lens culinaris*). *Plant Breed.* **2022**, *141*, 786–800. [CrossRef]
60. Gupta, S.; Das, S.; Dikshit, H.K.; Mishra, G.P.; Aski, M.S.; Bansal, R.; Tripathi, K.; Bhowmik, A.; Kumar, S. Genotype by environment interaction effect on grain iron and zinc concentration of indian and mediterranean lentil genotypes. *Agronomy* **2021**, *11*, 1761. [CrossRef]
61. Mannan, M.A.; Tithi, M.A.; Islam, M.R.; Al Mamun, M.A.; Mia, S.; Rahman, M.Z.; Awad, M.F.; ElSayed, A.I.; Mansour, E.; Hossain, M.S. Soil and foliar applications of zinc sulfate and iron sulfate alleviate the destructive impacts of drought stress in wheat. *Cereal Res. Commun.* **2022**, *50*, 1279–1289. [CrossRef]
62. Gorafi, Y.S.A.; Ishii, T.; Kim, J.-S.; Elbashir, A.A.E.; Tsujimoto, H. Genetic variation and association mapping of grain iron and zinc contents in synthetic hexaploid wheat germplasm. *Plant Genet. Resour. Characterisation Util.* **2018**, *16*, 9–17. [CrossRef]
63. Kallala, N.; M'sehli, W.; Hammi, K.M.; Abid, G.; Mhadhbi, H. Efficiency of antioxidant system in barrel medic (*Medicago truncatula*) genotypes and the orchestration of their key components under iron deficiency. *Crop Pasture Sci.* **2021**, *73*, 138–148. [CrossRef]
64. Kallala, N.; M'sehli, W.; Jelali, K.; Batnini, M.; Badri, M.; Nouairi, I.; Mhadhbi, H. Biodiversity within *Medicago truncatula* genotypes toward response to iron deficiency: Investigation of main tolerance mechanisms. *Plant Species Biol.* **2019**, *34*, 95–109. [CrossRef]
65. Ifie, J.E.; Ifie-Etumah, S.; Ikhajiagbe, B. Physiological and biochemical responses of selected cowpea (*Vigna unguiculata* (L.) Walp.) accessions to iron toxicity [Fiziološki in biokemični odziv akcesij kitajske vinje (Vigna unguiculata (L.) Walp.) na toksičnost železa]. *Acta Agric. Slov.* **2020**, *115*, 25–38. [CrossRef]
66. Sileshi, F.; Nebiyu, A.; Van Geel, M.; Abeele, S.V.; Du Laing, G.; Boeckx, P. Spatial variability of iron, zinc and selenium content in faba bean (*Vicia faba* L.) seeds from central and southwestern highlands of Ethiopia. *Plant Soil* **2022**, *473*, 351–368. [CrossRef]
67. Mahmoud, A.W.M.; Ayad, A.A.; Abdel-Aziz, H.S.M.; Williams, L.L.; El-Shazoly, R.M.; Abdel-Wahab, A.; Abdeldaym, E.A. Foliar application of different iron sources improves morpho-physiological traits and nutritional quality of broad bean grown in sandy soil. *Plants* **2022**, *11*, 2599. [CrossRef]
68. Nyiraguhirwa, S.; Grana, Z.; Ouabbou, H.; Iraqi, D.; Ibriz, M.; Mamidi, S.; Udupa, S.M. A genome-wide association study identifying single-nucleotide polymorphisms for iron and zinc biofortification in a worldwide barley collection. *Plants* **2022**, *11*, 1349. [CrossRef]
69. Mazhar, M.W.; Ishtiaq, M.; Maqbool, M.; Akram, R.; Shahid, A.; Shokralla, S.; Al-Ghobari, H.; Alataway, A.; Dewidar, A.Z.; El-Sabrout, A.M.; et al. Seed priming with iron oxide nanoparticles raises biomass production and agronomic profile of water-stressed flax plants. *Agronomy* **2022**, *12*, 982. [CrossRef]
70. Shareef, H.J.; Hzaa, A.Y.L.; Elsheery, N.I. Foliar iron and zinc nano-fertilizers enhance growth, mineral uptake, and antioxidant defense in date palm (*Phoenix dactylifera* L.) seedlings. *Folia Oecologica* **2023**, *50*, 185–195. [CrossRef]
71. Barhoumi, Z.; Atia, A.; Hussain, A.A.; Maatallah, M.; Alalmaie, A.; Alaskri, K.I.; Assiri, A.M. Effects of salinity and iron deficiency on growth and physiological attributes of *Avicennia marina* (Forssk.) Vierh. *Arch. Agron. Soil Sci.* **2023**, *69*, 2753–2766. [CrossRef]
72. Ndaba, B.; Roopnarain, A.; Vatsha, B.; Marx, S.; Maaza, M. Synthesis, characterization, and evaluation of *Artemisia afra*-mediated iron nanoparticles as a potential nano-priming agent for seed germination. *ACS Agric. Sci. Technol.* **2022**, *2*, 1218–1229. [CrossRef]
73. Alharbi, K.; Alshallash, K.S.; Hamdy, A.E.; Khalifa, S.M.; Abdel-Aziz, H.F.; Sharaf, A.; Abobatta, W.F. Magnetic iron–improved growth, leaf chemical content, yield, and fruit quality of chinese mandarin trees grown under soil salinity stress. *Plants* **2022**, *11*, 2839. [CrossRef] [PubMed]

74. Mnafgui, W.; Rizzo, V.; Muratore, G.; Hajlaoui, H.; Schinoff, B.D.O.; Mnafgui, K.; Elleuch, A. *Trigonella foenum-graecum* morphophysiological and phytochemical processes controlling iron uptake and translocation. *Crop Pasture Sci.* **2022**, *73*, 957–968. [CrossRef]
75. El-Sonbaty, A.E.; Farouk, S.; Al-Yasi, H.M.; Ali, E.F.; Abdel-Kader, A.A.S.; El-Gamal, S.M.A. Enhancement of rose scented geranium plant growth, secondary metabolites, and essential oil components through foliar applications of iron (nano, sulfur and chelate) in alkaline soils. *Agronomy* **2022**, *12*, 2164. [CrossRef]
76. Saudy, H.S.; El–Samad, G.A.A.; El–Temsah, M.E.; El–Gabry, Y.A.E.-G. Effect of iron, zinc, and manganese nano-form mixture on the micronutrient recovery efficiency and seed yield response index of sesame genotypes. *J. Soil Sci. Plant Nutr.* **2022**, *22*, 732–742. [CrossRef]
77. Salhi, K.; Hajlaoui, H.; Krouma, A. Genotypic differences in response of durum wheat (*Triticum durum* Desf.) to lime-induced iron chlorosis. *Plant Direct* **2022**, *6*, e377. [CrossRef]
78. El-Shehawi, A.M.; Rahman, M.A.; Elseehy, M.M.; Kabir, A.H. Mercury toxicity causes iron and sulfur deficiencies along with oxidative injuries in alfalfa (*Medicago sativa*). *Plant Biosyst.* **2022**, *156*, 284–291. [CrossRef]
79. El-Gioushy, S.F.; Ding, Z.; Bahloul, A.M.E.; Gawish, M.S.; Abou El Ghit, H.M.; Abdelaziz, A.M.R.A.; El-Desouky, H.S.; Sami, R.; Khojah, E.; Hashim, T.A.; et al. Foliar application of nano, chelated, and conventional iron forms enhanced growth, nutritional status, fruiting aspects, and fruit quality of washington navel orange trees (*Citrus sinensis* L. Osbeck). *Plants* **2021**, *10*, 2577. [CrossRef]
80. El-Monem, A.A.A.; Abdallah, M.M.S.; Bakry, B.A.; El-Bassiouny, H.M.S.; Sadak, M.S. Physiological role of iron chelators and/or arginine for improving yield and active constituents of roselle sepals. *Asian J. Plant Sci.* **2020**, *19*, 77–90.
81. González, M.R.; Hailemichael, G.; Catalina, Á.; Martín, P. Combined effects of water status and iron deficiency chlorosis on grape composition. *Sci. Agric.* **2019**, *76*, 473–480. [CrossRef]
82. Ben Abdallah, H.; Mai, H.J.; Slatni, T.; Fink-Straube, C.; Abdelly, C.; Bauer, P. Natural variation in physiological responses of tunisian *Hedysarum carnosum* under iron deficiency. *Front. Plant Sci.* **2018**, *9*, 1383. [CrossRef] [PubMed]
83. Wissal, M.; Nadia, K.; Haythem, M. Legumes: Model plants for sustainable agriculture in phosphorus and iron deficient soils. *Agric. Sci. Dig.* **2020**, *40*, 445–447. [CrossRef]
84. Garcia-Oliveira, A.L.; Chander, S.; Ortiz, R.; Menkir, A.; Gedil, M. Genetic basis and breeding perspectives of grain iron and zinc enrichment in cereals. *Front. Plant Sci.* **2018**, *9*, 937. [CrossRef] [PubMed]
85. Mc Inturf, S.A.; Khan, M.A.; Gokul, A.; Castro-Guerrero, N.A.; Höhner, R.; Li, J.; Marjault, H.-B.; Fichman, Y.; Kunz, H.-H.; Goggin, F.L.; et al. Cadmium interference with iron sensing reveals transcriptional programs sensitive and insensitive to reactive oxygen species. *J. Exp. Bot.* **2022**, *73*, 324–338. [CrossRef] [PubMed]
86. Kobayashi, T.; Nishizawa, N.K. Iron sensors and signals in response to iron deficiency. *Plant Sci.* **2014**, *224*, 36–43. [CrossRef] [PubMed]
87. Riaz, N.; Guerinot, M.L. All together now: Regulation of the iron deficiency response. *J. Exp. Bot.* **2021**, *72*, 2045–2055. [CrossRef]
88. Therby-Vale, R.; Lacombe, B.; Rhee, S.Y.; Nussaume, L.; Rouached, H. Mineral nutrient signaling controls photosynthesis: Focus on iron deficiency-induced chlorosis. *Trends Plant Sci.* **2022**, *27*, 502–509. [CrossRef]
89. Jones, M.P.; Dingkuhn, M.; Aluko, G.K.; Semon, M. Interspecific *Oryza sativa* L. X *O. glaberrima* Steud. progenies in upland improvement. *Euphytica* **1997**, *92*, 237–246. [CrossRef]
90. Heuer, S.; Miézan, K.M.; Sié, M.; Gaye, S. Increasing biodiversity of irrigated rice in Africa by interspecific crossing of *Oryza glaberrima* (Steud.) x *O. sativa* (L.). *Euphytica* **2003**, *132*, 31–40. [CrossRef]
91. Orr, S.; Sumberg, J.; Erenstein, O.; Oswald, A. Funding international agricultural research and the need to be noticed. A case study of NERICA Rice. *Outlook Agric.* **2008**, *37*, 159–168. [CrossRef]
92. Olivares Diaz, E.; Kawamura, S.; Koseki, S. Physical Properties of NERICA Compared to Indica and Japonica types of rice. *Agric. Mech. Asia Afr. Lat. Am.* **2018**, *49*, 68–73.
93. Matsuoka, Y.; Takumi, S.; Kawahara, T. Natural variation for fertile triploid F_1 hybrid formation in allohexaploid wheat speciation. *Theor. Appl. Genet.* **2007**, *115*, 509–515. [CrossRef] [PubMed]
94. Satyavathi, C.T.; Ambawat, S.; Khandelwal, V.; Srivastava, R.K. Pearl Millet: A climate-resilient nutricereal for mitigating hidden hunger and provide nutritional security. *Front. Plant Sci.* **2021**, *12*, 659938. [CrossRef] [PubMed]
95. Mittal, D.; Kaur, G.; Singh, P.; Yadav, K.; Ali, S.A. Nanoparticle-based sustainable agriculture and food science: Recent advances and future outlook. *Front. Nanotechnol.* **2020**, *2*, 579954. [CrossRef]
96. Babu, S.; Singh, R.; Yadav, D.; Singh Rathore, S.; Raj, R.; Avasthe, R.; Yadav, S.K.; Das, A.; Yadav, V.; Yadav, B.; et al. Nanofertilizers for agricultural and environmental sustainability. *Chemosphere* **2022**, *292*, 133451. [CrossRef]
97. Zhu, M.T.; Wang, B.; Wang, Y.; Yuan, L.; Wang, H.J.; Wang, M.; Ouyang, H.; Chai, Z.F.; Feng, W.Y.; Zhao, Y.L. Endothelial dysfunction and inflammation induced by iron oxide nanoparticle exposure: Risk factors for early atherosclerosis. *Toxicol. Lett.* **2011**, *203*, 162–171. [CrossRef]
98. Maher, B.A.; González-Maciel, A.; Reynoso-Robles, R.; Torres-Jardón, R.; Calderón-Garcidueñas, L. Iron-rich air pollution nanoparticles: An unrecognised environmental risk factor for myocardial mitochondrial dysfunction and cardiac oxidative stress. *Environ. Res.* **2020**, *188*, 109816. [CrossRef]
99. Küpper, H.; Andresen, E. Mechanisms of metal toxicity in plants. *Metallomics* **2016**, *8*, 269–285. [CrossRef]

100. Aung, M.S.; Masuda, H. How does rice defend against excess iron? Physiological and molecular mechanisms. *Front. Plant Sci.* **2020**, *11*, 1102. [CrossRef]
101. Saif, S.; Tahir, A.; Chen, Y. Green synthesis of iron nanoparticles and their environmental application and implications. *Nanomaterials* **2016**, *6*, 209. [CrossRef] [PubMed]
102. Tao, Z.; Zhou, Q.; Zheng, T.; Mo, F.; Ouyang, S. Iron oxide nanoparticles in the soil environment: Adsorption, transformation, and environmental risk. *J. Hazard. Mater.* **2023**, *459*, 132107. [CrossRef] [PubMed]
103. Kornberg, T.G.; Stueckle, T.A.; Antonini, J.A.; Rojanasakul, Y.; Castranova, V.; Yang, Y.; Wang, L. Potential toxicity and underlying mechanisms associated with pulmonary exposure to iron oxide nanoparticles: Conflicting literature and unclear risk. *Nanomaterials* **2017**, *7*, 307. [CrossRef]
104. Murgia, I.; Delledonne, M.; Soave, C. Nitric oxide mediates iron-induced ferritin accumulation in *Arabidopsis*. *Plant J.* **2002**, *30*, 521–528. [CrossRef] [PubMed]
105. Arnaud, N.; Murgia, I.; Boucherez, J.; Briat, J.F.; Cellier, F.; Gaymard, F. An iron-induced nitric oxide burst precedes ubiquitin-dependent protein degradation for *Arabidopsis* AtFer1 ferritin gene expression. *J. Biol. Chem.* **2006**, *281*, 23579–23588. [CrossRef] [PubMed]
106. Murgia, I.; Vazzola, V.; Tarantino, D.; Cellier, F.; Ravet, K.; Briat, J.F.; Soave, C. Knock-out of the ferritin AtFer1 causes earlier onset of age-dependent leaf senescence in *Arabidopsis*. *Plant Physiol. Biochem.* **2007**, *45*, 898–907. [CrossRef] [PubMed]
107. Briat, J.F.; Céline Duc, C.; Karl Ravet, K.; Gaymard, F. Ferritins and iron storage in plants. *Biochim. Biophys. Acta (BBA)* **2010**, *1800*, 806–814. [CrossRef]
108. Briat, J.F.; Ravet, K.; Arnaud, N.; Duc, C.; Boucherez, J.; Touraine, B.; Cellier, F.; Gaymard, F. New insights into ferritin synthesis and function highlight a link between iron homeostasis and oxidative stress in plants. *Ann. Bot.* **2010**, *105*, 811–822. [CrossRef]
109. Vigani, G.; Tarantino, D.; Murgia, I. Mitochondrial ferritin is a functional iron-storage protein in cucumber (*Cucumis sativus*) roots. *Front. Plant Sci.* **2013**, *4*, 316. [CrossRef]

MDPI
St. Alban-Anlage 66
4052 Basel
Switzerland
www.mdpi.com

Plants Editorial Office
E-mail: plants@mdpi.com
www.mdpi.com/journal/plants

www.ingramcontent.com/pod-product-compliance
Lightning Source LLC
LaVergne TN
LVHW070702100526
838202LV00013B/1015